Introduction to Rural Planning, 2nd Edition

Introduction to Rural Planning: Economies, Communities and Landscapes provides a critical analysis of the key challenges facing rural places and the ways that public policy and community action shape rural spaces.

The second edition provides an examination of the composite nature of 'rural planning', which combines land-use and spatial planning elements with community action, countryside management and the projects and programmes of national and supra-national agencies and organisations. It also offers a broad analysis of entrepreneurial social action as a shaper of rural outcomes, with particular coverage of the localism agenda and Neighbourhood Planning in England. With a focus on accessibility and rural transport provision, this book examines the governance arrangements needed to deliver integrated solutions spanning urban and rural places. Through an examination of the ecosystem approach to environmental planning, it links the procurement of ecosystem services to the global challenges of habitat degradation and loss, climate change and resource scarcity and management.

A valuable resource for students of planning, rural development and rural geography, *Introduction to Rural Planning* aims to make sense of current rural challenges and planning approaches, evaluating the currency of the 'rural' label in the context of global urbanisation, arguing that rural spaces are relational spaces characterised by critical production and consumption tensions.

Nick Gallent is Professor of Housing and Planning and Head of the Bartlett School of Planning at UCL. **Iqbal Hamiduddin** is Lecturer in Transport and Housing at the Bartlett. **Meri Juntti** is Senior Lecturer in Environmental Governance and Rural Development in the Department of Law and Politics at Middlesex University. **Sue Kidd** and **Dave Shaw** are both in the Department of Geography and Planning at the University of Liverpool, where Sue is a Senior Lecturer and Head of Planning and Dave is a Professor of Planning.

The Natural and Built Environment Series

Editor: Professor John Glasson
Oxford Brookes University

Introduction to Rural Planning, 2nd Edition

Economies, communities and landscapes

••

Nick Gallent, Iqbal Hamiduddin,
Meri Juntti, Sue Kidd and Dave Shaw

Routledge
Taylor & Francis Group

NEW YORK AND LONDON

First published 2015
by Routledge
711 Third Avenue, New York, NY 10017

and by Routledge
2 Park Square, Milton Park, Abingdon, Oxon OX14 4RN

Routledge is an imprint of the Taylor & Francis Group, an informa business

Library of Congress Cataloging in Publication Data
Gallent, Nick.
Introduction to rural planning : economies, communities and landscapes / Nick Gallent, Iqbal Hamiduddin, Meri Juntti, Sue Kidd and Dave Shaw. — 2nd edition.
pages cm. — (The natural and built environment series)
Includes bibliographical references and index.
1. Rural development—Great Britain. 2. Great Britain—Rural conditions. I. Title.
HN400.C6G35 2015
307.1'4120941—dc23
2014048380

ISBN: 9781138811447 (hbk)
ISBN: 9781138811454 (pbk)
ISBN: 9781315749280 (ebk)

Typeset in Stone Serif and Akzidenz Grotesk
by FiSH Books Ltd, Enfield

MIX
Paper from
responsible sources
FSC
www.fsc.org   FSC® C013056

Printed and bound in Great Britain by
TJ International Ltd, Padstow, Cornwall

"This is the definitive textbook on rural planning in the UK. It is up to date, relevant and clearly explained. The book provides critical perspectives on rural planning issues that will continue to ensure its relevance in a rapidly changing policy arena."
Richard Yarwood, Plymouth University

"*Introduction to Rural Planning* highlights stark changes in British rural planning since 2010. Reviewing longstanding rural themes, today's issues are impressively linked to broad trends, with new, intriguing insights. A snappy presentation gives a sharp, informative introduction to critical issues for the future governance, management and existence of rural Britain."
Keith Hoggart, Emeritus Professor of Geography, King's College London

"This book provides an invaluable resource for students and teachers of rural planning, environmental planning and rural geography. In this significantly updated and restructured second edition, the authors provide insightful and comprehensive analysis of rural challenges and the existing policy context, while offering clear prescription for more sustainable and resilient rural futures."
Mark Scott, University College Dublin

Contents

List of illustrations

Figures

Boxes

Tables

Notes on authors

Nick Gallent is Professor of Housing and Planning and Head of the Bartlett School of Planning at University College London. He is a Chartered Town Planner and a Chartered Surveyor. His research focuses on UK planning policy as it pertains to housing delivery and as it affects rural communities. He has conducted research for a wide range of funding bodies and is the author or editor of several previous works on planning, housing and communities, the most recent being 'Community Action and Planning' (2014, edited with Daniela Ciaffi).

Iqbal Hamiduddin is a lecturer in transport planning and housing at the Bartlett School of Planning, University College London. His research specialisms are in transport, housing and particularly the interface of transport and housing policies in residential design. His PhD thesis on the *Social Implications of Residential Car Reduction* was based largely on detailed comparative field research undertaken in different neighbourhoods of Freiburg. He has also investigated different elements of transport planning and housing delivery separately in a variety of projects for organisations including the Royal Institution of Chartered Surveyors, the Regional Studies Association and the European Union.

Meri Juntti is a senior lecturer in the Department of Law and Politics at Middlesex University. Her research focuses on agricultural and environmental policy decision-making and implementation in the European Union. She has worked on both EU and domestically funded research projects focusing on a number of EU member states and has authored publications on the discursive construction of environmental policy, the role of the socio-material context in differentiating rural policy outcomes and the nature and role of 'evidence' in the policy process.

Sue Kidd is Head of Planning in the Department of Geography and Planning at the University of Liverpool. Her research interests focus around integrated planning and management. She has a particular interest in exploring ways in which environmental considerations may be more effectively combined with economic and social concerns in policy making. Sue is a Chartered Town Planner.

Dave Shaw a Professor of Planning in the Department of Geography and Planning at the University of Liverpool. He is particularly interested in the way rural areas and rural places are being transformed and the role of planning in that process. He also participates in rural affairs and is currently the Chair of the Mersey Rural Leader programme, represents 'the rural' in the Liverpool City Region's European Investment Fund Subgroup and is a Trustee of the Community Forest Trust.

Preface

•••

Much has changed in Britain[1] in the seven years since the publication of the first edition of this book. The continuity of rural policy development under the Labour Government since 1997 was seemingly broken in the aftermath of the economic crises of 2008 and the subsequent election of a Conservative–Liberal Democrat Coalition Government in May 2010. Since then, there appear to have been fundamental changes to the statutory planning system in England, and although there seems to have been relative stability and continuity in other areas of policy affecting the countryside, dramatic public spending cuts have significantly reduced the capacity for state leadership and action. Between 2010 and 2012, the Coalition government executed a plan to localise and rescale planning in England. Its first move was to revoke regional strategies and dismantle the entire apparatus of regional planning. It then enacted key legislation in the form of the Localism Act 2011, which created a system of Neighbourhood Development Planning (NDP) and tasked local authorities with greater responsibility for strategic oversight and cooperation with neighbouring areas. A year later, a process of reducing planning policy in England to just 53 pages was completed with the publication of a National Planning Policy Framework, augmented with further technical guidance. These big changes to the planning system in England seemed to offer a cue to revise our *Introduction to Rural Planning*, so as to provide students with an up-to-date account and analysis of the system as it exists, paralleling what has been happening in England with insights from Scotland and Wales.

However, this is not our main reason for producing a new edition at this time. Although developments over the last five years have been promoted by some as a radical departure from former practice, the changes that have occurred are perhaps better viewed as a resurgence of long-standing 'post-modern' and 'neo-liberal' undercurrents affecting not just the UK but many other parts of the world as well. Even prior to the dramatic events of 2008, the Labour government had spent much of its period in power wrestling to achieve an acceptable balance between the promotion of local democracy and strategic control of planning and other areas of public policy. And all governments have faced the same dilemma since the 1970s, as society becomes more pluralistic generally and more diverse locally. Social pluralism has been manifest in a diversification of needs and political allegiance, making it difficult for governments to define and pursue a clear 'public good'. The response has been for the apparatus of representative democracy to try to connect more directly to communities, to devolve greater responsibility to the local state and downward to the people whose lives and neighbourhoods seem to be most directly affected by different policies, programmes and projects. There has been a gradual move to renew local democracy and thereby create a new political consensus in support of government. Some commentators have called this a 'governance shift' and others have referred to the desire to 'co-produce' public policy with its users or achieve a new politics of consensus. The existence of this changed reality has been acknowledged around the globe and it has affected not only land-use planning, but all areas of public policy design and intervention. Responses to a variety of challenges from economic development and transport delivery all the way to countryside management, renewable energy production and climate change adaptation have been devolved to local partnerships and community actors. Exogenous investment has given way to endogenous, localised enterprise, often built on collaborative approaches that seek to achieve broad consensus around particular resolutions.

But in the UK, government in the 1980s seemed to swim against this tide. Margaret Thatcher was not a fan of consensus:

> To me, consensus seems to be: the process of abandoning all beliefs, principles, values, and policies in search of something in which no one believes, but to which no one objects; the process of avoiding the very issues that need to be solved, merely because you cannot get agreement on the way ahead.

> (Thatcher, 1993)

By the late 1990s, however, a view that policy development at the top needed to reconnect to a grassroots of local communities at the bottom began to gain momentum, not only in the area of land-use planning (the focus of the reforms noted above), but in all areas of public service delivery and in approaches to rural development, broadly defined. Labour Governments after 1997 brought forward a number of Local Government Acts in quick succession, all aimed at devolving powers to the local state and promoting the idea of community leadership. A system of Local Strategic Partnerships was established, tasked to produce Community Strategies that would give the informal representatives of community networks and action groups a direct input into local policy, including policy determining land-use planning outcomes. But at the same time, Labour needed mechanisms with which to deliver against major development and infrastructure priorities. It sought to strengthen regional planning and eventually, in 2008, moved key responsibilities for major infrastructure to a new quango. England was washed over by a powerful tide of community rhetoric during the 2000s, but most communities were underwhelmed by government promises and saw instead the growing power of regional planning to determine local outcomes from distant regional offices. This failure alongside the events that followed the 2008 economic crisis gave the Conservatives the platform they needed to launch a withering attack on the 'steering centrism' of the Labour Governments and to embark on a significant slimming down of the state. They were, for example, able to portray the key tenets of labour planning – especially stronger regional control – as undemocratic. Their answer was to extend the idea of community leadership, empowering local people to take control of key aspects of the planning system as well as local service delivery: to instigate what they badged as a 'control shift'. The revocation of regional strategies, noted above, amounted to a significant rescaling of planning. But it also seemed to restrict government's control over local planning frameworks and outcomes as well as other policy priorities. How would government deliver the housing, the infrastructure and the economic development and growth that Labour tried to lever through regional planning and its various quangos? The answer came in two forms: first, the continued primacy of local plans and a National Planning Policy Framework that prioritised 'growth'; and second, a significant shift to market and community-based delivery of rural services and development as government funding was withdrawn.

In the context of greater social complexity in which all conceptions of the public good are contested, the 2010–2015 UK government walked a tightrope between promoting local consensus (as well as market- and community-based action) and retaining the capacity to take tough executive decisions and provide strategic direction. The intention in this second edition is to show, first, how the shift in the scale and mode of 'rural planning' (towards local action) is a long-term trend and, second, how the broader global challenges (for example, of climate change and resource scarcity) are now providing the bigger context for that planning. Britain is presented as a case study facing challenges that are shared with the global community of nations. Our objective is to lay bare the context for rural planning and its composite parts, revealing how a range of challenges are being addressed at scales from the supra-national to the very local. This has required a new overall structure and a full revision of each chapter, with some of the first edition chapters now gone and others inserted. The book has been completely rewritten and whilst it shares its title with the 2008 edition, that is where any resemblance ends. Each of the new chapters are now divided into two sections, the first dealing with the 'bedrock' of principles and key ideas; and the second with the shifting sands, and the narrative, of rural planning policy and practice in Britain. We hope that this structure will bring greater clarity to the material presented and help readers to understand how policy and practice has developed, and also how seemingly radical reforms are often veiled continuities.

Nick Gallent, Iqbal Hamiduddin and Meri Juntti, London
Sue Kidd and Dave Shaw, Liverpool
December 2014

Note

1   This book deals with rural planning in Britain, but the main case study presented is an English one. The majority of illustrative case studies are from England, though some are from Scotland and Wales. The chapters refer to Britain or the UK, depending on whether particular processes or events in England, Scotland and Wales are being discussed or reference is being made to the UK as a whole or as a unit for data reporting.

Acknowledgements

We have received a great deal of support during the production of this book, starting with the team at Routledge's New York office, Nicole Solano and Judith Newlin, who provided the encouragement needed to start the project and thereafter guided us during the journey to its completion. Colleagues at Liverpool University – this time, Suzanne Yee and Tinho Da Cruz – have again done a fantastic job of producing and redrawing a variety of line diagrams, figures and maps. We are indebted, as ever, to our families: they provide the love, support and distraction needed for these 'extracurricular' projects. Team Gallent comprises Manuela, Sue, Marta and Elena. Iqbal would particularly like to thank Sarah and Jenny for their support. Meri's immediate support comes from Yuki, Marika and Joonas, but thanks are also due to the extended Juntti family. Dave has been helped, as always, by Vicky, Andrew and Matthew. And last but not least, Sue's support comes from the ever-patient Paul, Will, Francis and Anna; and the writing of this book has been lightened by the arrival of Aiofe, already a very special granddaughter.

Finally, some dedications: first, this book is dedicated to the memory of Nick's dad Ron Gallent born in Birmingham in 1922, but later a resident of Offenham in Worcestershire and various locations in Warwickshire (Stratford-upon-Avon, then Warwick itself, and finally Wellesbourne). Like many, he was a man with urban roots and rural passions, who regularly accompanied his young son on rambles across seemingly endless fields in search of the perfect fishing spot. And second, the book is also dedicated to Iqbal's mother, Dorothy: a rural dweller for most of her life from her early years in the Cotswolds and later life in Alresford, Hampshire and from whom he has inherited a love of the countryside.

Abbreviations

....

ACRE	Action with Communities in Rural England
ACV	Asset of Community Value
ANC	Area of Nature Constraint
AONB	Area of Outstanding Natural Beauty
BDUK	Broadband Delivery UK
CAP	Common Agricultural Policy
CBA	Cost Benefit Analysis
CBD	Convention on Biodiversity
CIL	Community Infrastructure Levy
CLT	Community Land Trust
COP	Conference of the Parties
CPRE	Council for the Preservation of Rural England (now the Campaign to Protect Rural England)
CRC	Commission for Rural Communities
CRoW	Countryside and Rights of Way (Act)
DBIS	Department for Business, Innovation and Skills
DCLG	Department for Communities and Local Government
DGSF	Department for Children, Schools and Families
DECC	Department for Energy and Climate Change
DEFRA	Department for Environment Food and Rural Affairs
DETR	Department of the Environment Transport and the Regions
EA	Ecosystems Approach
EAFRD	European Agricultural Fund for Rural Development
EAGGF	European Agricultural Guarantee and Guidance Fund
EAGF	European Agricultural Guarantee Fund
EC	European Commission
EIA	Environmental Impact Assessment
ELC	European Landscape Convention
ES	Ecosystem Services
EU	European Union
FAO	Food and Agricultural Organisation (of the United Nations)
GIS	Geographic Information System

GDP	Gross Domestic Product
GM	Genetically Modified
GVA	Gross Value Added
HCA	Homes and Communities Agency
HLPE	High Level Panel of Experts (on Food Security and Nutrition)
HMA	Housing Market Area
ICT	Information and Communication Technologies
IGBP	International Geosphere-Biosphere Programme
IHDP	International Human Dimensions Programme on Global Environmental Change
IMBY	In My Back Yard
IPCC	Intergovernmental Panel on Climate Change
IRDA	Irish Rural Dwellers Association
JNCC	Joint Nature Conservation Committee
LAA	Local Area Agreement
LANR	Local Authority Nature Reserves
LEP	Local Enterprise Partnerships
LNG	Liquefied Natural Gas
LNP	Local Nature Partnership
LNR	Local Nature Reserves
LSC	Local Service Centre
LSP	Local Strategic Partnership
MAA	Multi Area Agreements
MAFF	Ministry of Agriculture, Food and Fisheries
MCA	Multi-criteria analysis
MES	Millennium Ecosystem Assessment
MTI	Market Town Initiative
NCA	National Character Areas
NDP	Neighbourhood Development Plans
NDO	Neighbourhood Development Orders
NDPB	Non-Departmental Public Bodies
NE	Natural England
NGO	Non-Governmental Organisation
NHS	National Health Service
NIMBY	Not In My Back Yard
NNR	National Nature Reserves
NPA	National Park Authorities
NPPF	National Planning Policy Framework
NPPG	National Planning Practice Guidance
MNR	Marine Nature Reserves
NUTS	Nomenclature of Units for Territorial Statistics
ODPM	Office of the Deputy Prime Minister
OECD	Organisation for Economic Co-operation and Development
PCT	Primary Care Trust
PES	Payment for Ecosystem Services

PINS	Planning Inspectorate
PPG	Planning Policy Guidance
PPH	Persons per Hectare
PPP	Polluter Pays Principle
PPS	Planning Policy Statement
PSA	Public Service Agreements
RCAN	Rural Community Action Network
RCC	Rural Community Council
RDA	Regional Development Agency
RES	Regional Economic Strategies
RP	Registered Providers
RPG	Regional Planning Guidance
RSPB	Royal Society for the Protection of Birds
RSS	Regional Spatial Strategies
SA	Sustainability Appraisal
SAC	Special Areas of Conservation
SEA	Strategic Environmental Assessment
SBSTTA	Subsidiary Body on Scientific, Technical and Technological Advice (to CBD)
SHLAA	Strategic Housing Land Availability Assessment
SFP	Single Farm Payment
SPA	Special Protection Area
SSSI	Site of Special Scientific Interest
TTWA	Travel to Work Area
UTF	Urban Task Force
UNEP	United Nations Environment Programme
VDS	Village Design Statement
WCRP	World Climate Research Programme
WFD	Water Framework Directive
WTO	World Trade Organisation
WWF	World Wildlife Fund

Part 1

Rurality, rural governance and planning

1 Introduction

•••

Background and aims

This book provides an introduction to rural planning in England, Scotland and Wales (henceforth 'Britain'), but with a primary focus on England. As an introduction, it offers a way in, for new readers, to the many different subjects and issues that define the field of rural planning and confront policy making and practice. Planning in Britain can seem to be in a state of perpetual flux: the institutional and governance frameworks in which its formal planning systems are situated are forever changing, and policy is interpreted and used differently from place to place. Moreover, the boundaries of 'planning' seem to be ever-expanding, embracing the programmes and actions of many different agencies and stakeholders – from the supra-national to the very local. Planning exists in many forms and at multiple levels and is very much an instrument of shifting political priority. Making sense of the 'system' and drawing out constants, or at least key principles, is no easy task. But such is the primary aim of this introductory text: to make sense of rural planning in Britain and to reveal the continuities and shifts in thinking that have delivered the approach to managing rural change that we have today, at the beginning of the twenty-first century.

It achieves this aim within twelve chapters, each structured in a broadly similar way. The chapters begin with discussions of the central questions, challenges or debates that might confront anyone wishing to intervene in (or merely understand) matters of rural development and change. The chapters end by looking at how these same questions, challenges and debates are expressed in Britain and have shaped development and planning outcomes. The first part of this book is concerned with perceptions and definitions of rurality and with the governance frameworks and debates in which planning is situated. The three parts that then follow propose that economic change (and, in the case of rural areas, a transition away from land-based activity to other forms of production and consumption) underpins all other social and environmental outcomes, which the various aspects of policy and planning are called upon to respond to. Hence the book's core structure is a simple one, moving from baseline economic and land questions, to issues of people and communities in the countryside, and then onto debates around landscape and environmental change and management. The social and environmental parts could easily have been reversed, without any disruption to the logic of economic drivers underpinning social and environmental outcomes. This is part of the basic premise of the book: to make sense of rural planning, it is necessary to understand the economic undercurrents of rural change, on the one hand, and the broader politics of governance and planning on the other.

This introductory chapter sets the broad parameters of rural planning debate, leaving detailed practices and outcomes to later parts. So, what is rural planning, what does it do and why do we need it? Planning, wherever it takes place, is very much defined by its context and by the problems that it is called upon to help resolve. This means that a logical place to start this 'introduction to rural planning' is to unpack its constituent parts: the 'rural' and the 'planning'. Arriving at some definition of each, which can be carried forward into later chapters, is the main aim of this introduction. Beyond that, it also addresses two sets of key questions.

The first set is generic, dealing with broad understandings of rurality (as the defining context), rural challenges, rural development and the scope of planning in the countryside. The second set is focused on Britain: how rurality is perceived and defined in the nations of Britain; how these perceptions colour understanding of rural problems and how these might be addressed through public policy and planning. The intention is to provide an introduction to both general 'rural' and 'planning' issues before drilling down into the narrative of rural change and planning debate in Britain. The chapter then ends with a navigational aid for the reader, in the form of a short sketch of the content of later chapters.

Part 1: Rurality, rural development and planning

Perceptions of rural change and continuity

In the opening chapter of his 2005 textbook on *Rural Geography*, Michael Woods challenges the reader to imagine a typical rural place. He contrasts likely European, and particularly English and Scandinavian, images with those from North America or Australia. He then asks the reader to populate their images with people, with economic activities – from farming to tourism – and with the problems, social and economic, that such places face (Woods, 2005a: 3). He concludes that perceptions of rurality are culturally specific – borne of different experiences and exposure to competing ideas communicated through television and other media – and may ultimately be vague and ambiguous. In some parts of the world, rural areas may be viewed nostalgically, being seen to represent a nation's roots in an agrarian past and its heritage. Elsewhere, they are backwaters, where essential development has not been able to achieve a secure enough foothold. The rural areas of some countries are protected as recreational spaces, with policy and planning emphasis placed on conservation and the promotion of leisure activities. But in other places, the countryside is an economic hinterland, which feeds the city and forms part of the essential infrastructure for urbanisation. The rural can either be seen in opposition to the urban, or as a less densely developed and populated extension of the city. Building on the work of Lefebvre (1970), Brenner and Schmidt (2011: 12) argue the case for 'planetary urbanisation' in which '[...] spaces that lie well beyond the traditional city cores and suburban peripheries [...] have become integral parts of the worldwide urban fabric'. This means that the '[...] idea of the "non-urban" [including the rural] appears increasingly to be an ideological projection derived from a long dissolved, pre-industrial geo-historical formation' (Brenner and Schmidt, 2011: 30). In other words, there is no rural, only an extended urban. This theoretical position is motivated by a desire to reset the boundaries of urban studies, and claim firstly that 'rural' is an outdated idea and secondly that spaces are functionally inter-connected. The view taken in this book is that inter-connectivity does not invalidate the claim of geographical specificity (see Chapter 11 for an extended discussion). There are of course difficulties with the rural label; it can denote many different things and there are inevitable differences from one rural place to the next. But if the term conveys particular problems and challenges then it would seem to retain some value. Planning, as we will see later, is a spatial undertaking, dealing with problems or managing opportunities that are place-based, stemming from the rise or decline of a particular economic activity and the consequent spatial misalignment between where people need to live and where they work or receive services. It is reasonable, we think, to talk of a 'rural problem' even if that problem is differently constructed from place to place.

In very general terms, the rural areas of many advanced industrial countries (the core focus of this book) are often described as 'post-productivist', meaning that their primary function is no longer centred on food production or other 'land-based activities'. The advent of global markets means that the food or timber is being produced overseas. This can mean that the relationship between urban and rural places has altered, and sometimes that production has been substituted with consumption, and significant numbers of people residing in towns and cities now see the countryside as a leisure space – a destination for day trips, longer holidays or the purchase of second homes – or at least a non-productive space from which to commute to urban jobs or retire to in later life. Going back to images of the rural: one popular one is of lowland farming landscapes; another is of recreation spaces, for rambling or more adventurous mountain trekking, inland boating and such like. But whether food is being produced or tourists attracted, the relationship with urban spaces and urban populations is strong. And it is often the case that these consumption and production functions co-exist: the countryside is 'multi-functional' or post-productivist with an element of retained production. This might suggest an economic adjustment that has produced a new balance, between the remnants of land-based activity and a new economy centred on tourism. However, economic change is sometimes accompanied by negative social consequences or environmental degradation.

Where there are strong consumption-driven counter-urbanisation pressures affecting rural areas, families and communities reliant on traditional rural jobs, or seasonal employment in tourism, may find it difficult to compete for housing. Service provision may change to suit the tastes and needs of newcomers and quickly a social 'reconfiguration' of the countryside takes place, marked by new social and economic inequalities. This may be accompanied by either a clamour for further development, which may affect the look and perhaps the 'quality' of rural places, or by a rejection of development and change (by conservative incomers), which denies existing residents the homes they need and accentuates the process of gentrification. But where this pressure is weaker and places are being abandoned, a different set of rural problems will arise. Depopulation of the countryside is selective, depriving areas of its youngest and most ambitious people. Areas experience demographic aging and their needs change, but it becomes difficult for national or local governments to service these needs as the local tax base collapses (and other voluntary or community actions may become the only means of supporting vulnerable residents). Much depends on wider economic forces and the changing relationship between rural and urban areas, the physical connection between the two and their proximity. The problems of rural areas are often different in more affluent compared to poorer countries. And within the same country, the nature of problems varies depending on distance from urban cores. Reference is often made to the challenges of urban-fringe, accessible rural, or near-urban areas, compared to those that are more 'deeply' rural, remoter and perhaps isolated.

The different images of rural areas – ranging from the lowland pastoral idyll to the mountainous uplands of Europe; and from the prairies and vast farming belts to the wildernesses of North America – are often infused with a general awareness of the difficulties that economic restructuring has brought, triggered by industrialisation in the nineteenth century and running alongside urbanisation, and a flight to towns, in the first half of the twentieth century. Those same difficulties have been modified, in many instances, by counter-urbanisation – and a flight from towns – since the later twentieth century. In some places, this means a loss of core economic activity and an emptying of rural places, with little or no replacement migration. Elsewhere, it means diversification into non land-based economic activities and occasionally significant levels of migration. Rural areas may face abandonment; or stagnation and decay; or a replacement of former economic activity that brings profound social consequences. Likewise, they may find themselves locked into a new relationship with urban cores, or occupying a distant economic periphery where, without external assistance, they are consigned to gradual economic and social decline.

the countryside and the nature and governance of planning in rural areas. That same attribute also regularly frames discussions of rural development: how development should be promoted and managed, and how 'rural development' might be viewed as a broader set of goals, encompassing economic and community dimensions.

Rural development

Discourses and strategies for the promotion (and management) of rural development are framed by the 'either/or' context of depopulation /repopulation, by the reality of the rural comprising spaces of production or consumption (or something in between) and by the important structural and governance characteristics noted above. Woods (2011: 131) suggests that 'the aims of rural development [...] are relatively simple: sustainable economic growth and improved living conditions, bringing areas up to national standards of development, and ensuring that regions are attractive places to live and able to contribute positively to the national economy'. This aim has given rise to a push for rural 'modernisation' based on the assumption that, through a range of interventions, areas that are technologically and economically laggard can be encouraged to catch up with those that are more advanced in these respects. Such modernisation has involved different sets of interventions in the countries of the global north and global south and these have been critiqued by Woods (2011: 32–39). Modernisation involved looking at the problem of rural development from a top-down vantage point; the industry of rural areas – farming, for example – was contrasted with the industry of urban centres. The latter was modern, mechanised, and reliant on a smaller pool of highly skilled labour. The idea of 'modern farming' then took root, as highly mechanised, based on factory-principles, using the latest technological innovations in crop types and pesticides, and requiring far less labour. Traditional, low intensity farming was spurned in favour of this modern vision. Governments around the world then set about investing in new farm infrastructure and technology; and some even planned their farming industry centrally. But by the late twentieth century, the over-production of food and its wastage, along with evidence of environmental damage, led to a critique of and turning away from the modernisation paradigm. Thinking on rural development gradually became more localised and less top-down, shifting to ideas of local endogenous development with communities leading on a range of development initiatives.

This changing view of promoting and managing rural development aligned with what was increasingly thought to be a key feature of rurality: a strong social communitarianism that, conveniently, sat very comfortably with neo-liberal thinking affecting governance and planning from the 1980s onwards. A prior emphasis on top-down planning, on structural advancement (through high-level investment in infrastructure) and on modernisation as developmental convergence, gave way to: endogenous, local development;

community initiative and innovation; the mobilisation of social capital; and a divergence of solutions that celebrated local distinctiveness by, for example, commercialising local produce and tradition. These big shifts in approaches to rural development paralleled similarly dramatic changes in approaches to planning. Greater acknowledgement of the inherent complexity of the social world, and therefore of planning problems (Rittel and Webber, 1973), as a result of growing social pluralism from the 1960s onwards, is marked today by a retreat from top-down planning and a greater emphasis on seeking resolutions to social and developmental problems through engagement with social entrepreneurialism, innovation and action (Gallent and Ciaffi, 2014b). Both development and planning paradigms have been shifted by the same communitarian and neo-liberal undercurrents.

This book is primarily concerned with aspects of development that connect to planning, including the community development that is needed if planning is to become an effective instrument of local democracy, promoting solutions to development dilemmas that balance competing interests and win broad support. Planning provides a means and a tool for promoting and managing development with a view to delivering not only sustainable economic growth, but sustainable places.

The scope of rural planning

Rydin (2011: 12) has recently defined planning as '[...] a means by which society collectively decides what urban change should be like and tries to achieve that vision by a mix of means'. There is no reason why this same definition cannot be applied to rural change, and the idea of achieving a vision by a 'mix of means' – many well outside the traditional boundaries of land-use control – has particular resonance in the countryside. Planning, whether undertaken in urban or rural areas, was once very much a public sector endeavour. Many European countries established comprehensive land-use planning systems after the Second World War and charged the local state with responsibility to orchestrate and deliver reconstruction on a grand scale. But this model has since given ground to greater 'interaction between the public and private sectors' (Rydin, 2011) and more indirect means of steering and facilitating development agendas, which are often set locally and embrace the visions and priorities of different communities and stake-holders. There has been a backlash against public planning and professionalism in recent decades, and also against the 'distant administration' of urban and rural change and development (see above). In 1973, Rittel and Webber (p.155) observed that '[...] we've been hearing ever-louder public protests against the professions' diagnoses of the clients' problems, against professionally designed governmental programs, against professionally certified standards for the public services'. These protests have arisen because definitions of the 'public good', defined centrally by experts, are often disputed.

In rural areas, there is often protest against the new road that will bring 'necessary' external investment, against the modernisation of farming practices that will see a reduction in agricultural labour or the planting of GM crops, and against the new housing and industrial units that will bring new people and new jobs, and are hailed as being in the 'public interest'. Such public interest justifications for top-down intervention, and planning, seem to challenge the right of communities to set their own goals and pursue their own interests, leading to development outcomes that look to have been imposed. Today, public planning with its emphasis on intervention is often contrasted with a liberal worldview which holds that '[...] society consists, or should strive to consist of an association of free individuals, detached from imposed duties and obligations and free to form their own aspirations and interpretations of the good life' (Sage, 2012: 267). More strident neo-liberalism in the 1980s went further, seeking to emancipate not only individuals but also markets from regulatory influence, rejecting planning in its various forms. However, neo-liberalism tends to substitute professional paternalism with market tyrannies and often with the inequitable distribution of private goods. Therefore, between top-down public planning and rampant neo-liberalism, there is an opportunity space for alternative forms of interactive planning. It was in this space that 'collaborative planning' took root in the 1990s (Healey, 1997; Innes and Booher, 2010), inspired in part by ideas of communicative rationality (Habermas, 1984) and in part by communitarian theory, which posits that 'individuals belong to and participate in a wider group (or groups) of common interest and shared goals' (Sage, 2012: 267).

Planning rooted in communitarianism seems likely to relegate politically motivated and professionally formed ideas of the 'public good' behind community interest, believing that it is at the level of the community (or village) that needs can most clearly be understood and solutions to social or development problems formulated. In many countries, there has been a journey from post-war comprehensive public planning to various levels of collaborative or communitarian planning. In the countryside, this journey has often been easier as the rural 'way of life' seems to lend itself to social entrepreneurialism and action (Moseley, 2000), and public planning has often struggled to deliver its own interpretation of the public good in areas of low population density. However, although there is a different governance context for planning in the countryside – one that is often more flexible, more communitarian and more networked – public policy and planning have continued to exert a top-down influence on development, and on environmental and social agendas. Hierarchical authority (whether exerted by national government, by regional or provincial authorities, or by officially appointed agencies) provides government with a means of achieving strategic development priorities and mitigating the risk of local self-interest threatening the realisation of broader community and social goals.

Today, rural planning comprises a composite (see Table 1.1) of core public planning functions (primarily the control of development), spatial or territorial planning that brings together a wider range of public and private stakeholders (but remains public sector led; Albrechts, 2004: 747), community action and planning in various forms, countryside management (which deals with the spaces beside or between settlements and is often led by farming interests) and the programmes and projects spearheaded by government departments and agencies or by local groups (Bishop and Phillips, 2004: 4). Rural planning, wherever it happens, is broad in scope; it is undertaken by numerous agents; it comprises a range of initiatives, interventions and local actions; and it goes well beyond mere land-use planning particularly when ideas of countryside management and the role of rural areas delivering multifunctional benefits are added into the mix.

Table 1.1 The scope of rural planning

Components	Functions
Public land-use planning	• national policy; • strategic planning for infrastructure and housing; • development (settlement) planning; • land-use control and other regulatory functions.
Spatial or territorial planning	• area visioning; • co-ordination of service investments; • co-ordination of all public/private and third sector initiatives.
Community action and planning	• campaigning and lobbying; • voluntary control of services; • support for community development and social infrastructure; • community visioning; • interfacing with public and spatial planning activity.
Countryside management	• farming and stewardship; • strategies and actions that focus on the spaces besides or between physical development; • strategies for renewable energy, mineral extraction or 'fracking'.
Other projects and programmes	• governmental and pan-national directives and programmes; • departmental or agency-based (sectorial) projects around health, education, transport and so on; • development agency interventions; • private sector (industrial) programmes and initiatives.

Part 2: Rural planning in Britain

Many of the issues introduced above are reconsidered in greater detail in later chapters. The intention now is to use these various framing ideas – around rurality, rural development and the scope of rural planning – as anchors for understanding the rural narrative and challenges, development priorities, and the governance and scope of planning in Britain.

Perceiving and defining the British countryside

Satsangi *et al.* (2010) reject the notion of a 'British countryside', arguing that Britain's countrysides are very different, in terms of how they have been socially constructed and also in terms of the challenges they face. It is often the case that the rolling countryside of lowland, pastoral England dominates thinking on rurality; and the idea of the 'rural idyll' (Bell, 2006) is extended to cover Britain as a whole. This idea – or 'romantic myth' (Satsangi *et al.*, 2010: 10) – is rooted in William Blake's 'green and pleasant land' and in the later poetry and imagery – including the works of Wordsworth and Constable – that romanticised the English countryside as a place of retreat and as a 'repository' of cultural values (Gallent *et al.*, 2003). But the experience and the imagery has been very different in Scotland and Wales. Central to the

Figure 1.1

Village and fields, Great Chart, Kent

'Highland myth' and to Scottish rurality is the history of clearance and crofting; the malevolent exercise of power and the displacement of rural people from the land. This is often juxtaposed with natural grandeur, with the hardiness needed to survive in the Highlands and the past injustices that need correcting. The Welsh experience is very similar. Although some lowland and coastal areas seem to share characteristics with lowland England, the Welsh-speaking heartland (Cloke *et al.*, 1997) has its own distinctive identity, rooted in a history of displacement associated with land enclosures. Both the Welsh language and nationalist fervour have traditionally been strongest in the deeper rural parts of the country. These images of Britain's countrysides are reflected today in quite different planning problems and debates. Much of rural England is a place of retreat, which puts pressure on housing and resources. And the English often venture into Wales and Scotland (and indeed into mainland Europe) because of this shared cultural image of the countryside. In Scotland, the history of forced eviction has given impetus to land reform and far greater acceptance of communities' rights to develop land. Wales sits somewhere in the middle; its own battle against malevolent landlordism was not quite so bitter or transformative, but still there is a sense of a culture under threat from new patterns of rural consumption.

But these perceptions, or social constructions, of rurality are frequently contested and certainly unbounded. Significant effort has been expended on defining and delimiting rural areas across Britain. This effort is premised on there being distinctive challenges for public policy and programmes in rural areas. A general narrative of change in the countrysides of many advanced industrial nations was suggested above; the detail of change in Britain is introduced below. Broadly, there have been both national and global shifts in the focus, location and mode of economic production since the industrial revolution, changing the function of Britain's countrysides and bringing a mechanisation of surviving farming activity. These processes caused a significant decline in the rural population during the twentieth century. Although the number of people living in the countryside (i.e. the countryside or countrysides of Britain) was stable between the mid-nineteenth century and the 1960s (Allanson and Whitby, 1996), its relative share of population plummeted as cities grew during the nineteenth century. There have since been some increases – owing to a range of counter-urbanisation pressures – but four-fifths of Britain's population now resides in urban areas (see Table 1.2). Lower population densities, the slow decline of the conventional farming sector, and more recent population movements are the drivers of a number of social and environmental challenges that confront the various aspects of rural planning listed in Table 1.1.

That the countryside is different, and subject to a particular set of pressures, has prompted periodic efforts to map rurality. Paul Cloke's Index of Rurality (Cloke, 1977) brought together a range of key statistics – on population density, percentage economic activity in primary industries and commuting patterns – to generate a spatial index. Added to these core indicators were data

Table 1.2   Population change in rural Britain, 1951 to 2008 (Satsangi *et al.*, 2010: 60 [where various issues with these data are noted])

| | England and Wales | | Scotland | |
	Rural population	% of total population	Rural population	% of total population
1951	8,193,000	19.7	1,529,506	30
1961	8,954,000	19.4	1,528,723	29
1971	10,568,000	21.7	1,516,936	29
1981	11,320,000	23.1	715,342	14
1991	10,073,963	21.2	538,341	11
2001	10,460,900	20.1	880,268	17
2005/6	10,606,815	19.9	n/a	n/a
2008	n/a	n/a	954,009	18

Note: n/a = not applicable.

on age structure (because rural areas tend to be demographically older as a result of selective out-migration), household amenity, housing occupancy rates and distance to larger settlements. His first index drew heavily on the 1971 Census. An amended version that appeared nine years later (Cloke and

Figure 1.2

Vernacular house and fields, Bilsington, Kent

Edwards, 1986) used statistics from the 1981 Census, but was also more self-critical, concluding that 'rurality' is not merely an amalgam of key data: it is not possible to capture the lived experience of the countryside in a single figure. Since the 1980s, researchers have tended to shun purely functional delimitations of rural areas, accepting instead that 'rural is what people recognise as rural' (Troughton, 1999). Perhaps more pragmatically, Hoggart (2005) has argued that despite the claim of functionality in the statistical profiling, delimitations of rurality often neglect the functional connectivity between town and country. The economic reality in much of Britain is that rural areas are dependent on cities and vice versa. They are joined within a single labour or housing market, or supply people with complementary opportunities – e.g. jobs, housing and recreation. Despite these debates, there is now a generally accepted statistical (and functional) delimitation of rural England (and equivalents for Wales and Scotland). This was first produced by a team of researchers from the University of Sheffield and Birkbeck College in 2004 on behalf of a number of English and Welsh government departments and agencies. It is a delimitation that emphasises population density and the concept of 'sparsity'. The starting point was a division of England and Wales into hectare grid squares and the association of each square with a '[...] particular settlement type: dispersed dwellings, hamlet, village, small town, urban fringe and urban (>10,000 population)' (Countryside Agency *et al.*, 2004: 4). By bringing together the concept of settlement sparsity with population data from the 2001 Census, the team were able to conclude that of the 20 per cent of people living in identifiably rural areas, nearly half reside in small towns within the rural–urban fringe (i.e. close to cities) whilst the other half reside in villages or in smaller hamlets, clusters of a few homes or in isolated dwellings.

An updated mapping of this index, drawing on 2011 Census data, is shown in Figure 1.3. It provides a visualisation of darker urban places and lighter sparse or less sparse rural places. If it is accepted, broadly, that rural areas are different and that planning – broadly defined – may need to be tailored to rural situations, then the map indicates where a 'rural' approach might be needed. But it says nothing about the local challenges. These are more likely to be captured by 'typologies' of rural places that attempt to highlight critical spatial differences. Terry Marsden has argued that typologies should try to capture the increasingly '[...] polyvalent rural scene and regulatory structure' (Marsden, 1998: 107). His own typology (Marsden *et al.*, 1993), focusing on economic opportunity and power is outlined in Chapter 3. Another more recent one, also described in that chapter, draws on a selection of economic data and rurality indices to construct a series of descriptors that infer key employment and social challenges (Lowe and Ward, 2007). Some, for example, are said to be 'deep rural', economically fragile and unattractive to migrants. Others are 'dynamic' and often the hinterlands of knowledge-hubs (i.e. university towns) or 'retirement retreats' with a specific set of service challenges. Figure 1.4 shows this particular typology, although it is discussed more fully in Chapter 3.

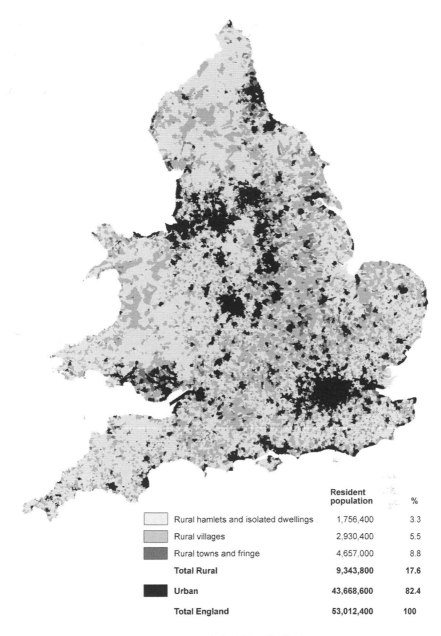

	Resident population	%
Rural hamlets and isolated dwellings	1,756,400	3.3
Rural villages	2,930,400	5.5
Rural towns and fringe	4,657,000	8.8
Total Rural	**9,343,800**	**17.6**
Urban	**43,668,600**	**82.4**
Total England	**53,012,400**	**100**

Source: 2011 Census, Rural-Urban Classification
Contains National Statistics data ©Crown copyright and database right 2015
Contains Ordnance Survey data ©Crown copyright and database right 2015

Figure 1.3

Rural–urban classification, showing population figures and splits for England only, 2011

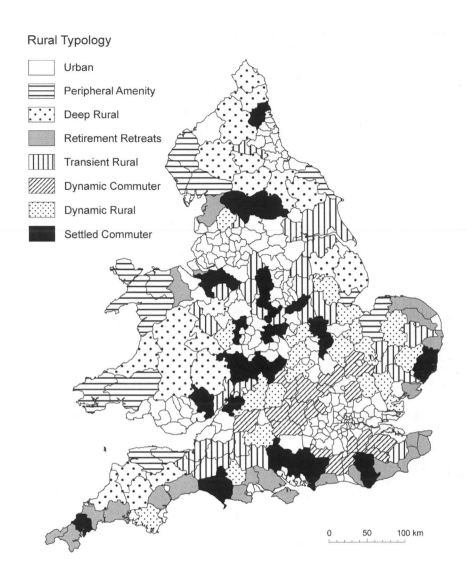

Rural Typology

- ☐ Urban
- ☰ Peripheral Amenity
- ⋮⋮ Deep Rural
- ▨ Retirement Retreats
- ⦀ Transient Rural
- ▨ Dynamic Commuter
- ⋮⋮ Dynamic Rural
- ■ Settled Commuter

0   50   100 km

Figure 1.4

A typology of rural England and Wales, 2001
Source: Lowe and Ward, 2007

These typologies pinpoint some of the challenges that parts of the countryside face and in doing so they also capture something of the narrative of rural change in Britain, particularly the economic transition, population shifts, and consumption and service pressures. These are of course dealt with in detail in the chapters that follow, but they are now briefly introduced.

The narrative of rural change in Britain

It was noted above that rural narratives in the global north tend to follow the sequence of depopulation, abandonment and poverty, or repopulation, social transformation and inequality. Britain has had both of these narratives. Much of the countryside suffered relative depopulation until the 1950s; and some of it then enjoyed or endured a period of repopulation. A fundamental economic transition, marked in particular by changes in farming during the twentieth century, has underpinned this narrative shift. Before 1939, the British farming economy seemed to be in state of terminal decline; support for it was not a political priority and it faced stiff competition from foreign grain and livestock producers. The country was increasingly dependent on imported food. But the Second World War highlighted Britain's lack of 'food security'. A wartime blockade of British ports, along with attacks on the Atlantic convoys bringing supplies from North America, caused severe shortages. Farmers intensified production and led a 'dig for victory' campaign aimed at increasing the supply of basic foodstuffs. Their role in feeding the population during the war years gave them a 'moral account' from which to draw, after the war ended in 1945. Support for British farming and future food security became key post-war priorities, delivered through the Agriculture Act 1947, which established a system of revenue support and capital grants for farmers. The result was rapid mechanisation and a post-war emphasis on maintaining a productive countryside. An intensification of farming followed, which became mixed with new economic activities from the 1960s onwards, as recovery from post war recession and austerity gave rise to higher incomes and greater economic confidence: more people started taking holidays in the countryside and some even invested in rural property. But still, the intensification of farming on an industrial scale (and model) continued and was given further impetus by Britain's accession to the European Economic Community in 1973 and consequently by access to the system of grants administered under the Common Agricultural Policy (CAP). However, subsidy-driven production and over-production, along with its environmental outcomes, ultimately led to a turning away from this 'modernisation' paradigm (see above) towards more localised and endogenous models of development. Economic recovery from the 1960s brought new people and new opportunities to rural areas, and eventually triggered economic diversification centred on consumption and new modes of localised production; organic, 'slow' and so on.

The British countryside became economically and socially differentiated after the initial post-war period of intensified farming activity (Marsden *et al.*, 1993). This transition has been explored at length by Newby (1979) who points to uneasy co-existence between new rural residents and traditional farming groups from the 1960s, with the former acquiring ever-greater economic and political power. New forms of consumption – including the purchase of second homes, lifestyle downshifting, retirement and commuting from rural areas to urban jobs – generated new social pressures and tensions in the countryside, which were exacerbated by the scarcity of rural resources (principally housing) that resulted from strong post-war comprehensive land-use planning, which tended to release land for development in key settlements but not in small village locations (Gallent, 2009). This resulted in rural gentrification (Gallent and Robinson, 2011) and a widespread social reconfiguration of the countryside (Hoggart and Henderson, 2005). Many newcomers, with their income and wealth derived from urban jobs, found that the countryside delivered the promise of the rural idyll. But others, reliant on low-paid farming jobs or seasonal employment associated with tourism, came to endure a number of disadvantages: centred on employment and income, education (because of fewer training opportunities in rural locations), housing (because of its scarcity, their lower incomes, and competition from 'adventitious purchasers'; Shucksmith, 1981), health (especially stress-related illness), retail (owing to likely distances from the nearest shop or post office) and transport (or more broadly 'accessibility' because of disinvestment in public transport and high levels of car-dependency) (Commission for Rural Communities, 2006c). Over the last few decades, the countryside has become socially diverse, with many competing (and sometimes conflicting) interests and needs. It has become a more complex place for rural planning, whereas it was once marked out by long-standing social stability and homogeneity (Martin, 1962).

The environmental and landscape narrative of the British countryside follows the economic lead. Britain's landscapes are a product of human contrivance; they are man-made. And in the post-war period, they were very much a product of the intensive farming described above. In lowland England, the post-enclosure 'patchwork quilt' of relatively small fields was disrupted by the removal of hedgerows as farms 'industrialised': larger farm machinery, such as combined harvesters, required bigger open fields. In many upland landscapes, visitor pressure increased during the same period, introducing new threats and new management challenges. But there are two principal themes in the environmental narrative. The first is diversity: Britain's countrysides range from more rugged uplands to chalk downlands; from the semi-wild (but never wholly 'natural') to the heavily farmed; and from the relatively remote to the rural–urban fringe. The second theme is landscape change, tied to the fortunes of farming. In the farming-centred rural planning debates immediately after the Second World War, it was acknowledged that the cultivation of crops and keeping of livestock had been instrumental in making the landscape over at least a thousand years, and without the continuation of such activity – and the 'stewardship' that farming provided – the iconic rural landscape would be lost. Yet this rationale for farming began to unravel as over-production in the 1970s and 1980s started to change the landscape and threaten economic diversity, particularly diversity into tourism.

These three underpinning narratives provide the backdrop to our focus on economy and land, people and communities and landscape and environment in Parts 2, 3 and 4.

Debating rural development in Britain

What does development mean in rural Britain? It is important to unpack the many dimensions and understandings of development as it provides a significant focus for planning. Different sorts of development are promoted, occur, and are steered and managed by different aspects of planning and public policy. It was noted in the first part of this chapter that rural development is often equated with economic modernisation and growth; it is a transition from being backward to catching up, orchestrated by the state or some high-level agency. There was a strong focus on modernisation after the war, first promoted by the Agriculture Act 1947 and then by the Common Agricultural Policy a few decades later. But this emphasis on centrally steered modernisation has since given way to more localised forms of private enterprise and social entrepreneurialism, aimed at realising development that is appropriate to local circumstances and serves community needs rather than national ambitions, though it is hoped that the former will deliver against the latter. There has been a shift, in broad rural development terms, from the exogenous and general to the endogenous and specific. This same shift has been witnessed across much of Europe and the global north. Planning can facilitate and manage this type of development through its land-use, spatial, community-based, countryside management, or programmes/projects dimensions (see Table 1.1); the British characteristics of which are examined below.

But development has other important meanings in Britain. As well as suggesting an economic transition, it also points to the physical outcomes that underpin or are consequential to that transition; hence the development of new infrastructure (particularly roads), industrial units or additional housing. These physical things can be called development or development outcomes. They are often contested – in location, form or quantity – as rural residents have different needs and therefore varying interest in achieving these outcomes. Again, Rittel and Webber (1973) have observed that in an increasingly pluralistic society, there is no such thing as the uncontested public good. Whilst some groups will see new housing as essential; others will reject it as unnecessary. This seems to be particularly true in the English countryside, where gentrification has taken hold in many locations and where many new residents have no direct interest in new housing, sometimes fearing that it will have a detrimental impact on village character or on existing property values.

Understandings of development commonly focus on the physical. Regulatory planning views it only as operations on, under or over land that result in land-use changes or material changes to existing infrastructure or buildings. In England, it is asserted in the Ministerial Foreword to the National Planning Policy Framework (NPPF) that 'development means growth' (DCLG, 2012: i) and is about earning a living, housing a rising population and not stagnating. Although not specific to rural areas, it suggests a focus on the physical and on having more. Planning is assigned a role in facilitating that growth, and regulating it to achieve 'sustainable development', defined simply as 'change for the better' (DCLG, 2012: i). This slight broadening of the definition points to softer development outcomes. Cadman and Austin-Crowe (1991) have defined development as the transformation of land and property, leading to the replacement of one use with another that is more profitable or *socially valuable*. The idea that development can or should have social value, at scales from the local to the global, is critically important. Development that delivers social value has a positive (or neutral) environmental impact, helps support sustainable economic wellbeing (and not necessarily growth), and concerns itself with the way in which communities function and are sustained. Specifically, it may mean delivering low-impact or zero-carbon housing (Williams, 2013); supporting a diversification away from intensive farming towards local craft industries through the conversion of buildings; and the protection and enhancement of community assets and 'interactional infrastructure' (Kilpatrick *et al.*, 2014). 'Community development' is a crucial aspect of the rural development agenda. It is not only achieved through the development or protection of physical things, such as pubs or village halls, where people can meet and interact – but these things are important.

Planning has an important role in promoting 'change for the better'; it is able to work across the environmental, economic and social dimensions noted here; and the 'mix of means' (Rydin, 2011) that constitute planning, in urban and rural areas, all have roles in promoting and steering that change.

Means and scales in rural governance and planning

Rural planning is a composite of five key components: public land-use planning; spatial planning; community action and planning; countryside management; and a range of other projects and programmes. These are the mix of means through which change is managed and delivered. Throughout this book, these different means are shown to produce particular outcomes and to interact in certain ways. And each operates across a specific band of scales. Land-use planning in England has recently been localised, as we shall see below. Spatial planning is a broad, multi-scalar activity with multiple dimensions. Community action and social entrepreneurialism is highly localised. Countryside management is broad and specific, operating at a landscape scale but also through local initiative. And other projects and programmes can be initiated by a wide range of actors at many scales. These components are briefly introduced here and examined in greater detail in the next chapter.

Comprehensive **land-use planning** in England and Wales was introduced through the Town and Country Planning Act 1947. That legislation, and the system of local plan-making and development control that it created, was a hundred years in the making. The pressure to exert greater political control over 'market processes' began to build in the middle of the nineteenth century as rapid industrialisation and urbanisation generated a number of problems centred on poor public transport, uncoordinated infrastructure provision, bad housing, and declining public health. At first, these problems were tackled through private philanthropy and the projects of the great Victorian social reformers: Ebenezer Howard's 'Garden Cities' – planned new towns on the edge of London – were perhaps the first comprehensive attempt to deal with the challenges of urbanisation, or at least demonstrate the potential of good land-use planning. Legislation for the whole country was then rolled out in 1909. The Housing, Town Planning, Etc. Act tasked local authorities with the preparation of 'schemes' to suggest how land might be used for new housing development. But – and very critically – without the power to control development (or interfere with the right to develop land) local authorities could only suggest how land might be used, but had no direct influence over that use. The right to develop remained a private right and until that changed, there could be no 'comprehensive' planning.

This would be a massive change requiring a huge catalyst. That catalyst came in the form of the Second World War. National mobilisation delivered greater 'acceptance of the need for economic planning' (Simmie, 1994: 2) and the scale of reconstruction that would be needed after the War galvanised support for stronger land-use control. In 1944, the wartime government published its planning White Paper – 'Control of Land Use' – which set future planning the goals of advancing public interest, protecting private property, and balancing social, environmental and economic objectives. The 1947 legislation which then followed was largely concerned with steering and controlling land-use change. It handed two key responsibilities to 'planning authorities' (County and Borough Counties at that time, generally covering rural and urban areas respectively): the drawing up of a 'local plan' and the subsequent control of development, in line with the objectives and policies of that plan.

In short, authorities were to engage in plan-making and development control. They did so, for the next 20 years, in a strategic vacuum. Then, in 1968, government legislated for the creation of a higher-tier of 'Structure Planning': essentially strategic plans produced by County Councils (with development planning dropping down to district councils in most rural areas). These started to appear in the 1970s and identified broader infrastructure needs and major housing sites. Local plans needed to be in conformity with the Structure Plans, and so began a process of lifting decision-making away from elected local authorities. This process continued with the advent of regional plans in the 1990s and then their 'legal' incorporation into local plans in 2004. A greater proportion of planning outcomes were being determined at a level ever-more remote from affected communities. And all this was happening whilst society became more plural (see above) and needs more diverse and fragmented. The upward rescaling of land-use planning appeared out of kilter with the social and community pressures that should, perhaps, have resulted in a downward rescaling. There were signs in the early 2000s that government acknowledged this tension, but it was only when a Coalition of the Conservatives and Liberal Democrats replaced the incumbent Labour Government in 2010 that a more radical downward rescaling of planning began.

But of particular note here is that the 1947 system should have been about more than land-use control. Bruton (1974: 15) credited the system with other '[...] limited and unstated social and economic objectives' including the delivery of (spatial) equalities of opportunity, urban renewal, the promotion of good transport infrastructure, and the conservation of natural resources and heritage. Immediately after the war, planning was involved in the delivery of comprehensive development in the New Towns, but as the volume of public development was eclipsed by growing private enterprise, the system's narrower control function seemed to become its main *raison d'être* (Gilg, 2005). There was great disappointment in the operation of land-use planning, and by the 1990s the need for the system to embrace 'broader goals or outcomes in terms of more sustainable development' rather than very narrow outputs – e.g. a specified number of new homes in a given area, without really thinking about how these might contribute social value – became a dominant discourse in planning (Nadin, 2007: 48).

This discourse centred on the idea of **spatial planning** (Tewdwr-Jones *et al.*, 2010), concerned with the co-ordination of various policies, programmes or actions that deliver against the objectives agreed by different partners. Land-use decisions and actions are only one possible means through which spatial planning ambitions might be realised. Others could include the co-ordinated investments of development agencies and their private partners, or the decisions over the siting of schools or medical centres taken by a County Council or local Primary Care Trust (PCT). Building on the earlier work of Kunzmann (2000) and Healey (1997), Albrechts (2004: 747) notes that whilst spatial planning is generally public-sector-led, it should be seen as a broader 'socio-spatial process through which a vision, actions, and means for implementation are produced that shape and frame what a place is, and may become'. This is very much what spatial planning became under the Labour Government during the 2000s. A series of Local Government Acts established Local Strategic Partnerships (LSPs comprising different local authority departments, business and third sector partners and community groups), the need to develop a 'Community Strategy' for an area – essentially a 'vision [of] what a place is, and may become' – and a requirement for partners to sign a Local Area Agreement (LAA) committing them to take the concerted actions needed to deliver that vision. Planning under Labour became 'spatial planning': broad in its concerns and partnerships; integrative in its approach;

extensive in its territorial reach; multi-sectoral and professionally inclusive. However, spatial planning seemed, during the 2000s, to morph into something highly nebulous, punctuated by endless acronyms and framed by complex local government and partnership frameworks which seemed to some community actors to be opaque and impenetrable (Gallent and Robinson, 2012a). But worse still, community representation was thought to be weak and confined to quite high-level networks or representative bodies (e.g. associations of community councils). Despite the laudable intention to broaden the system's scope and reach, Labour's reforms appeared to further bureaucratise and jargonise the planning process.

The term 'spatial planning' is omitted from the National Planning Policy Framework for England (DCLG, 2012). It has been erased entirely from all government policy and web resources. Although spatial planning continues to be a reality – different agencies are today co-ordinating their programmes and actions under the umbrella of Local Enterprise Partnerships (LEPs) – it is rarely mentioned. There is a new focus in the planning system in England; not on the co-ordinated activities of agencies and experts but on the power of communities to take greater control of planning decisions and local outcomes.

Spatial planning is perhaps tainted by the rescaling of statutory planning that happened under Labour. Hence, there is an emphasis today on planning for communities, by communities. This has been packaged by the Coalition inside a wrapper of localism and neighbourhood planning. But it builds on an inclination towards **community action and planning**. This has always existed but has gathered pace since the 1960s (Gallent and Ciaffi, 2014b). Disillusionment with the remoteness of public planning, whether land-use or spatial, and its seeming disregard for local circumstances have been key drivers of community action. And this action has been particularly important in rural areas or in urban locales seemingly forgotten by public investment or unattractive to private enterprise. Although neighbourhood planning at the level of parishes has now become a key feature of planning activity in the countryside, since the Localism Act 2011, it is in fact rooted in more than half a century of village-level planning (Owen and Moseley, 2003: 445) and over the last 20 years, '[...] there has developed a pattern in which the planning system has given some form of formal "credence" to plans produced at the community level, be they parish plans or village design statements' (Bishop, 2007: 343). Although the contribution of communities to steering and shaping change in the countryside extends well beyond formal or even informal planning, it is perhaps worth briefly tracing the evolution from parish to neighbourhood planning. The original parish plans were in fact simple 'village appraisals' undertaken in the 1970s which catalogued various local attributes (community assets, key socio-economic data), often with the help of support groups or others with a keen interest in the vitality of rural communities (Gallent and Robinson, 2012a: 40). These evolved, by the 1990s, into open-ended Parish Plans that went beyond the cataloguing of problems or statistical profiling, arriving instead at wish-lists of changes and interventions. Their authors were often motivated by a desire to influence land-use plans. The messages were simple: don't close the post-office; build more affordable housing; protect this or that site from development as it is a valued community asset; etc. and so on.

A rural White Paper at the turn of the millennium (DETR and MAFF, 2000) gave further impetus to these plans by suggesting that 'communities could play a much bigger part in their own affairs and shaping their future development' (DETR and MAFF, 2000: 145). More specifically, it offered advice on the form of Parish Plans (DETR and MAFF, 2000: 146 and 150) hence giving them some degree of official endorsement. A sense that this type of community plan was of value, not only because it might link to local land-use planning but also because it assisted in community development, grew during the 2000s and the possibility of better integrating them with local plans was soon being highlighted by the Countryside Agency (2004: 14). Yet ambitions towards integration or alignment can obscure the fact that community planning is rooted in a very different model of government from all of the public-led planning described above. Whilst 'strategic frameworks are prepared mainly through the processes of representative democracy – by elected bodies working through formal democratic structures as part of their continuous representative responsibilities' community action and planning is '[…] undertaken mainly through participative democracy – by self-identifying groups of articulate residents pursuing their own objectives' (Owen *et al.*, 2007: 55). Neighbourhood Development Plans, examined at various points in later chapters, can be seen as an extension of the Parish Plans that evolved in the 1990s. But they are subject to a greater degree of 'instrumental framework' (their general conformity to local plans needs to be demonstrated) to ensure a tie-in between the ambitions of community actors and the strategic priorities of local authorities. That said, a different model of governance underpins the production of these plans and drives broader community actions, often in opposition to public policy. What happens at the community-scale is now being accepted as an important and formal component of rural planning.

The components listed above focus, principally, on human settlement or physical aspects of rural areas. Equally important for these areas is **countryside management**, dealing with the spaces on the edges or in between settlements. Environmental and landscape changes affecting rural areas have a variety of global, European and national policy drivers, with policy frequently attempting to frame the countryside as a multi-functional space, in which opportunities to enhance biodiversity, deliver climate change adaptation and mitigation, improve water quality and alleviate flood risk should go hand-in-hand with other forms of development. At the same time, the traditional landscapes of the countryside, with their inherent biodiversity, are viewed as a precious resource for future generations. Exploiting and protecting 'nature' have become the twin goals of countryside management, though this aspect of rural planning often struggles to reconcile beneficial exploitation – from conventional farming to energy crops and from wind farms to fracking – with protection aimed at maintaining current amenity: e.g. access to rural land, aesthetic value and quiet enjoyment. It is countryside management, in broad terms, that needs to bring together different interest and users of rural spaces, often attempting to reconcile projects such as onshore wind farms – that have a huge climate change mitigation potential – with community and societal interests in maintaining countryside character. The role that the countryside can play in addressing global challenges makes it an increasingly contested space in which global goals and local ambitions conflict. Today, a broad range of actors are involved in countryside management, from communities and local interest groups, to farmers and institutional land-owners, all the way to government agencies and supra-national bodies which set the directives and frameworks that determine the nature of that management.

The final component, linking strongly to countryside management, comprises a mix of **programmes and projects** (and framing directives), initiated at different scales by a broad range of actors. At the level of the European Union (EU), various grant schemes operated under the auspices of the Common Agricultural Policy have had a profound impact on the development of rural areas, focusing initially on agricultural production and later on a wider set of environmental goals. The EU's 'INTERREG' (inter-regional) Structural Programmes have channelled large sums of money into rural areas judged to need significant investment in their infrastructure. These programmes (and the projects taken forward) have been very much top-down, but other EU initiatives – including various rounds of the LEADER (Liaison Entre Actions de Développement de l'Économie Rurale) programme – have lent support to local initiative, funding partnerships that promote 'integrated, high-quality and original strategies for sustainable development'. It is also the case that 'Europe' frames aspects of British planning and conservation policy through various EU Directives: for instances, the Birds Directive (1979) requiring the designation of Special Protection Areas (SPA) and the Habitats Directive (1992) leading to the designation of Special Areas of Conservation (SAC). Some EU programmes support projects that are defined and delivered by a range of national ministries and agencies. But these national bodies also run their own rural initiatives.

In England, the Department for Environment, Food and Rural Affairs (DEFRA) operates a number of grant schemes that directly promote enterprise and rural training, amongst many other things. Other 'non-rural' ministries affect the planning of rural areas either through the exercising of direct planning powers (in the case of DCLG) or through specific programmes that promote or fund training, schools and healthcare. Furthermore, a number of Executive Non Departmental Public Bodies (NDPBs) or Executive Agencies, attached to DEFRA or DCLG (in England) or the Welsh or Scottish Governments operate a number of rural programmes or undertake different projects that impact on rural areas. These include, or included, Natural

England (with various agri-environmental or conservation projects), the Commission for Rural Communities (closed in 2013), the Environment Agency (preoccupied recently with flooding and now undertaking the dredging of the Somerset Levels), the Homes and Communities Agency (HCA, with its responsibility for housing and regeneration), the Forestry Commission, the Planning Inspectorate and many other agencies and executive bodies which are introduced later in this book. Together, the various high-level bodies – all organised nationally – co-ordinate a number of interventions that affect parts of the countryside. Other agencies often take forward these interventions at a local level, including Rural Community Councils (RCC) with their focus on community development and the Registered Providers (RP, perhaps better known as housing associations) of social housing funded by HCA grant or developer contributions levered by local authorities. The list of agencies and groups – public, voluntary and private – likely to be undertaking projects in rural areas is extensive. Some are community spin-offs, and these might include land trusts. Others operate across districts or counties, and include the RCCs. And others may be part of the new economic governance of the countryside, including Local Enterprise Partnerships, which are increasingly seen as sub-regional successors to the Regional Development Agencies dismantled by the Coalition Government after 2010.

It is extremely difficult to set fixed boundaries around rural planning. Decisions and actions are being taken at many levels and by a broad spectrum of actors. There are many influences on the trajectories of development and change in the countryside, though the separation of planning into five main areas seems to capture the breadth of this influence, and these areas are returned to in each of the chapters that now follow.

A navigational aid

This opening chapter has provided both an overview of generic concerns in rural planning – from the question of what is 'rural' to issues of development and the scope of that planning – and a meta-narrative of change and intervention in Britain's countrysides. In the next eleven chapters, that narrative is developed further. Chapter 2 begins by exploring the broadening scope of rural planning across England, Scotland and Wales and its embeddedness in the changing structures of governance. Planning has been subject to dramatic rescaling over the last few decades, in response to significant socio-political pressures and as part of a broader rescaling of the state. Those pressures are largely economic in origin, with changes in the productive function of the countryside triggering key transformations and the claim that 'society has been completely urbanised' (Lefebvre, 1970). In Chapters 3 and 4, how urban–rural relationships have changed and how it might be argued that the rural areas have been 'urbanised' is examined. This analysis underpins two parallel narratives. The first is concerned with People and Communities:

Chapter 5 looks at the nature of rural society and communities, unpacking the inclination towards communitarianism and community-based planning introduced in this opening chapter. Chapter 6 then looks at the challenges – specifically accessibility and equality of opportunity – that those communities face, before Chapter 7 considers how social entrepreneurialism – built on a foundation of community governance and social relations and capital – contributes to addressing at least some of these challenges. This first narrative then ends with an examination of housing supply and pressure, contrasting normative planning responses that are centrally steered with those grown from communities, and now framed by ideas of 'localism'. The second narrative is concerned with the rural landscape and environment; with the economic and social determinants of change, and with planning's multiple influences on both the natural and built countryside.

All the narratives are of course unfinished so it seems appropriate that the book should end with speculation on the paths ahead; on how the countryside, and the influences on its development, may change during the course of this century.

Summary

- This textbook aims to make sense of rural planning in Britain and to reveal the continuities in thinking that have delivered the approach to managing rural change that we have today.
- Some writers have questioned the validity of differentiating between urban and rural areas, but the latter have their own representations, functions and political-economies; they present planning with unique challenges.
- The principal challenge is one of 'development', which is a process, a desired goal and an outcome. It is grounded in political discourse, contested and ultimately managed by planning, broadly defined.
- Rural planning has five components; land-use control; spatial planning; community action and planning; countryside management; and multi-level programmes and projects.
- Britain's countrysides have experienced episodes of depopulation and repopulation, at different degrees in different places. These countrysides have transformed over the twentieth and into the twenty-first centuries. Significant economic and social changes have occurred, delivering big challenges for planning and also bringing into question the appropriateness of rural modernisation and top-down interventions.
- The current role and functions of Britain's countrysides is highly complex and contested, meaning policy interventions are debated as many of the dilemmas facing the countryside can be described in Rittel and Webber's terms as 'wicked' problems.
- The scope of planning in Britain and its embeddedness in new scales of governance has become a critical discourse.

Key readings

- Woods, M. (2005) *Rural Geography: Processes, Responses and Experiences in Rural Restructuring*, Sage: London.
  (This book provides a broad overview of rural issues with examples of development or planning dilemmas from around the world.)
- Woods, M. (2011) *Rural,* Routledge: London.
  (Key readings on representations of the countryside and on rural development.)
- Rydin, Y. (2011) *The Purpose of Planning*, Policy Press: Bristol.
  (Offers critical reading on the purpose of planning with a chapter dealing with the countryside.)
- Satsangi, M., Gallent, N. and Bevan, M. (2010) *The Rural Housing Question: Communities and Planning in Britain's Countrysides*, Policy Press: Bristol.
  (Looks at representations of the countrysides of Britain and at the use of data to delimit the countryside in its appendix.)
- Cloke, P., Marsden, T., and Mooney, P. (eds) (2006) *Handbook of Rural Studies*, Sage: London.
  (Broad encyclopaedia of rural issues from an international perspective.)

Key websites

- DEFRA: www.gov.uk/government/organisations/department-for-environment-food-rural-affairs.
  (See also Chapter 2 for a full description of DEFRA's function.)
- DCLG: www.gov.uk/government/organisations/department-for-communities-and-local-government.
  (See also Chapter 2 for a full description of DCLG's function.)

2 Approaches to rural planning

....

A changing scene

This chapter revisits the narratives of rural change introduced in Chapter 1, linking these with shifting rural planning paradigms. As noted previously, *context* – economic, social and environmental – is critically important for the development of all forms of policy intervention and determines the nature of planning responses brought to bear on a range of rural challenges. Part 1 provides a brief and eclectic overview of changing approaches to rural planning, set within the context of economic and social change and shifting political ideologies and views on the role of the state. This is a greatly simplified and partial account but it conveys an important message: that the forces shaping rural areas and rural planning approaches are multi-dimensional, closely interrelated and constantly evolving. Part 2 then begins by mapping these trends against the British experience and highlights a second major theme – that while rural planning approaches are increasingly embedded within the wider international scene, at the same time they are shaped by country and locally specific contexts, and great diversity in experience is evident. That diversity of experience is now framed by the many agencies – supra-national, national, regional and local, and also governmental or non-governmental – that create the institutional context for rural policy and strategy. These are examined later in Part 2, before a final consideration of statutory planning in Britain, which builds on the introductory remarks from Chapter 1. Many structures and processes are introduced in this chapter, with their impacts on dealing with different rural challenges becoming apparent in later thematic chapters.

Part 1: Forces for change and shifting rural planning paradigms

Economic change

The changing economic context of rural areas has been a central factor influencing the evolution of rural planning over the past 100 years or so and there is no doubt that a radical reshaping of rural areas has taken place, driven to a large degree by wider patterns of modern economic growth.

From the late nineteenth century through to the 1960s, the industrialisation of Western economies had far-reaching consequences for rural areas. On the one hand, it brought new technologies, expanding urban markets and reducing transportation costs that transformed the character of agriculture – traditionally the core of the rural economy. Supported, particularly in the post Second World War years, by public policy with a productivist view of the rural economy, it enabled increased mechanisation and specialisation of farming, accompanied by a growing capital intensity of agriculture and major improvements in agricultural productivity and output, but also dramatic falls in agricultural employment. On the other hand, in many rural areas, industrialisation ushered in an era of rural decline as non-agricultural economic activity and employment increasingly centred on towns and cities and the forces of cumulative causation and economic agglomeration came to the fore.

By the 1960s, 1970s and 1980s, however, a new dynamic in rural economies was being detected in the form of rural manufacturing and service sector growth. There was increasing evidence that the economic structures of urban and rural areas were converging, at least in terms of the balance of employment. This sparked much debate as to whether a fundamental rural restructuring was taking place in tandem with the development of a post-industrial or post-Fordist economic and social model (Cloke and Goodwin, 1992; Hoggart and Paniagua, 2001). It was argued that new information and communication technologies (ICT) were eroding the locational advantages of urban areas for manufacturing businesses, which were declining anyway in many Western economies. Instead service sector growth associated, for instance, with rural tourism, recreation and residential development was occurring in response to a growing trend of urban people seeking respite from the stresses of urban living. It was at this time that a new 'consumptionist' view of the rural economy began to emerge. It was evident, however, that the extent of this restructuring was spatially variable, with those areas with strongest physical or psychological connections with urban areas experiencing the most significant change (Terluin, 2003). For more remote and peripheral localities, agricultural employment remained central and the pattern of rural economic decline seemed more entrenched than ever.

Table 2.1 overviews this changing economic scene with reference to shifting employment patterns in some of Europe's rural uplands between 1850 and 2000. It is interesting to note that the economic trends described

Table 2.1 The changing employment structure of selected European uplands (Collantes, 2009: 311)

	Agricultural employment (%)		When did non-agricultural employment rise above 50% (⇧) and 75% (↑)?			
	1980/2	2000/1	1850–1913	1913–1950	1950–1980	1980–2000
Scottish Highlands	11	4	⇧50	↑75		
Swiss Alps	9	5	⇧50		↑75	
French Alps	8	3		⇧50	↑75	
Italian Alps	15	4		⇧50	↑75	
Apennines	36	8			⇧50	↑75
Cordillera Cantábrica	42	15			⇧50	↑75
Spanish Pyrenees	21	9			⇧50	↑75
Spanish inland ranges	41	15			⇧50	↑75
Cordillera Bética	55	25			⇧50	⇧50

above not only varied spatially but also temporally, reflecting the diverse national contexts and different economic trajectories of countries and their regions. In Britain (represented here by the Scottish Highlands) for example, it can be seen that the shift towards non-agricultural employment was well underway before the start of the twentieth century, whereas it occurred later in other European countries.

The experience of the 1970s, 1980s and 1990s has set the context for discussions about the character of the rural economy in the early years of the twenty-first century. Two key patterns are apparent. The first, noted in Chapter 1, is the structural convergence in many Western countries between urban and rural economies. The second is the spatial diversity of rural economies. Regions within the same countries often have vastly different economic structures from one another.

The general trend of convergence between urban and rural economic structures is suggested by rural development statistics from the European Union (see Figures 2.1 and 2.2), which reveal the growing dominance of the tertiary (service) sector in all types of region and the diminishing contribution that primary industries (e.g. agriculture, forestry fishing, mining/quarrying) make to local economies, even in predominantly rural regions. However, other statistics point to considerable variation in economic experience between rural regions. A recent study of rural development patterns in EU countries – 'EDORA' – highlighted four key types of rural economy. The study emphasised the need to look beyond simple 'production'/'consumption' dualisms. While agrarian-based economies (the first type) still exist in some parts of the EU (mainly in eastern and southern regions), elsewhere a

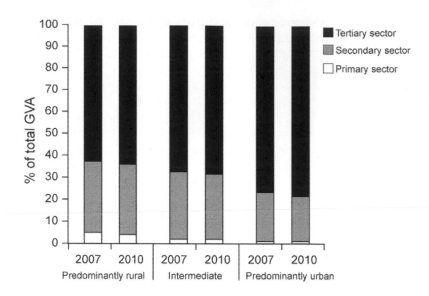

Figure 2.1

Structure of the economy by branch of activity in the EU Regions, 2007 and 2010
(% of total gross value added)
Source: European Commission, 2014

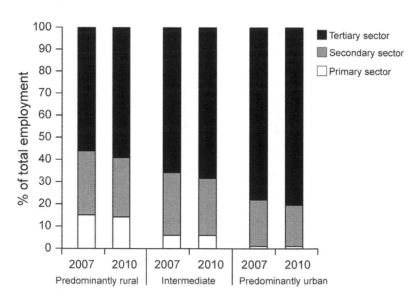

Figure 2.2

Structure of the economy by branch of activity in the EU Regions, 2007 and 2010
(% of total employment)
Source: European Commission, 2014

patchwork of other economic characteristics is evident. A large number of rural regions demonstrate consumption-countryside characteristics (the second group) where countryside public goods, environmental or cultural assets, or local produce underpin an economy based upon 'consumption' activities enjoyed by urban visitors, such as leisure, recreation and hospitality. But the study also identified two other distinct rural economy types centred on other forms of rural diversification, and suggested that these deserve more attention when formulating new approaches to rural policy and planning. In some rural areas economic diversification from agriculture is largely based around secondary manufacturing industries (type three), while in others it relates to the tertiary/private service sector (type four). This last category seems to be particularly associated with the most accessible rural areas. Figure 2.3 shows the spatial distribution of these different rural economic types within the EU.

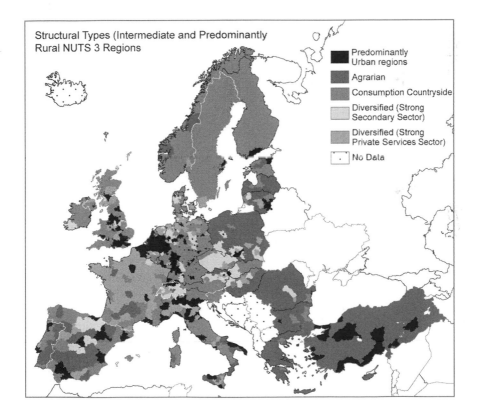

Figure 2.3

Economic structure typology of European rural regions, 2010
Source: EDORA, 2010a: 6

Development trends and social change

There is a close intertwining of economic and social change in rural areas as countries pass from being agricultural to industrial and then post-industrial societies. Development patterns and associated shifts in population are an expression of these complex relationships, and broad trends are often explained with reference to a four-stage model of urban development (Hall and Hay, 1980; van den Berg *et al.*, 1982). The model suggests a transition from urbanisation; through suburbanisation and counter-urbanisation; to re-urbanisation as a final state. Each phase involves a changing population dynamic between urban and rural areas. The following account presents a rural perspective on the model, its implications, and a broad timeline of change from a Western country viewpoint.

As we have seen, 1850 to 1950 was the period when the USA and many European countries made the transition to become industrialised economies and experienced a phase of rapid urbanisation as industrial production and employment concentrated in towns and cities and prompted mass migration from the countryside. This was accompanied by significant rural depopulation. As towns and cities grew and urban populations became more established however, the forces of suburbanisation became apparent. This was a marked trend in many Western countries between 1930 and 1960. In this phase, some more affluent urban dwellers were able to exercise choice about where they lived and increasingly looked for less polluted and less densely developed settings away from the congested urban core, but still within relatively close reach of urban centres. The processes of suburbanisation expanded the footprint of towns and cities as urban development spread to nearby countryside and suburban uses displaced agriculture as the main land use in urban hinterlands.

But as early as the 1950s, the driving forces behind a new phase of counter-urbanisation (sometimes referred to as 'dis-urbanisation' or 'de-urbanisation') began to emerge. These included, the growth of road transport and motorway infrastructure, new forms of goods handling such as containerisation, as well as rising land costs within established urban areas and the growing land intensity of manufacturing, all of which supported a more dispersed pattern of industrial development. These factors also undermined the appeal of the traditional industrial heartlands, which, by the 1970s and 1980s, often seemed to be in terminal decline. This was also a time when many Western countries witnessed the growing importance of more footloose service-sector industries where companies were increasingly able to include lifestyle factors in location decisions. It is in this context that rural areas began to experience a new era of population growth, and a more dispersed pattern of urbanisation across the wider countryside began to emerge. Such trends were well established in the USA by the end of the 1960s, but were not evident in Nordic countries until the beginning of the 1990s and are a very recent phenomenon in Eastern Europe (Hosszu, 2009).

Despite early predictions of an inherent tendency towards urban decline, evidence from the 1980s revealed that re-urbanisation of once shrinking cities was in fact occurring in parts of northern Europe including Germany, the UK, Denmark and the Benelux countries (Cheshire 1995). However, this phenomenon was by no means uniform, and at that time seemed to favour medium-sized cities that offered a quality urban environment – expressed in low crime and congestion rates – and a vibrant higher education sector. The shift to a service-based economy was seen as a central factor in this development. Although service industries were often relatively mobile, they wanted to be close to their customers and have access to an extensive and specialised pool of labour. And for some service sectors – such as finance – agglomeration in urban areas remained critical. Other social factors were thought to be reinforcing this re-urbanisation trend, including more women going to work, smaller family sizes and the increasing number of all-adult households, all of which meant that good quality and cheap public transport, alongside easy access to a broad range of job opportunities, were perceived as increasingly important. These factors tended to favour urban areas or nearby rural locations with good access to jobs and transport.

Although there is some variation in observed patterns between different countries and regions, a long-term underlying trend of urban population growth and rural decline is evident and this is forecast to continue even in the more 'mature' capitalist economies of Europe and North America (United Nations, 2014) (see Figure 2.4). This urbanisation manifests itself in both the dispersal of activity to newer cores and in the re-urbanisation of older cores, as predicted by the model above.

Although typical rural land uses – agriculture, forestry and 'nature' – dominate across Europe (when compared to 'artificial' urban uses) (see Figure 2.5), there are now fewer regions that are described as predominantly rural (Figure 2.6) in which half or more of the population are living in areas with fewer than 150 inhabitants per square km. Much of England, for example, could now be described as 'urbanised countryside' in which the majority of people live in larger villages and towns, which presents rural planning with a very particular type of challenge. That is not to say, however, that there are no extensive areas of dispersed population within this generally urbanised countryside, which will demand a very different planning approach.

It is perhaps wrong to view urbanisation, suburbanisation, decentralisation and re-urbanisation as a linear model but rather as a range of development possibilities that may be occurring alongside each other and varying spatially and temporally. In fact, an increasing diversity and divergence in the trajectory of rural areas has been highlighted by the recent EDORA study, which showed that many EU rural regions are in fact growing faster than the EU and national averages in both economic and population terms, while others – particularly in the east – are declining economically and losing population at a rate and to an extent that raises serious concern (see Figure 2.7). Comparing Figures 2.3 and 2.7, it can be seen that 'agrarian' and

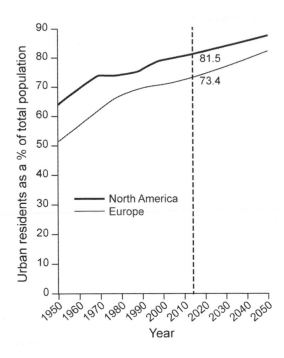

Figure 2.4

Urban and rural population as proportion of total population, 1950–2050
Source: United Nations, 2014: 8

'diversified' (secondary sector) regions are generally the poorest performing socio-economically while those regions classified as 'consumption countryside' or 'diversified' (private services) tend to be high performers and are anticipated to continue to accumulate economically and in population terms in the future. This mixed picture is not just a feature of the EU. It is also apparent within the wider group of OECD countries. An analysis of urban and rural development statistics between 1997 and 2010 revealed that rural rather than urban regions saw the highest average growth in GDP per capita and in productivity during this period. However, whilst those rural regions closest to cities were often extremely dynamic, the OECD also drew attention to the stark and growing differences in experience between rural regions and expressed particular concern over the declining fortunes of remoter areas (Garcilazo, 2013).

Analyses of the factors producing differences in performance now regularly emphasise that all regions (urban and rural) are subject to common external forces of change associated with globalisation and global trade liberalisation. Supported by developments in new technology, many areas have experienced

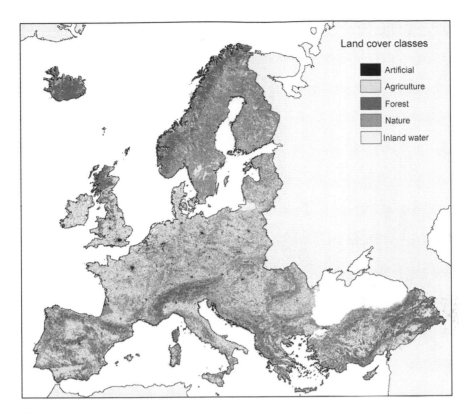

Figure 2.5

Land cover, 2006
Source: European Commission, 2014: 2

a growth in 'connexity' – that is, new interrelationships and interdependencies, between individuals, economies and wider society, which are realised through new technology and reflected in an international flow of money, goods, people and ideas (Mulgan, 1998). With these external factors accepted as common, there is renewed interest in the significance of local factors as being key determinants of contrasting socio-economic experiences. Why, despite shared 'connexity', do some areas out-perform others? Traditionally, explanations have focused on 'hard' factors: availability of natural resources, transport and accessibility, and land availability. These all continue to be important, and recent statistics confirm that OECD member states are all experiencing net migration to their amenity-rich rural areas (Brown, 2010). However, there has been recent focus on the ways in which an area's store of human, social and cultural capital – alongside institutional capacities and different styles and structures of governance – may help explain development

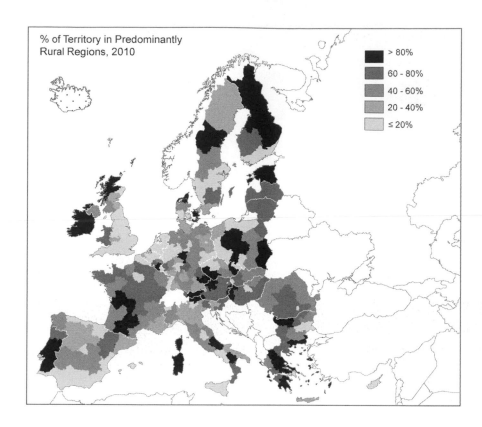

Figure 2.6

Percentage of territory in predominantly rural regions, 2010
Source: European Commission, 2014: 3

outcomes (Terluin, 2003; EDORA, 2010a; Garcilazo, 2013). The importance of social capital as a driver of rural development was introduced in Chapter 1. A fuller discussion is provided in Chapter 5.

Before leaving this brief introduction to contemporary socio-economic change within rural areas, attention is drawn to a notable trend in their demographic structure, which is shared by the majority of rural areas in Western countries: rural populations are ageing. This is not only a product of the general ageing trend that is being experienced in many countries. The rate of ageing is greater in rural areas and is associated with a mix of factors that vary spatially, but include increased longevity and *in situ* ageing, chronic low fertility, and the migration of retiring older households to rural areas. But most worryingly, demographic ageing is a result of the exodus of many younger people from rural areas; they leave for a combination of educational and employment reasons, causing a rural 'brain drain' (Brown, 2010). Many

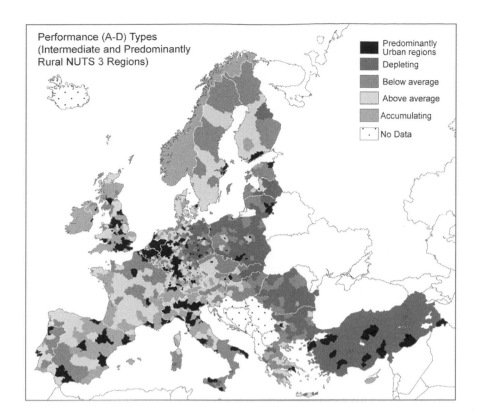

Figure 2.7

The differential performance of rural regions
Source: EDORA, 2010a: 6

areas are being socially and demographically reconfigured by the combination of rural retirement and the loss of younger, more ambitious and potentially entrepreneurial people. Some of the reasons for younger people leaving, and its economic impacts, are discussed in Chapter 8.

Shifting political perspectives and changing approaches to rural planning

The developments outlined above have produced their own ideological reactions and political responses, often manifest in changing approaches to rural planning.

Industrialisation and what has been referred to as the 'Fordist' period of economic development was a feature of many Western economies until the post Second World War years. This was accompanied by a complex mix of liberal *laissez-faire* politics and by a commitment to free trade in some countries such as the United Sates and the UK, and national protectionism in others. At this time, historic central and local government structures were evolving in response to rapid population and urban growth. However, growing popular discontent with some of the social consequences of liberalism and industrialisation stimulated the emergence of socialism in the late nineteenth and early twentieth centuries. This, combined with the economic crisis of the 1930s and with the social and economic upheavals of two World Wars, gave rise to a period of stronger state intervention and social democratic politics in the post-1945 years, as affected countries sought to recover from the trauma of war and adjust to the attendant social and economic transformations. In many countries, this new socio-political reality resulted in the development of a 'welfare state' of some description, in which central and local governments – driven by a commitment to equality, social justice and redistribution – became the providers of a wide range of public services and state regulation of many other aspects of life was established. The modern systems of 'town and country' or 'land-use' planning that emerged at that time, were part of this trend, often aiming to extract maximum public good from the development of private land.

Numerous national policy frameworks for rural areas came into being, and these tended to be sectoral in character, emphasising agricultural modernisation as the cornerstone of rural development. In line with political concerns over inequality, regional economic planning also began to emerge, which sought to assist declining rural areas and stem population loss through the development of new industries and other initiatives. Local-level planning (for example, 'town and country planning' in Britain) was in its infancy and often focused on managing patterns of urban development in the countryside in support of the productivist view of rural areas, underpinned by continuing concerns over food security in the wake of the Second World War. Statutory planning systems were, however, emerging as an important factor driving the differentiation of rural economies and rural experiences as contrasting approaches emerged built on local politics (Marsden *et al.*, 1993).

By the late 1960s, increasing social differentiation (manifest in growing affluence for some, and in greater individualism, producing more mixed and 'plural' societies) resulted in a questioning of the post-war order. This, in turn, led to a critique of modernist rationality, underpinning the social welfare model and the notion that states should control and deliver the bulk of public services. One particularly powerful 'post-modern' perspective became manifest in neo-liberal politics, which rejected state intervention and emphasised the need to promote market and personal (individual) freedoms. The state was too clumsy an instrument to deal with the growing diversity of needs and situations, particularly in the broader context of globalisation and internationalisation in which the ability of states or localities to 'go it alone' was being eroded. Only by breaking down borders (representing obstacles to the natural functioning of markets) and through collaboration could those states and localities address contemporary challenges. It was in this context that the European Union emerged during the latter half of the twentieth century, expressing a mix of emergent neo-liberal thinking and retained social democratic ideals. Through the opening up of free trade within the Common Market, Common Agricultural Policy (CAP) subsidies and a range of environmental and other directives, the EU increasingly became a potent force influencing the pattern of rural change within European member states.

By the early 1980s, the spread of neo-liberalism had radically altered post-war thinking on the role of the state, from being a direct provider of public services to that of an enabler of provision by other sectors. Emphasis was also being put on stimulating the growth of private enterprise, which, it was thought, would henceforth resolve numerous public service dilemmas. Running alongside this promotion of the state as facilitator and deregulation aimed to promoting market-led solutions was a burgeoning critique of sectoral approaches to government and calls for more 'joined up' working, between different partners, by multi-sectoral agencies, and across the public–private divide. As a consequence, a more collaborative style of *governance* emerged supporting a partnership approach to the development and delivery of public policy, involving a range of public, private and third sector organisations. This reinterpretation of the role of government, rooted in the social differentiation emerging from the 1960s onwards, and the move to a mixed-mode governance model has also been associated with a 'rescaling' of the state and increasing emphasis on bottom-up, sometimes community-led, development, regulatory and service delivery models. Lange *et al.* (2013) argue that this combination of governance shift and rescaling, is a trend that has continued into the twenty-first century. It has been manifest in the shift from exogenous to endogenous rural development (and a new 'rural paradigm'; see Table 2.2; see also Chapter 1), a turn away from the centrally orchestrated 'welfare state' to local welfare models (including that of the 'welfare city' – see Van der Pennen and Schreuders, 2014), and a rise in community action grounded in communitarianism, which is viewed by some as a more inclusive space between neo-liberalism and state-led social welfare (Gallent and Ciaffi, 2014b).

A clear narrative emerged from 1945: there was a journey from state-led, through market-led, to community-led interventions in a range of policy areas. This ideological and political narrative is interwoven with the socio-economic changes affecting rural areas, and supported periodic shifts in the nature, focus and purpose of statutory planning. Broadly, modern planning started off as a tool of the welfare state, in particular facilitating the large-scale delivery of public housing. This was true in France, Italy, Britain and many other parts of Europe. That welfare model faced growing criticism from the 1960s and large 'modernist' housing projects fell out of favour by the 1970s. A decade later, neo-liberalism was sweeping many parts of Europe, with planning reined back to allow unfettered market access to different service and development areas. But deregulation created room for a range of different

Table 2.2 The new rural paradigm (OECD, 2006)

	Old approach	New approach
Objectives	Equalisation, farm income, farm competitiveness	Competitiveness of rural areas, valorisation of local assets, exploitation of unused resources
Key target sector	Agriculture	Various sectors of rural economies (e.g. rural tourism, manufacturing, ICT industry, etc.)
Main tools	Subsidies	Investments
Key actors	National governments, farmers	All levels of government (supra-national, national, regional and local), various local stakeholders (public, private, NGOs)

solutions to market failure and to the inadequacies of public welfare, meaning that from the 1960s/1970s, harder and softer forms of neo-liberalism have been accompanied by rescaled and community- or third-sector-led actions, including more localised forms of land-use planning. Part 2 of this chapter now turns to consider how the trends discussed above have been reflected in the British experience of land-use planning for rural areas.

Part 2a: Rural planning in Britain

Pre-war rural trends

Britain was the world's first industrial nation, experiencing the shift from an agrarian to an industrial economy much earlier than most other countries. As well as driving a very early decline in agricultural employment, industrialisation also brought with it new economic opportunities for the countryside. Mining and quarrying were concentrated there and a whole industry grew up around the transportation of raw materials from country to town (Gilg, 1976; Murdoch, 1996). However, these early successes tended to be short-lived in many areas. Britain was at the centre of world commerce, and these emergent rural industries found it difficult to compete against the overseas suppliers of cheap imports. Likewise, new industries that might have remained in more rural areas, quickly sought the locational advantages of towns and cities. It is against this backdrop of declining rural employment (most acute in the more remote rural areas of Scotland, Wales, the Lake District, the Northern Pennines, East Anglia and the South West) that the Development Commission (renamed the Rural Development Commission in 1987) was established under the Development and Road Improvement Fund Act 1909. The Commission was responsible for administering a fund that was designed to facilitate the development of agriculture and rural industries, forestry, land drainage and reclamation, rural transport, harbours, internal navigation and fisheries. The creation of the Commission represented a first significant step towards state intervention in rural economic and social affairs. In the years leading up to the Second World War, its activities broadened to include the training and advisory services needed to revive rural craft-based businesses (Countryside Agency, 1999).

Despite these efforts, during the first part of the twentieth century, the disparity between the economic trajectory of London and the South East and much of the rest of Britain widened. Declining employment was not just a feature of rural areas, but also spread to the traditional manufacturing heartlands of the North, South Wales and Scotland, leading to mounting public concern over the social, economic and strategic consequences of the concentration of employment and population growth in and around London. This was the focus of the influential Barlow and Scott Reports (1940 and 1942). As in other countries, the climate of war also brought into sharp focus the strategic disadvantages of reliance upon imported primary goods, particularly food and timber and there was growing concern about the weak state of British primary production, the pace of urban expansion and loss of countryside around London from both a production capacity and an amenity perspective. It was in this context that a political consensus developed that the state should play a more active role in directing economic affairs.

Pre-war rural planning

The emergence of 'town and country planning' as an activity of government was one reflection of this trend; and a useful starting point for understanding how the embryonic planning system in Britain responded to the economic and social problems of rural areas is the analysis presented by Ebenezer Howard in his classic text 'To-morrow: A Peaceful Path to Real Reform' (Howard, 1898: see Figure 2.8).

Although Howard's analysis was mainly a diagnosis of *urban* ills, it also included a powerful account of the rural economic and social problems of the period, which included agricultural decline (land lying idle), long working hours, low wages, rural unemployment and consequent rural depopulation ('deserted villages'). However, his proposed marriage of town and country, through the creation of Garden Cities, did on the face of it do little to address the economic and social plight of rural areas. As Cherry comments '[...] with hindsight, we can appreciate today that this was a planning solution which was a form of urban exploitation of the countryside' (Cherry, 1976: 6). A slightly more charitable view is presented by Abercrombie (1959). He saw Howard's model of Garden Cities (see Figure 2.9), in which both urban and rural workers would be housed, as key to protecting the integrity of the

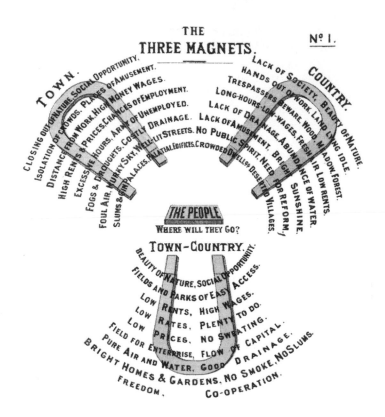

Figure 2.8

Ebenezer Howard's analysis of country life
Source: Howard, 1898: 9

surrounding agricultural belts and for stemming the outflow of rural population. As he states '[...] the farm worker instead of living his isolated life is to have the benefits of a social existence in large centres of population situated conveniently for access to his work' (Abercrombie, 1959: 178).

However, the spirit of Abercrombie's subsequent remarks perhaps more accurately reflect the focus of countryside planning at the time:

> It is the last thing that is necessary for the preservation of the countryside that it should be depopulated; there is nothing more antagonistic to the beauty and seemliness of typical country than the untidiness of uncultivated ground, unless it be the indiscriminate scattering over its face of discordant urban elements.
>
> (Abercrombie, 1959: 178)

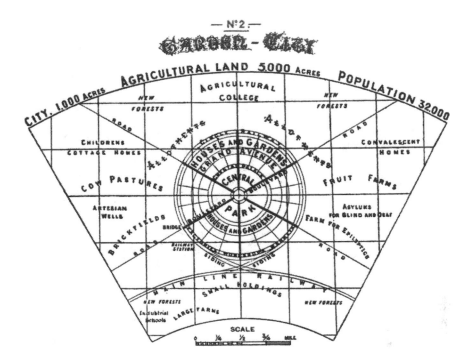

Figure 2.9

Ebenezer Howard's Garden City Model
Source: Howard, 1090: 13

These remarks reveal a central concern with the aesthetics of open countryside, a belief in the key role of agriculture in providing appropriate environmental stewardship (and employment) for rural areas and the identification of urban concentration and containment as a core feature of good planning for the countryside. All of the ingredients of future rural planning are present in this statement.

Indeed, it is this view of rural affairs that is reflected in the planning legislation of the inter-war years. The first Act of Parliament to recognise a rural role for planning was the Town and Country Planning Act 1932, which extended the scope of planning schemes to cover all land (urban and rural) and the objects of planning to cover the preservation of places of natural interest or beauty (Abercrombie, 1959). Further planning powers to limit urban encroachment of the countryside were provided under the Restriction of Ribbon Development Act 1935. In both cases, the economic and social dimensions of rural life were only indirectly addressed, though the containment of economic activity and the concentration of population, perhaps inadvertently in support of rural services.

The productivist era (1945–mid 1970s)

During the Second World War (1939 to 1945), many rural areas in Britain experienced a marked revival in their fortunes. Government initiated a 'dig for victory' campaign in order to reduce reliance on imported food. New factories, including munitions plants, were located in areas with surplus labour and many military establishments were built well away from towns and cities that were the target of German bombing campaigns. These initiatives reflected a hasty adoption of many of the recommendations set out in the wartime Barlow and Scott Reports (into the decentralisation of industry and land utilisation) and paved the way for a transformation in the government's approach to rural areas in post-war years. A Cabinet Office report provides a useful overview of what this entailed (Box 2.1).

It was within this political context that rural areas in Britain entered into their 'productivist' era marked by strong state intervention in which government policy was explicitly concerned with the preservation of rural land for agricultural production and with increasing the productivity of primary industries (Halfacree, 1999). Agriculture was the main focus of attention, but forestry, fishing and mineral production also received varying levels of government support.

The Agricultural Act 1947 set the scene for a 'second agricultural revolution', which saw fundamental changes in the nature and structure of the farming industry in Britain (Walford, 1999). The overall aim was to increase agricultural production whilst keeping food prices low. The net effect of the introduced policy measures was to drive the processes of concentration of ownership and intensification and specialisation in production (Walford, 1999), resulting in spectacular growth in agricultural output. Increased production did, however, come at a cost. Contrary to the government's initial expectations, growth in output was accompanied by continuing decline in

Box 2.1   Government's productivist view of the countryside (Cabinet Office, 1999: 5)

At that time [i.e. in the 1940s], the government had a clear view of what the countryside was for, and could command a wide consensus of support for this view. It saw agriculture as the primary function of rural areas and, therefore, as their key economic sector; and it viewed agriculture's primary role as to ensure security of food supply. It assumed farmers, as the stewards and shapers of the countryside, could be relied upon to protect the quality of the rural environment. It assumed the shift in population from rural to urban areas would continue, and therefore that the greatest threat to the environment would be around the fringes of urban areas. And it believed that rising output from agriculture and other primary sectors in rural areas, combined with universal models of public service provision, would generate sufficient wealth to tackle rural poverty.

agricultural employment as farmers increasingly mechanised their activities. Between 1945 and 1965, North and South Wales, the South East and parts of the Midlands all saw a 50 per cent decline in agriculture jobs and all areas experienced losses of 35 per cent or more (Gilg, 1976).

It is against this backdrop of continuing employment decline in primary industries that government renewed efforts to stimulate new job opportunities in other sectors of the rural economy. This intervention took a number of forms. These included the designation of Development Areas in localities suffering from high unemployment and consistent population decline. Initially established in the 1930s, the name, coverage and focus of government activity related to Development Areas went through numerous modifications in the years following 1945 (Gilg 1976). At first, most Development Areas were urban; but by the mid-1960s 40 per cent of Britain's land area was covered by these designations, including *all* the remote rural districts. Within the Development Areas a number of forms of government support were available including financial assistance for new capital investment and staff training and tax subsidies.

The Development Commission often worked hand-in-hand with Local Authorities and increasingly initiated 'grass roots' action to stem rural employment decline. The pioneering Mid Wales Industrial Development Association established in 1957 is a noted example of early partnership working in rural regeneration (Garbett-Edwards, 1972, Pettigrew, 1987). Supported by the five mid-Wales County Councils, the Association was charged with taking direct action to create and maintain jobs with the objective of stemming the outflow of population and attracting people from elsewhere to live and work in the area. This and similar initiatives such as the Highlands and Islands Development Board and the Pennines Rural Development Board (Gilg, 1976) made significant contributions to the economic and social fortunes of the rural areas they covered. They achieved notable successes in bringing new manufacturing industries to outlying rural districts and supporting the growth of service industries, including those related to tourism, which, as we shall see, was an increasingly prominent source of rural employment in post-war years (Dower, 1972). Equally, through experimentation and innovation they encouraged greater sophistication and professionalism in approaches to rural economic development and marketing. In particular, they began to highlight the case for a more locally-sensitive approach to rural development and the valuable role that regionally-based agencies could play in addressing economic and population decline (Pettigrew, 1987).

The productivist era and planning

But how did the post-war system of 'town and country' planning interact with these developments? Perversely, it could be argued that the interaction was quite strong and that the Town and Country Planning Act 1947 put in place a planning system that was entirely consistent with the government's productivist ambitions for the countryside, which aimed to preserve so called 'white land' for the use of agriculture so that it could freely develop into a strong export-oriented economic sector. Agriculture and forestry were exempt from the planning controls created by the 1947 Act, and therefore the vast majority of land in rural areas was not of direct concern to the new County Planning Authorities (with their newly assigned duty to draw up development plans and make development control decisions in light of plan policy). The focus of their attention was directed instead to the villages and towns, but even in this context, development plans were required to minimise the impact of development on good agricultural and forestry land and to cause minimal damage to farm units. As Travis (1972: 188) puts it: '[...] it appears to have been assumed that the 1947 Agricultural Act and government control by subsidies, incentives, etc., could achieve both the best economic and spatial arrangement for the countryside'. Moreover, the Scott Report, which informed much of the approach of the 1947 Act, saw the beauty and pattern of the landscape to be a direct result of soil cultivation and therefore perceived no antagonism between productivist agriculture and landscape values (Allanson and Whitby, 1996). This understanding of agriculture as the bedrock of rural communities also pervaded the focus of planning attention in rural settlements in the early post-war years, which tended to centre on questions of amenity and conservation rather than nurturing the economic or social wellbeing of rural communities (see Box 2.2).

It was not until the 1960s that planners began to show a wider interest in rural economic and social affairs and an appreciation of issues related to rural decline. These concerns were most typically reflected in the identification of a settlement hierarchy of towns, small towns, and 'key villages' or 'king villages', which were defined as key service centres and 'growth points' (Cloke, 1979, 1983). Travis (1972) suggests that these might have been more accurately described as 'holding points' because their chief rationale was to stem the strong trend of population and economic decline that continued to be a feature of many rural areas in this period. The development of this approach reflected Christaller's Central Place Theory, which held that economic efficiencies could be achieved by nucleating public services and infrastructure and maintaining suitable levels of support population, which in turn could then provide more viable locations in which to retain existing employers and attract new businesses (Christaller, 1933). The application of this idea is examined in Chapter 7. A further development of this approach was the creation of new 'Central Places' in remote rural areas, such as the Moray Firth Urbanisation Programme in Scotland and the development of Newtown (designated under the New Town Act 1946) in mid Wales. These were planned as major new regional employment centres, designed to stem the exodus to the cities, particularly among the younger sections of the rural population.

Barlow Report (1940): concerned with the 'distribution of the industrial population', this report led to both a concern for the development of new towns and the distribution of industry across Britain. It resulted in the New Towns Act 1946 and the Distribution of Industry Act 1945.

Scott Report (1942): concerned with 'land utilisation in rural areas' contained proposals for the protection and advancement of agricultural interests. It set the tone for post-war agricultural policy, led directly to the Agricultural Act 1947 and also resulted in a special status for agriculture in the land-use planning system, with many agricultural uses exempt from planning control.

Uthwatt Report (1942): led to the creation of the comprehensive planning system, development control by County and Borough Counties, and the first of a series of attempts to capture 'betterment' (land value uplift) through the granting of planning permission. Resulted in the Town and Country Planning Act 1947.

The consumption countryside (1970s–2010)

This search for new sources of rural employment beyond the primary sector was illustrative of a shift in thinking about rural areas that gathered momentum in Britain in the 1970s, and heralded the beginning of a consumption-oriented era for the British countryside (Halfacree, 1999). This shift reflected a number of developments. These included: fading national security imperatives to maximise home production of primary goods, mounting concern about the environmental damage being caused by intensive agriculture and about growing food surpluses, and increasing questioning of the validity of continuing government aid and other special support for farmers, as Britain weathered a series of economic recessions and public-sector purse strings began to tighten. Running alongside these issues, a new realism about the relative importance of farming as a source of rural employment began to emerge. Continuing decline in the agriculture workforce was paralleled with job growth in other sectors. Measures to attract new manufacturing saw success in many rural areas over the latter part of the twentieth century, an experience that was in marked contrast to that of urban areas over the same period. Studies revealed that together with the improvements in transport, communications and utility networks, 'quality of life' issues proved to be an important factor influencing location/relocation decisions (Murdoch, 1996). The rural service sector also saw job growth. Recreational and tourist use of the countryside, which blossomed in the 1950s and 1960s (Dower, 1972), was responsible for much of this growth. This reflected a number of significant social trends in the post-war era including: rising affluence, increased mobility, shorter working hours, improved education and the growing influence of the media in developing wider awareness of the countryside (White, 1976). There was an increase in the 'marketing value' of the so-called rural idyll, a crucial factor in the shift to a consumption oriented view of the countryside. The popularity of rural areas as leisure destinations was paralleled by what Cloke and Goodwin (1992) described as service class in-migration, which occurred alongside amenity-driven migration expressed through retirement to the countryside and the purchase of second homes (see Chapter 8). After many years of population decline, many rural areas, including even some remote districts, began to experience a steady influx of population from the 1970s onwards and, as Halfacree (1999: 69) so aptly describes, this presented a major challenge to the old productivist regime:

> The crisis in productivism is not just of significance to the agricultural population, as it also signals a crisis for all those implicated in the produc-tivist era. It suggests the hegemonic domination of rural areas and rural society by agriculture, which of course dates back further than 1945, is itself coming to an end. Hence, post-productivism signals a search for a new way of understanding the countryside. New interests and actors are coming on the scene in an attempt to create a rurality in their image.

The spatial significance of these developments were revealed above in Figure 2.3 which shows that today virtually all rural areas in Britain can be classified as 'Consumption' countryside or 'Diversified' (private service) regions which Figure 2.6 reveals to be experiencing economic and population growth which is forecast to continue. However, as we will see in later chapters, beyond the general trends and statistics lies a complex picture of local variation in the economic and social circumstances of rural areas and this has prompted a number of planning responses that aim to be sensitive to the differentiated rural experience.

It is also important to note that during this post-productivist period, there have been raised expectations concerning access to the countryside for leisure and recreational purposes. Because so much public money had been, and was continuing to be, given to support farming there was a growing expectation that public access rights should enhanced. This has given rise, in numerous areas, to conflicts surrounding who the countryside is for, its central function, and what sorts of activities might be appropriate or perhaps banned from different parts of the countryside. This often led to questions of countryside management rather than land-use or statutory planning *per se* although often there is overlap between these elements, as noted in Chapter 1.

The consumption countryside and planning

Despite the transition described above, the need to protect and preserve the countryside from change has remained a central objective of planning, often

---

reinforced by powerful political elites who relocate to rural areas and wish to realise their particular visions of what the countryside should be (Marsden *et al.*, 1993). Hence, planning policy, manifest for instance in national designations (such as National Parks or Areas of Outstanding Natural Beauty (AONBs)) and other strategic planning tools (such as Green Belts – see Chapter 10) have sought to protect the countryside from development. Development restraint has also been reinforced through the continuation of the generic settlement policy, described above, which prescribes a hierarchy of towns or key settlements. Any new development is frequently concentrated in these designated settlements and, as a corollary, strict restrictions apply to new development proposals outside of these centres (Cloke, 1979, 1983), extending to a general presumption against development in the open countryside. Any development that is eventually permitted needs to fit into the rural landscape and satisfy strict design criteria, which often leads to escalating costs.

In the face of consumption pressures, town and country planning has increasingly become a means of preserving and protecting the enduring characteristics of the countryside. Pressure to do so has come from new migrants who wish to protect the lifestyles that they have bought into and preserve landscape amenity – including the open, rolling vistas – and the character of villages. Throughout the consumption period, migrants have often achieved this by gaining influence or control over parish or community councils and local authorities, becoming a potent political force. As a consequence, development has become increasingly restricted, demand for property has increased, and sections of the 'local' population have been priced out of the housing market (Chapter 8), which has underpinned a pattern of rural gentrification (Chapter 5). The restrictive approach to development has endured for half a century, leading to what Taylor (2008) has described as a 'sustainability trap': by restricting growth to a few key settlements, planning denies other places the things they need – housing and basic services – to thrive, creating 'inherently unsustainable settlements'. This may result in the local workforce being displaced up the settlement hierarchy but then having to back-commute 20 miles to run a local shop. Alternatively, the viability of that shop, and other basic services, may simply be lost, resulting in closure and the loss of vital community infrastructure.

These challenges within the 'consumption countryside' are largely rooted in economic and social change (see Chapters 3, 4 and 5 for detailed discussions). There are now a wide variety of claims as to the legitimate function of rural places and the relative emphasis that should be placed on farming, on landscape protection, on the preservation of village character, on the need to provide new employment opportunities, on leisure uses and so on. Because of the consumption patterns that now characterise the countryside, rural society is far more heterogeneous than it once was, suggesting that local solutions are needed to local problems: innovations that can address the complexity of different situations and competing needs. But a wide variety of interests and agendas are assembled in the countryside, giving rise to debates over the scale and nature of rural governance. Governance debates are infused with uncertainties around what the countryside is for and whose agenda should frame future policy and action: should it be for people ('settled' or 'newly arriving'?) who live in the countryside and who may have certain needs and expectations, or for visitors who may well have different priorities, or indeed for the state, which takes a more strategic view as to the opportunities different uses of rural land present? Such dilemmas are readily apparent in recent and ongoing debates over the impact of new and renewable energy resources generated or accessed from the countryside. The impact of onshore wind turbines or solar panels, as renewable energy sources, are often located in rural settings with arguably important landscape implications. There are ongoing debates surrounding the intensity of agricultural production, often livestock-based, and the need for good animal welfare versus satisfying the demand for cheap food. Similarly in National Parks, which were originally intended to promote the 'peaceful enjoyment' of the countryside, there are growing debates as to what type of development might be appropriate in these contexts (see Chapter 10) given the competing needs of communities, farmers and visitors.

During the first decade of the current century there was a re-awakening of the debate surrounding what the focus of rural policy should be, often framed around ideas of 'evidence', 'users', 'visions', 'co-ordination', and 'delivery', with a focus on 'rural governance' based upon partnership approaches, delivering some degree of consensus, and community leadership and involvement. The current Rural Strategy, produced in 2012, adopts a very people-centric perspective (see Box 2.3), prioritising economic growth delivered through rural businesses and in support of communities, who are able to take greater control of their lives and surroundings through the 'localism' agenda. This idea that local communities should have a greater say in both shaping the scale of local development and taking more responsibility for managing local assets of 'community interest' has been given renewed emphasis by the Localism Act 2011, which has important implications for the scale at which statutory planning is delivered – see below.

There is currently much excitement as to whether this really represents a devolving of power and responsibility to the local level, often below the level of local government, or whether rural policy will continue to be driven by overriding national priorities. One of the great debates at the current time, in rural policy in general and land-use planning in particular, is the effect that social capital (see Chapters 1 and 5) will have on the ability and ambitions of communities to use new powers, and how this might further advantage some rural areas over others. The Localism Act provides communities with the opportunity to take greater responsibility for planning decisions and take control of different community assets. But inevitably, some communities will possess a deeper pool of the skills and resources needed to take advantage of such opportunities, meaning there will be spatial variability in the benefits

Box 2.3   Rural Statement 2012 (DEFRA, 2012: 2)

This Rural Statement is intended to underline our commitment to Rural England. It reflects our vision of successful rural businesses and thriving rural communities in a living, working countryside, and is based around three key priorities:

- **Economic Growth** – we want rural businesses to make a sustainable contribution to national growth;
- **Rural Engagement** – we want to engage directly with rural communities so that they can see that Government is on their side; and
- **Quality of Life** – we want rural people to have fair access to public services and to be actively engaged in shaping the places in which they live.

that localism brings. What is clear, however, is that rural places are very diverse and increasingly need to respond to specific place-based opportunities and challenges. Yet, despite an apparent localisation of rural policy delivery, the design of policy – and the framing of its delivery – is occurring at numerous levels and within a variety of different bodies above the point of delivery.

So far in this chapter, we have considered how planning – before and after the introduction of the comprehensive system in 1947 – adapted to, and reinforced, the narrative of economic and social change in the countryside. In the remaining part, an overview of the agencies framing (and sometimes directly delivering) rural policy is provided, followed by an account of the current planning system and its recent localisation. These final discussions serve as markers for the detailed analyses provided in later chapters.

Part 2b: Agents of rural policy design and delivery

Rural policy and its implementation are increasingly shaped by local action, as will be seen in later chapters. However, this action is framed by forces and agents well above the point of implementation. It is important to appreciate the roles and responsibilities that key agents have in shaping rural places, where locally-specific opportunities and priorities are determined. There is also a complex relationship between the statutory planning system examined above, which seeks to regulate new development in rural areas, and much broader rural policy and funding activities, which aim to address specific social, economic or landscape issues. The main agents involved in the 'composite' rural planning introduced in Chapter 1 are listed in Table 2.3.

Table 2.3 Agents involved in rural planning in England (excluding third sector)

Level	Key governmental bodies	Non-governmental bodies
European	• Directorate Generals responsible for Agriculture, Environment, Regional Development, etc.	
National	• Department for Communities and Local Government; • Department for Environment, Food and Rural Affairs; • Department for Business Innovation and Skills; • Department of Energy and Climate Change.	• Planning Inspectorate; • Housing and Communities Agency; • Natural England; • Environment Agency; • Forestry Commission; • National Park Authorities.
Sub-national	• County Councils	• Local Enterprise Partnerships; • Local Nature Partnerships.
Local	• Local Authorities; • Parish Councils.	

The European Union

The European Union plays an important role in moderating rural change (Duhr *et al.*, 2010) in two distinct ways. First, various funding mechanisms continue to direct significant amounts of financial resource towards Europe's rural areas. The most significant has been the Common Agricultural Policy (CAP), which is examined in Chapter 4. Since 2005, this has comprised two 'pillars'. The first pillar, otherwise known as the European Agricultural Guarantee Fund (EAGF), provides direct support to farmers engaged in good environmental land management practices and who deliver a minimum level of animal welfare. Payments promote particular ways of managing farmed land and reward the provision of new environmental services. The second pillar, or European Agricultural Fund for Rural Development (EAFRD), is a smaller pot of money from which grants are distributed either to assist with efficiency and productivity (e.g. by supporting on-farm diversification) or to support wider development activities, with funds channelled through the LEADER+ programme.

European policy has long focused on the support of farming, but it has also concerned itself with the 'structural' framework for development. European 'Structural (Investment) Funds' are available to help Europe's lagging regions catch up with average economic performance. The current funding round runs from 2014 to 2020 and is focused on the promotion of 'smart specialisation': this essentially means exploiting local development opportunities in tourism, regional food or some other specialist asset that is key to realising endogenous development ambitions. Europe's 'less developed regions' (formerly known as Objective 1 areas) are the focus for Structural Fund investment. These have a GDP that is less than 75 per cent of the European average, and in the UK currently include Cornwall, the Isles of Scilly, West Wales and the South Wales Valleys. Lower priority is given to 'transition regions', which had GDP above 75 per cent of the European average but less than 90 per cent. The Structural Funds are divided between two pots: the European Regional Development Fund, earmarked for capital projects such as infrastructure investment, and the European Social Fund, available for capacity building.

Besides these European funding mechanisms, several key 'directives' oblige EU member states to take actions in support of key conservation or environmental objectives. Notable amongst these are the Birds Directive 1979, the broader Habitats Directive 1992 and the Water Framework Directive 2001. The first of these required the designation of Special Protection Areas (SPAs) for nesting birds; the second required the designation of Special Areas of Conservation (SAC); and the third sought to reduce pollution in rivers, particularly by restricting the discharge of agricultural chemicals. These three directives alone restrict how land can be used in certain locations and set limits on farming practice. More broadly, key areas of agreed European policy will have profound impacts on how rural land is used in the future. For instance, energy policy ties member states to fixed goals for increasing national energy supplies from renewable sources – to 20 per cent by 2020 and 27 per cent by 2030 – and will drive investments in onshore wind and solar energy, both highly contested rural land-uses.

Central government, executive agencies, and non-departmental public bodies

Whilst there have been periodic name-changes and portfolio adjustments, there are now four main government departments ('ministries') in England with remits that impact significantly on rural areas and policy: dealing with communities and local government; environment, food and rural affairs, business innovation, and energy and climate change. The administrations in Scotland and Wales have devolved responsibility for these areas. The Scottish Government identifies topics, from 'built environment' through to 'people and society', as being within its jurisdiction; and it also has a specific focus on 'farming and rural issues'. The Welsh Government has parallel responsibilities and a particular rural focus on 'environment and countryside'. Arrangements vary, but a division of responsibilities – into topics or departments – is common to different parts of Britain. In England:

- The **Department for Communities and Local Government** (DCLG) was created in 2006 and is concerned with local government issues, planning, housing, social inclusion and community engagement. Its ministers (led by Eric Pickles, the Secretary of State) have been the main proponents of 'localism', which has been largely concerned with the relocation of powers away from the old regions to local authorities. It led on the drafting of the Localism Act 2011 and has overseen – via a dedicated group within DCLG – the establishment of a system of Neighbourhood Planning in England. Its civil servants continue to monitor the creation of Neighbourhood Fora and the drawing up of Neighbourhood Development Plans. This is also the department that owns the planning guidance given to local authorities in England: previously a series of thematic 'planning policy statements' dealing with different areas of planning practice and now the single National Planning Policy Framework published in 2012. Together with more technical advice contained in the National Planning Practice Guidance, this aims to provide a comprehensive view of the planning system and a means for local authorities to interpret and execute their legislative responsibilities. Beyond all of this, DCLG is also the national body responsible for administering the European Structural Investment Funds described above.

- The **Department for Environment, Food and Rural Affairs** (DEFRA) was created in 2002, combining functions from the Department of the Environment (DoE) and the Ministry of Agriculture, Food and Fisheries (MAFF). It has a broad remit including acting on behalf of rural communities, promoting sustainable food and farming and protecting the countryside, including its biodiversity. It previously worked through the Commission for Rural Communities (CRC) and Natural England, both created in 2006. The CRC was disbanded in 2012 but Natural England was retained. CRC responsibilities for data gathering, rural proofing (see Chapter 7) and rural advocacy were taken into DEFRA, which also administers European funding under the CAP.

- The **Department for Business, Innovation and Skills** (DBIS) is responsible for facilitating business growth across England. It therefore has a role in promoting and supporting business growth within rural areas. It is responsible for the management of various national growth funds (currently the 'Growing Places Fund' created in 2011) which is then locally administered by Local Enterprise Partnerships (see below), and finally.

- The **Department of Energy and Climate Change** (DECC) has responsibility for energy security (planning to ensure continuity of supply for business and domestic users), action on climate change and promoting the use of renewable energy. As well as being concerned with generic energy issues, it leads on government's strategy for renewable energy, including wind and solar power in rural locations, with a view to achieving EU targets.

These key departments fund and are, in turn, supported by a number of independent agencies and non-departmental public bodies (NDPBs). The most significant of these for rural areas are listed below. It is worth noting that the number of these 'quangos' has been greatly reduced over the last few years. The Coalition Government was elected in 2010 on a pledge to cut the deficit in public finances. It reduced the budgets of all government departments and removed funding, partially or totally, from some NDPBs. It was noted above that the CRC was closed in 2012, becoming a high-profile victim of the government's 'bonfire of the quangos'. But others continue to play prominent roles in rural policy, design and delivery. These include:

- The **Planning Inspectorate** (PINS) provides independent advice to the Secretary of State for Communities and Local Government on planning matters. Through its appointed planning inspectors, it tests the 'soundness' of local plans to ensure that they have been developed in accordance with planning procedures and in conformity with the National Planning Policy Framework. It also tests their underlying evidence base; for example, whether enough land has been allocated to meet housing demand or whether policies framing the procurement of community gain through planning permissions are economically viable (see Chapter 8). PINS provides advice on 'called-in' applications, where the Secretary of State decides to directly determine the outcome of a planning proposal because of its scale, impact or significance. It also adjudicates on planning appeals where an applicant seeks to overturn a refusal by a planning authority to grant planning permission.
- The **Homes and Communities Agency** (HCA) is government's key housing and regeneration agency within England with a budget of roughly £4 billion for the 2012 to 2015 period. The agency distributes grants to the 'registered providers' of affordable housing (see Chapter 8). This function is performed by the devolved governments in Scotland and Wales.
- One of the oldest set of non-executive bodies funded by DEFRA is the thirteen **National Park Authorities** (NPAs) in England and Wales. These were established by the National Parks and Access to the Countryside Act 1949 and are tasked to conserve and enhance the natural beauty of the designated areas and promote the enjoyment and under-standing of the special qualities of the Parks by the public, whilst fostering the social and economic well-being of local communities within those Parks. They act as the designated planning authorities. Three further National Parks were created in Scotland after 2000 following the National Parks (Scotland) Act of that year.
- **Natural England** was created in 2006 and is funded by DEFRA. Its role is to protect and improve, and encourage public enjoyment of, England's natural environment.
- The **Forestry Commission** now works exclusively in England, although there are bodies in Scotland (Forestry Commission Scotland) and Wales (Natural Resources Wales) that fulfil a similar function. Funded by DEFRA, the Commission manages forestry in its own right and supports planting

programmes undertaken by others. It also plays a key role in the protection and enhancement of woodland, and finally.

- The **Environment Agency,** created in 1995, is concerned with pollution control and prevention, waste management, managing open water bodies, and protecting people from flooding. Many of these functions have important rural planning implications. For example, this is the agency responsible for implementing the Water Framework Directive and preparing catchment-wide river management plans.

Whilst the functions of many of these agencies have remained fairly constant in recent years, the way they operate has changed dramatically. Because of the recent cuts to the budgets of these agencies, they now frequently work with and through other groups including voluntary bodies. Some of their core work is now undertaken directly by the government departments they serve, though they continue to have a visible presence and an impact on how the countryside is planned.

Sub-regional governance and local government

Devolution to Wales and Scotland, and also to London, means that government in Britain might appear to be less centralised than it once was. However, the apparatus of regional government in England – Regional Assemblies and Development Agencies – has been dismantled by the Coalition Government since 2010. These regional bodies were established by the first Labour government after 1997 with the aim of developing approaches to planning and urban regeneration that would be more aligned to regional needs (Allmendinger and Tewdwr-Jones, 2006; Cullingworth and Nadin, 2006). Nothing has yet replaced the regional planning function, centred on the Assemblies, but government has attempted to grow economic governance – one of the remits of the development agencies – from the bottom-up, by inviting local authorities and local business groups to create voluntary partnerships which have been tasked to determine economic priorities and set the direction of business and jobs growth within their areas. There are currently 39 **Local Enterprise Partnerships** (LEPs) in England, which have developed their own 'growth strategies' using allocated monies from government's 'Growing Places' fund. These LEPs have now taken on responsibility for European Structural fund investment.

The LEPs are a relatively new construct and it is perhaps too soon to assess their contribution to local economic governance and development. However, some concerns have been raised over their capacity to deliver against their plans. Funding has been limited and the success of each partnership depends on the skills and resources of local partners, which vary from place to place (Shutt *et al.*, 2012). There are early signs, however, that LEPs are working well in many urban areas where there is an existing institutional infrastructure on which to build. In more rural areas, however, partnerships seem more patchy and fragmented, leading to concerns that LEPs may not work well for the countryside (Pemberton and Shaw, 2012; Henderson and Heley, 2014).

Unrelated to the LEPs, are the new **Local Nature Partnerships** (LNPs) in England. These were promoted by government through the Natural Environment White Paper 2011. There are currently 48 LNPs in England. They are constituted in a similar way to the LEPs, comprising '[...] a broad range of local organisations, businesses and people who aim to help bring about improvements in their local natural environment'. They are intended to be partnerships which '[...] work strategically to help their local area manage the natural environment. They aim to make sure that its value, and the value of the services it provides to the economy and the people who live there, is taken into account in local decisions, for example about planning and development'.

Although the LEPs and LNPs have different remits, they represent a common approach to policy delivery: one that is constituted on local partnerships and is intended to signal a shift from top-down imposed frameworks to local arrangements based on local policy and interest groups coming together to produce tailored responses to the particular challenges they face. The approach can be viewed as an extension of changes to the **local government** function since 1997. The incoming Labour government of that year brought with it a 'managerialist' approach to public service delivery, seeking more 'joined up' government that would be delivered not only by the traditional public sector but through partnerships with voluntary and private organisations. Labour brought forward a series of Local Government Acts which opened up service delivery to other bodies, requiring local authorities to seek and deliver 'best value' through partnerships and by contracting others to deliver key services. This evolved into a 'local strategic partnership' approach, with authorities leading consortia of business, voluntary and community groups and seeking to co-ordinate investments and actions so as to realise a particular set of development goals for an area – set out in a 'Community Strategy' (Morphet et al., 2007). Arguably, the Coalition has continued this approach, but created partnerships with more focused remits. It has done this in a context of fiscal austerity, asking local government to do 'more with less'. But today, this partnership approach is increasingly associated with a 'localist' agenda. There has been a rescaling of land-use planning, described below, that has seemingly concentrated more power in local authorities, largely by removing the regional planning apparatus. Greater responsibility has been handed to local partners, including community groups, to innovate solutions to local challenges. It is within this local governance context that changes to the statutory land-use planning function should be viewed.

Part 2c: Localising statutory ('land-use') planning

The deeper roots of statutory planning were considered in Chapter 1, and how approaches to planning have tracked the narrative of rural change was considered in Part 2a of this chapter. Statutory planning is only one part of the composite 'rural planning' that is examined in this book. But it is nevertheless an important part. The intention here is to provide an overview of the planning system as it currently exists. It was noted earlier that statutory planning splits into two key functions: the drawing up of land-use plans and taking decisions against those plans. This is the system that was created in 1947, with responsibility for managing it handed to local authorities (Counties or Borough Counties at the time). Those authorities comprise politicians who take the decisions and officers, whose role it is to advise politicians on the technical merit of those decisions or their conformity with planning law. Politicians, however, have always had the 'discretion' to weigh officer advice against other material considerations including the weight of local opinion in favour or against a particular plan or decision. Planned intervention in matters of land-use is one area of public policy that local authorities (and other tiers of government) have responsibility for. The system comprises (a) an institutional structure of competent bodies and (b) a set of instructions to those bodies, underpinned by law, as to how plans should be drawn up and decisions taken. The most recent and comprehensive overview of the system is provided by Cullingworth *et al.*, 2015.

Because land-use planning is just one part of the rural planning composite, it is given only summary treatment here, although key aspects of the system are unpacked in later chapters. The overview provided below centres on Table 2.4. This summarises the institutional structure (columns) and the form of instructions given to competent authorities ('national' row) following key structural changes made after Labour came to power in 1997.

Statutory planning: the national level

Planning is the responsibility of a single ministry in England: The Office of the Deputy Prime Minister until 2006 and the Department for Communities and Local Government thereafter. Contained within this ministry is a team of civil servants who normally lead on the drafting of planning legislation and guidance. That team is led by the Chief Planner for England. But the head of the ministry is the Secretary of State for Communities and Local Government, who has call-in powers over local planning decisions and is advised by PINS. During the 17 years represented in Table 2.4, there have been three fundamental shifts in the planning system: the creation of the regional apparatus outside government offices in 1998; the abolition of Structure Plans at the County Level following the Planning and Compulsory Purchase Act 2004; and the revocation of regional spatial strategies following the Localism Act 2011. There has been a 'hollowing out' of the middle tiers of planning, resulting in a system that today has only national and local parts. But at the top level, there is no guiding national plan. Guidance on how planning law should be interpreted by lower-tier competent authorities has been produced since 1988, but this is aspatial, setting out the intentions and aspirations of

Table 2.4 Scales of planning/planning reform in England

	1998/9–2004	2004–2010	2010–2015
National	Office of the Deputy Prime Minister (ODPM) (until May 2006)	Department for Communities and Local Government	Department for Communities and Local Government
	National Planning Policy Guidance (PPGs)	Gradually, Planning Policy Statements (PPS) replaced Planning Policy Guidance (PPG)	National Planning Policy Framework (NPPF) 2012
Regional	Government Offices RDAs/RAs	Government Offices RDAs/RAs	Nil – after 2012
	REGIONAL PLANNING GUIDANCE	REGIONAL SPATIAL STRATEGIES	
County	County Councils	Nil	Nil
	STRUCTURE PLANS		
Local	Local Planning Authorities	Local Planning Authorities	Local Planning Authorities
	DEVELOPMENT PLANS	LOCAL DEVELOPMENT FRAMEWORKS	LOCAL PLANS
Parish/Community	Parish/Community Councils	Parish/Community Councils	Parish/Community Councils
	VILLAGE DESIGN STATEMENTS/PARISH APPRAISALS/PARISH PLANS	VILLAGE DESIGN STATEMENTS/PARISH APPRAISALS/PARISH PLANS	NEIGHBOURHOOD PLANS AND ORDERS

government and how responsibilities should be locally discharged. The first comprehensive set of guidance comprised the 'planning policy guidance' notes or PPG. These were not legally binding, but authorities needed to have regard to them when preparing development plans. They were thematic and numbered: PPG3, for example dealt with housing and PPG7 with 'the countryside'. The PPG were replaced with a set of broadly similar 'planning policy statements' (PPS) after 2004: PPS3 still dealt with housing, but PPS7 was renamed 'sustainable development in rural areas'. A more dramatic shift occurred in 2012 when all twenty-five PPS were replaced with a single National Planning Policy Framework for England (NPPF), later supplemented with National Planning Practice Guidance (NPPG) designed to help authorities interpret and implement the now broader instructions coming from the national level.

Statutory planning: the regional level

The role of regional planning was clarified and strengthened after 1998 under the first Labour government led by Tony Blair. Regional Planning Guidance (RPG) was to be produced by the new Regional Assemblies and published by the Government Offices for each English region. County Structure Plans (produced from the 1970s onwards) needed to be in general conformity with the RPG. This system was swept aside after 2004 when the Labour government abolished Structure Planning (under the 2004 Act) and created Regional Spatial Strategies (RSS), which would henceforth be a component of Local Development Frameworks. These were essentially local plans that needed to be in compliance with, rather than just in general conformity with, the regional plans. This increased the power of regional planning, most notably because the regions were now driving the housing growth agenda, setting long-term targets and annual building rates, which needed to be included in the local plans. This all changed, however, after the Localism Act 2011 when the Coalition Government scrapped the RSS. This meant that outside of London there was nothing between the national and local planning tiers, and because there was no overriding national plan that local authorities needed to be compliant with, planning had essentially been localised to its lowest tier.

Statutory planning: the county level

The Counties (and County Boroughs, often covering urban areas) had been the original planning authorities empowered to produce plans and make development control decisions against them in 1947. A new distinction between structure (strategic) and development planning after 1968 divided responsibilities between County Councils and lower-tier District or Borough Councils (sometimes later fused under 'unitary' arrangements). But at the beginning of the period described in Table 2.4, County Councils retained their Structure Planning function. This function ceased after 2004. Although the Counties kept a range of responsibilities, including for highways and minerals, they no longer produced strategic Structure Plans after this date.

Statutory planning: the local level

Throughout the period, development planning duties have been discharged by local authorities, usually district councils in rural areas. It is at that level that local plans are drawn up. These are the basis of development control decisions. They are the plans in which land is allocated for different uses including, for example, new housing. The role of planning authorities has been a constant, though their power arguably weakened under Labour (as government sought to influence local outcomes by strengthening the role of regional planning) and then grew as the Coalition reasserted the primacy of local plans and removed the regional tier. As noted above, the way in which local authorities operate has evolved in recent decades. In the 1980s, the New Right governments (led by Margaret Thatcher) thought that authorities should be the 'strategic enablers' of actions taken by others. Hence, their housing development role, for instance, was largely removed and they were tasked to work through the voluntary sector and concern themselves with the wider functioning of local housing markets with the aim of satisfying as much need as possible without direct government spending. Under New Labour (led by Tony Blair) the 'managerialist' approach to local government also resulted in authorities positioning themselves within networks of local partners, pursuing broader 'place shaping' aims (see Lyons, 2007) through the actions and investments of those partners. This approach was continued under the Coalition (led by David Cameron), which established a number of new partnership structures (see above) and, more generally, continued efforts to promote local empowerment through the devolution of planning functions to the neighbourhoods and parishes below this statutory 'local level'.

There has been a significant degree of continuity in the reform of local government and the promotion of community interests under both right and left wing parties, though national politicians have inevitably sought to steer local actions in a strategically determined direction. Labour did this in the 1990s and 2000s by promoting the power of regional planning; the Coalition has promoted the role of PINS, allowing it to intervene in local matters where there is too great a divergence from national priority. It has also weighted the NPPF in favour of development as a counter-balance to local objection.

Statutory planning: the neighbourhood level

Throughout this book, there is regular consideration of how 'community action and planning' shapes rural policy and development outcomes. This was presented as a key dimension of the composite rural planning described in Chapter 1. Although the Coalition is now closely associated with community-based localism, there has been a long journey to formalised planning at a 'neighbourhood' level, which started in the 1970s. The early stages of this journey were charted in Chapter 1. They are also revisited and linked to growing 'social pluralism' in Chapter 5. Parker (2014), however, has identified three more recent stages, leading up to the establishment of 'neighbourhood planning'. The first stage (1995 to 2001) was characterised by the drawing up of often fragmented and disjointed village appraisals. These were patchy, though in some instances they alerted the local level to particular community aspirations and were found to be valuable for development control. The second stage (2001 to 2011) saw an increase in the production of more standardised Parish Action Plans and Village Design Statements. These were often found to be of greater value to local planning and in some instances were integrated into local plans. The third stage (from 2011 onwards) has been the establishment of a system of statutory Neighbourhood Development Plans which, it has been suggested, provide an opportunity for 'local people to get the right types of development for their communities (DCLG, 2012, Para. 184). These were legislated for in the Localism Act 2011. They are produced using a prescribed process and are voted on by local people in parishes or designated urban neighbourhoods. If 50 per cent of registered voters agree that the plan is right for an area, and that plan is aligned with the policies contained in the local plan (which retains primacy), then the Neighbourhood Development Plan (NDP) is 'made' and gains formal status in the development control process, still operated by local authorities. So far, NDPs have tended to focus attention on local design matters, the siting of housing, the identification of local assets and the promotion of specific land-use changes through associated development orders. By September 2014, more than 1,000 neighbourhoods or parishes in England were preparing plans but only 28 had been approved at local referenda (DCLG, 2014a). There is now much activity at the 'statutory' neighbourhood level and some of this is picked up in later chapters, especially in Part 3 of this book.

Conclusions

This chapter has unpacked several key aspects of rural planning. The first is the way in which statutory planning is both rooted in rural change, and especially the production to consumption transition, and the way in which the actions of planners and planning authorities have tended to reinforce particular development trajectories. Statutory planning across Britain, but especially in England, has been restraint-driven for much of the post-war period, initially in support of production but later because restraint, aimed at preserving amenity and property prices as well as the rural landscape, became the objective of powerful elites moving into and consuming the countryside. The second aspect unpacked was the institutional landscape of rural planning, from the European Union down to local actors. EU membership is an important feature of rural planning in Britain. Another is the move to partnership-based governance centred on local authorities. There has arguably been a 're-sectoring' and rescaling of policy delivery since the 1980s that was accelerated in the 1990s/2000s and is now expressed through new cross-sector partnership arrangements and a more visible role for communities in all areas of public policy, and notably in statutory planning. The final aspect to be unpacked was the statutory planning system itself, which has been subject to significant institutional changes since the late 1990s, resulting ultimately in a framework that has been 'hollowed' out in the middle, localised and more strongly connected to the machinery and aspirations of local governance. The aim has ostensibly been to achieve a more open planning system that more regularly delivers against local needs. The performance of the system, and wider rural planning, is examined in later chapters.

Summary

- At the European level, it is possible to identify and track the economic and social changes (and especially demographic changes) impacting on rural areas. These reveal key economic transitions and, very broadly, point to a distinction between rural areas of production and consumption.
- Whilst that distinction is evident spatially across Europe, differentiating more advanced and more laggard regions, it is also a key component of the changing narrative of the British countryside.
- Planning approaches in Britain have tracked three phases in rural development: a pre-war phase marked by concerns over depopulation and the 'discordant' countryside that would result, and then two major phases of intensive post-war production followed by a transition towards consumption, both underpinning distinct approaches to rural planning.
- Throughout the modern period, since 1945, there have been numerous changes in the 'institutional' environment in which planning occurs. This institutional environment has scales and actors: the scales cascade down from the European Union (which Britain acceded to in 1973) to the local level, and the actors comprise a range of governmental and non-governmental bodies, all with different resources that can be brought to bear on rural issues.
- Statutory planning (approaches to which are tracked in this chapter) is an important component of the 'composite' rural planning with which this book is concerned. It comprises actors and norms, which operate at different scales and are responsible for formulating policy and strategy and for producing local plans and taking decisions against those plans.

Key readings

- Cullingworth, B., Nadin, V., Hart, T., Davoudi, S., Pendlebury, J., Vigar, G., Webb, D. and Townshend, T. (2015) *Town and Country Planning in the UK*, 15th Edition, Routledge: London.
  (This provides a comprehensive and up-to-date account of the British planning system.)
- Duhr, S., Colomb, C. and Nadin, V. (2010) *European Spatial Planning and Territorial Co-operation*, Routledge: London.
  (Provides an excellent overview and analysis of planning across and between the European member states.)

Key websites

The following web-sites give official accounts of how government in the UK works, the current function of neighbourhood planning in England, the role of DEFRA and the role of DCLG:

- www.gov.uk/government/how-government-works
  (Explains how UK national government is organised.)
- www.gov.uk/government/policies/giving-communities-more-power-in-planning-local-development/supporting-pages/neighbourhood-planning
  (Provides access to resources related to Neighbourhood Planning.)
- www.gov.uk/government/organisations/department-for-environment-food-rural-affairs/about
  (Explains the responsibilities and organisation of DEFRA.)
- www.gov.uk/government/organisations/department-for-communities-and-local-government/about
  (Explains the responsibilities and organisation of DCLG.)

Part 2

Economy and land

3 The economic transition

••

Contingent and global rural economies

For many years, cities across the Western world have often been seen as the drivers of economic growth and rural areas, conversely, as serene, timeless, traditional and perhaps under-developed, or as Howard Newby (1988: 1) put it '[...] a source of peaceful certainty in an ever changing and modern world'. But scratch away at the surface and it quickly becomes apparent that rural areas are changing rapidly and their economic structures are becoming evermore 'urban'. It is sometimes the case that, owing to population dynamics and the flow of entrepreneurial individuals to the countryside (attracted by perceived quality-of-life factors), rural areas have become drivers of economic growth. Overall, there is a tendency in literature to problematise rural economies, seeing them as economically dependent on, or subservient to, much stronger urban economies. Rural areas more generally are framed as 'thin markets' as a result of lower population densities and their apparent inability to attract investment. Whilst there is undeniable truth in these presentations of the rural, there is also increasing recognition of the vital role played by rural areas globally, in delivering food security and in tackling major planetary challenges such as climate change. Moreover, it is acknowledged that rural areas are not inert economic spaces; rather they are host to a great deal of new and diverse activity. They are not merely a source of food for cities and 'the rural is no longer the monopoly of farmers' (Van der Ploeg *et al.*, 2000: 393). Rather, rural economies are shaped by the state, by large businesses, by 'un-embedded' finance capital (which is foot-loose, and not tied to the land) and also by communities themselves alongside an array of place-specific assets. The economic performance of rural areas is determined by 'connexity' (see Chapter 2) to global opportunities and also by the behaviour of middle-class migrants, who have become the new consumers and local investors in the countryside. Therefore, viewing the 'rural' as a residual pastoral periphery of nearby 'urban centres' obscures the complexity of economic drivers and relations that provide agency, not just to one homogeneous rural economy, but to various different and peculiar rural economies that are contingent on various specific local, regional and global characteristics and opportunities.

This chapter and the next together form the rural economy section of this book and they aim to provide the reader with a critical understanding of the recent evolution of rural economies globally and more particularly in Britain. Agriculture and its economic alternatives are examined more closely in Chapter 4 while this chapter charts the changes in rural economies more broadly and introduces the key tenets of the prevalent theoretical and policy approaches pertaining to rural development, building on the introductory comments made in Chapter 1.

Part 1: Defining and theorising the rural economy

Delineating rural economies

Whilst the general nature of 'rurality' is contested (see Chapters 1 and 11), there is greater consensus around the definition of a 'rural economy'. The concentration of economic activity in a particular place provides a straight-forward means of delineating dense urban economies from sparser rural economies (Gray, 2014). However, it should be observed that density of economic activity does not strictly correlate with growth performance, and many rural regions demonstrate faster growth than urban ones (OECD, 2011). The majority of rural typologies use the density of population to distinguish different types of rural region, or to set urban areas apart from those that seem to be obviously rural in a more general sense. The OECD defines 'rural communities' as those with fewer than 150 inhabitants per km^2 and divides these into a three-part typology: predominantly rural regions where more than 50 per cent of the population lives in rural communities; significantly rural regions where 15 per cent to 50 per cent of the population lives in rural communities; and predominantly urbanised regions with fewer than 15 per cent of inhabitants living in rural communities.

Parkin (1998, cited in Marini and Mooney 2006: 93) poses five questions that might be used to differentiate a rural from an urban economy:

1   What goods and services are produced?
2   In what quantities are these goods and services produced?
3   How are goods and services produced?
4   Where are goods and services produced?
5   Who consumes those goods and services?

Marini and Mooney (2006) suggest that Parkin's questions help tease out the distinctive *tendencies* of rural economies, such as their emphasis on land-based goods including agricultural products and environmental services, produced through a mix of traditional and advanced technologies in seasonal production cycles. They further suggest that rural production tends to occur in the open air, and is consequently more exposed to the impacts of climate and more reliant on local markets. It is certainly the case that, in the past, rural economies were very much a function of the rural environment and landscape; and that that landscape has in turn been shaped over many centuries by the presence of 'traditional' economic activities (Hoskins, 1955). However, industrialisation 'de-traditionalised' many Western rural economies and resulted in a greater mix of activities as well as movement away from farming. It was noted in Chapter 2 that industrialisation, concentrated around growing cities, generated a need for raw materials – including coal – that could often be found in rural locations. Hence, near-urban rural areas became loci for new extractive industries. This role was retained in some parts of the world. But elsewhere, including in Britain, such industries regularly fell prey to cheaper imports, leaving an imprint of economic decline on the rural landscape.

The questions posed by Parkin offer one means of understanding and demarcating rural economies, in an attempt to reveal what makes an economy identifiably rural. Marini and Mooney (2006) point to a range of popular approaches. In the US a range of policy challenges have been identified that seem to be more concentrated in rural areas: dependency on 'unearned income' in the form of benefits or pensions; the extent of public (federal) landownership and the extent, distribution and persistency of poverty within the population. These have all been incorporated into assessments of economic rurality. In Britain, and drawing on data from the 2001 Population Census in England and Wales (using Local Authority Districts as the spatial unit), Lowe and Ward (2007) have constructed a typology of rural economies comprising seven area types, each representing specific and tangible social and economic characteristics and the various ways that these manifest themselves. Having removed 'urban districts' from the Census database, data for more than 100 variables were subjected to a factor analysis enabling the identification of 15 critical, uncorrelated variables, central to shaping different types of rural economies. These variables, acting as drivers, opportunities and challenges on which rural economies are contingent, are listed in Box 3.1 below.

Similarly, the EDORA project that studied the development opportunities and challenges facing rural areas in Europe (see Chapter 2) identified the major processes characterising rural economies as those of ongoing economic diversification; commodification of rural assets such as cultural and environmental values; agricultural specialisation and intensification; and migration (rural exodus as well as in-migration). Transformations in public service provision are also a significant influence on rural economies, especially where austerity-driven marketisation of services has led to a search for benefits of scale (EDORA, 2010a), causing a concentration of key activity in hubs or service centres (see Chapter 7).

Further details of the EDORA analysis were provided in the last chapter,

Box 3.1 Delimiting rural economies: Key variables

- number of residents per 100 hectares;
- percentage change in resident population 1991–2001;
- proportion of population aged 65 and over, 2001;
- percentage change in total employment 1991–2001;
- rate of economic growth 1991–2001;
- average total income 2000–2001;
- proportion of knowledge workers to total economically active 2001;
- percentage change in knowledge-sector employment 1991–2001;
- proportion of managerial and professional workers to total economically active 2001;
- number of hotel and restaurant businesses;
- proportion of employment in agriculture, hunting and forestry relative to total economically active;
- proportion of households with two or more cars 2001;
- net commuting 2001 (GB = 100);
- number of national heritage sites per 1,000 sq. km 2002; and
- index of tranquillity 2001 (GB = 100).

with Figure 2.3 highlighting a broad demarcation of European rural economies into agrarian, consumption and diversified (secondary sector or private sector) classes. Such classifications, or the general understanding of economic processes and challenges that they provide, have become a framework for public-policy intervention, ranging from broad top-down investments to local capacity building. These are examined later in this chapter.

Understanding rural economies: key theoretical approaches

A range of economic theories can help policy-makers and planners understand how rural economies are likely to evolve and respond to various interventions. While an in-depth examination of these is beyond the scope of this book, this section will outline some of the most significant theoretical approaches that inform the wider discourse and support particular responses on the ground. Economic theories tend to treat the rural as a spatial or territorial unit below the national level, identifying spatial units of analysis and/or intervention. The NUTS classification (Nomenclature of Territorial Units for Statistics) is now the standard approach in Europe. This is used to divide up the economic territory of the EU for the purpose of socio-economic analysis, comparison, and the targeting of different policy interventions. NUTS provide a template (NUTS 3 being the smallest regions used for diagnosis) and ancillary analysis – for example, EDORA – can identify rural economies, but Gray (2014) points out that there is no tailored 'rural economics'. The theory applied is of a general nature, dealing at a micro-economic level with the behaviours of individuals, firms or sectors, and at the macro-economic level, with 'spaceless, aggregate measures of activity' (Gray, 2014: 33).

Theories focus attention or cast light on various aspects of an economy, how it functions and the predictable ways in which that function will alter in different contexts or in light of external forces. Theories of agglomeration, for example, draw attention to the impacts that household density or the clustering of certain types of business will have on the viability of services, the concentration of labour and the positive knock-ons for ancillary businesses and suppliers (Puga, 2010). This is an important theory in rural areas, supporting a range of interventions and planning approaches, from the promotion of key service centres (including market towns) to the effective barring of development in smaller settlements owing to restrictive planning (see Chapters 2 and 7). More broadly, export-based economic models help explain regional growth by examining the external linkages that enable an exporting sector to derive wealth from extended markets and therefore support a range of dependent sectors across knowledge, innovation and the service industries. Territorial advantage is rooted in those linkages rather than in local resources, or in opportunities for agglomeration (Grossman and Helpman, 1990), although the spatial distribution of skilled labour remains a key driver of that advantage, giving exporting sectors the means to specialise and innovate. This is often identified as a challenge for rural economies. Building the skills-base requires local investment (for example, in education) or the import of skilled labour: but investment is attracted to areas of greatest potential yield or, in the case of public services, where each pound, euro or dollar, delivers greatest leverage. Gray (2014) argues that rural regional economies are often path-dependent, with structural and institutional context determining the scope for export-led growth. The extent to which they are reliant on primary production relative to secondary (manufacturing and other value-added activities) and tertiary (service-based) seems to signal future economic trajectory, resulting in the growing 'agrarian'/'diversified' division across Europe.

Besides these local and regional structural insights, economic opportunities are increasingly propelled by the forces of globalisation and its associated flow of goods, money and people. Gray (2014) uses ideas of a 'new international division of labour' to explain how and why multi-national companies disperse research and development functions, headquarters and routine production across different regions and countries, situating them in such a way as to exploit advantages in terms of labour, resource and regulatory costs. This model can explain why some rural areas, with their unskilled and low-wage workforces, may become targets for international investment, although such investment will not necessarily lift those areas out of relative poverty. Business relocation, therefore, may have an ambiguous effect on economic

Figure 3.1

A fruit and vegetable processing plant in Almeria, Spain: sorting peppers, mass-produced locally, for export

wellbeing. Likewise, investors in rural hinterlands adjacent to regional centres often maintain or establish strong economic ties with suppliers and support services in those centres, meaning that potential benefits do not become embedded and are not retained, or at least there is limited local knock-on (Courtney and Errington, 2000). Leakage of activities and opportunities from rural areas seems to be a particular problem with manufacturing investment (Courtney and Errington, 2000) as many specialist suppliers locate in nearby cities; and also with the retail sector, with out-of-town shopping providing few opportunities for the rural workforce and taking business away from small retail operators (Guy, 1998).

The argument that information and communication technologies (ICT) have extended the reach of globalisation and resulted in a more networked society (Castells, 2004) with profound social and economic consequences is now commonly heard. Relationships between people and between businesses and their customers are less constrained by place and by distance. Competitive advantage is now gained by those able to exploit new techno-logical opportunities. In the 'network society' that we now find ourselves, innovation and the capacity to adapt to changing market demands are vital to achieving growth and competitiveness. This requires a business model that Castells (2004) describes as 'the network enterprise', which comprises companies that are, first, divided into looser networked segments rather than rigidly structured and, second, linked via the global financial market. This is a very general view of the right arrangements for business success. In rural areas, it translates into greater 'connexity' (see Chapter 2) because whilst research, development, marketing and branding are becoming critical to the production of economic value, all of these functions can be outsourced to small and medium-sized consultancies located anywhere in the world. The critical locational question is not about materials but about amenity and quality of life: where do people wish to locate and work? The answer in some Western countries has been the accessible countryside, to which there has been a steady counter-urbanisation flow since the 1960s. Gray (2014) therefore argues that connexity will drive the relocation of smaller, creative enterprises to rural locations given their particular lifestyle attractions.

Connexity is a potentially ubiquitous driver of growth. But whether or not it triggers that growth, and the spatial patterns that are eventually observed, will be determined by the presence of key *assets* (human, natural, social, financial and physical) and *capabilities* in different locations. It is also the case that *structures* (institutional, legal, political, and cultural) will have a decisive effect on outcomes (Schoones, 2009). Agency and structure are at work everywhere, the former providing the capability (through social capital and entrepreneurial culture, which are increasingly seen as important determinates of endogenous development outcomes; see Chapters 1 and 5) and the latter creating the context. Many structural determinants might be identified as critical to encouraging and supporting entrepreneurial action, including the extent to which land-use planning (embedded in political and regulatory frameworks) helps or hinders innovation. Does it, for example, permit the creation of spaces needed for creative business start-ups? But another important structure in relation to connexity is access to broadband and the elimination of internet 'notspots' (the inverse of more desirable 'hotspots'). Rural broadband access and speeds are examined in Chapter 7.

Whilst it is broadly acknowledged that interaction with nearby urban areas continues to impact on rural economic opportunities, not least by providing a market for rural goods and providing a source of consumption-led investment, the emergence of a regionalised patchwork of rural economies with varying growth trajectories and challenges reflects the mixed ability of regions to connect to more distant markets, diversify and innovate, and attract/retain new entrepreneurs. The interplay of assets, capabilities and structures helps to explain the mixed fortunes and the differentiation of rural economies revealed by performance data. In the remainder of this chapter, the changing nature of Britain's rural economies is examined and explained using some of the concepts introduced above.

Part 2a: The development of the rural economy in Britain

Of the approaches and processes outlined above, migration, commodification (and consumption) and a subsequent emphasis on the service sector as an employer (and GVA producer) are significant drivers of economic diversification in the British countryside. As the EDORA maps presented in Chapter 2 show, the rural NUTS 3 regions, with the exception of some areas on the English East Coast and Wales, are predominantly performing above the EU average in economic terms (EDORA, 2010a). Regarding structural characteristics, according to the EDORA classification, British rural NUTS 3 regions are mainly 'consumption countryside' where economic diversification is underpinned by the provision of countryside public goods and diversification into a range of new activities, such as food processing, recreation and tourism; or 'diversified' where the market-based services sector is emphasised. As suggested in Chapter 2 and noted above, the significance of 'connexity' – the increasing interconnectedness (rather than centralisation) of economic actors as well as the mobility of people and goods – is therefore clearly evident in the British rural economy.

The rural as an economic hub

Producing high value goods and services

A recent report by DEFRA (2014a) reveals that a disproportionate quantity of Gross Value Added (GVA) is generated in rural compared with urban areas. GVA is a measure of the contribution to the economy of each individual producer, industry or sector. In simple terms, it is the value of the amount of goods and services that have been produced, less the cost of all inputs and raw materials that are directly attributable to that production (see DEFRA 2014a). At the last population census, undertaken in 2011, there were 53 million people living in England: 82.4 per cent were resident in predominantly urban areas, and the remainder – 17.6 per cent – in predominantly rural locations, as defined by DEFRA.[1] During the previous year, 19 per cent of GVA (£211 billion) was generated in the predominantly rural areas and 12 per cent from significant rural areas (£137 billion). These figures compared with 25 per cent of national GVA from London and 43 per cent from the other predominantly urban areas (DEFRA 2014a). Productivity levels are, on average, much higher in London than elsewhere. Nevertheless, after London, significant rural areas have the highest productivity *per job* relative to the English average. Furthermore, a proportion of the urban GVA is likely to be generated by individuals who work in cities or major towns but live in the countryside.

All this perhaps flies in the face of the traditional narrative; that cities are the drivers of growth and, in some way, the custodians of national economic wellbeing. In an increasingly 'urbanised' world (see Figure 2.4, Chapter 2) it seems inevitable that the policies set out by governments will emphasise urban economic needs. One manifestation of this is the promotion of networked city regions, thought to be central to the growth paradigm. Interconnectivity between urban and rural areas (as hubs and spokes) remains important in development discourse, but there is increasing concern that the focus on cities is leading policy-makers to overlook the specific, distinctive and diverse needs of rural areas, not to mention the clear contribution of many to economic growth (Pemberton and Shaw, 2012; Henderson and Heley, 2014). Figure 3.2 shows which industry sectors are the most significant GVA producers in rural and urban economies. In general terms, the structures of these economies today appear very similar.

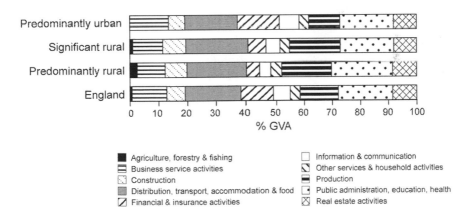

Figure 3.2

Percentage breakdown of Gross Value Added by industry and by local authority classification in England, 2010
Source: DEFRA, 2014a: 34

High employment, low salaries

Drawing on employment data, the Commission for Rural Communities' State of the Countryside Report for 2010 shows how rural and urban employment structures are converging (CRC, 2010: 106). Mirroring the EDORA project's findings, according to the CRC the property and business services sector was the largest employer in rural areas, providing 16.1 per cent of all jobs. In England's urban areas, the sector employs 20 per cent of the active population. Together with retail, hotels and catering deliver another 15 per cent of total employment in rural England; and education, public administration and other services – together with the health sector – provided exactly one quarter of jobs in 2009 (CRC, 2010: 106). While manufacturing (13.1 per cent), agriculture (7.5 per cent) and mining/quarrying and utilities (5.1 per cent) remain significant – the latter two mostly in remote rural areas – it is fair to say that the employment structure of rural England evidences (at least for the time being) the post-war narrative shift, retold in Chapter 2, from production to consumption. This story may develop further in the years ahead; health, wellbeing and food security may all be affected by a combination of growing global demand for resources and the impacts of climate change on food production systems.

Food remains important; not just for survival – as fundamental as that is – but also as an economic product. The UK 'Food 2030 Strategy' suggests that instead of viewing farming (i.e. direct production) in isolation, it should be considered as part of a much bigger 'agri-food' sector – comprising production, distribution, processing and selling to consumers through shops, supermarkets, cafes and restaurants – which, in 2007, was worth £80.5 billion to the UK economy, making it the single largest manufacturing sector (HM Government, 2010). For the agri-food sector, rural–urban linkages are critical; all activities from production through to consumption occur across networked space. It is also worth noting that although the proportion of total employment in agriculture is low in rural areas, in terms of business stock it dominates in the smallest settlements (CRC, 2010). Nevertheless, agriculture is unlikely to become a significant source of employment opportunities except for unskilled seasonal workers. This issue is picked up again in Chapter 4.

In terms of other sectors, whilst there are broad similarities in industrial structure between rural and urban areas, there are also key differences. Manufacturing is proportionately more significant in rural than urban areas. But the former are out-performed in the fields of financial services (which are clustered in urban locations, including London and Leeds) and in information and communication businesses. These are the industries that have greatest reliance on connexity, underpinned by not just good but excellent ICT infrastructure. The 'notspots' are very significant for these industries, as are difficulties in finding deep pools of skilled labour. And whilst rural areas are big producers of GVA, average business turnover per employee is significantly higher in urban (£180,000 per annum) compared to rural areas

(£118,000 per annum) with those located in sparse settings generating just £76,000 in turnover per employee. This is a function, in part, of the size of local markets, which is an important determinant of turnover for a great many businesses. And where turnover is lower (despite GVA) potential earnings for employees is also reduced. In the major urban areas, average earnings were £26,900 per annum in 2012 compared with £21,400 for significant rural areas and £19,900 for those living in the Rural-80 authorities (DEFRA, 2014a). These statistics are not easy to unravel. Earnings link to turnover and not to GVA, and lower earnings relate to part-time employment (see below) in businesses constrained by their thinner and smaller markets and a preponderance of seasonal, farm-based employment. But these features of rural employment are overlaid with other complex patterns; of commuting to urban jobs; of high turnover opportunities in creative industries, subcontracting work from city-based companies; and of course the general differentiation of rural economies.

Part time jobs

Despite the impact of the recent recession on small and medium-sized enterprises, employment rates (the proportion of the working aged population in employment) are higher in rural (at 75.2 per cent) compared with urban areas (70.1 per cent). Employment rates are, however, misleading.

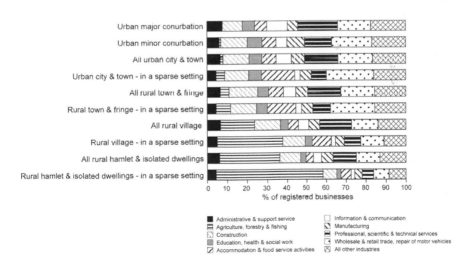

Figure 3.3

Percentage of registered businesses (single site or headquarters) in England, 2012/13
Source: DEFRA 2014a: 46

The reality is that employment data are recorded at the home rather than the work address, and many people recorded as employed are in fact commuting to jobs in towns. Other employment data are, however, indicative of the patterns of work in the countryside. For example, there is a slightly higher proportion of people working part time in rural areas compared with urban ones; men are slightly more likely to work full-time (89.6 per cent compared with 88.4 per cent), but women are significantly more likely to work part-time (48.7 per cent compared with 41.2 per cent), and thus the data are skewed in favour of part-time employment. These gender differences might be explained in terms of counter-urbanisation choices, with one partner in a dual household opting to work part-time as the price of moving out of a town or city to capture the lifestyle or amenity benefits of a rural location. DEFRA (2014a) also points out that the 'gendering' of employment patterns also relates to work availability, accessibility to jobs (by public or private transport) and an individual's (usually a woman's) caring responsibilities, all of which conspire to change the nature of work opportunity.

Numerous but small businesses

The suggestion that rural areas are becoming hubs of economic activity is based on a key piece of evidence: that there are significantly more businesses per 10,000 inhabitants in predominantly rural areas compared with predominantly urban ones. But there is an important caveat in this evidence: that most rural businesses fall into the category of small and medium-sized enterprises ('SMEs' with fewer than 250 employees) and the majority of these are micro-enterprises (employing fewer than 50 people). The majority of larger businesses with a workforce of more than 250 are urban-based (61 per cent of those employed in registered businesses). The figure for rural-based businesses is 27 per cent. The big companies (e.g. in banking or financial services) or institutional employers (government or universities) are in the cities whilst in rural areas, the only big employers are in manufacturing. Hence the urban/rural split in number of employers per business is a mirror to the national industrial structure, skewed in favour of services (private and public) rather than manufacturing. Other major activities in rural areas, including farming, generally have smaller workforces or engage casual labour. Smaller businesses therefore tend to provide the economic lifeblood for rural areas.

Prior to the economic downturn that began in 2007, the rate of business start-ups per 10,000 inhabitants was generally higher in rural than urban areas. This can be explained with reference to the consumption narrative: more people heading to the countryside to spend their money on rural products or rural accommodation, and also a vibrant service sector dealing with property investment and all the legal services that support that investment. But after 2007, a slow-down in consumption had a dramatic effect on small rural businesses, and the recovery in more recent years appears

Figure 3.4

A small rural enterprise: The 'Woolpack' pub in Stroud

to be slower and more difficult to achieve in those same areas. In 2005, the rate of business start-up (per 10,000 inhabitants) was 50 in rural areas compared to 47 in urban areas. This dropped to 36 in 2009 (compared to an urban figure of 40) before recovering to 40 in 2012 (compared to 47 in urban areas (DEFRA, 2014a).

Many small businesses are tourism related, producing local goods or providing accommodation or information for visitors. Tourism as a sector is thought to offer huge growth potential for rural economies. Many of the activities shown in Figures 3.2 and 3.3 have tourism dimensions, and many of the customers for rural products and services are weekend or seasonal visitors. Today, tourism features as a reference point in many rural economic and planning strategies, being seen as both an opportunity (see the remainder of this chapter and the next) and a pressure to be managed (see Chapter 10). Visit England's current Rural Strategy Action Plan targets a 5 per cent per annum growth in income from tourism between 2010 and 2020. Tourism often builds on or complements traditional activities such as farming, forestry and fishing and is examined further in Chapter 4.

The statistical evidence from DEFRA discussed in this section paints a picture of rural England in which agriculture, if not entirely ripped out, has a decreasing role in producing economic value and delivering employment opportunities. However, whilst direct farm incomes can be modest, the agri-food industry, broadly defined, is hugely significant. And agriculture retains a highly significant role in producing many of the supporting goods on which the consumption countryside relies. Moreover, farming is a varied sector associated with complex patterns of diversification and 'pluriactivity'. Although the EDORA research cited earlier showed that population growth and retention in much of the European countryside is a result of proximity to urban centres, agriculture still provides a demographic anchor in remoter areas, drawing migrants to farming jobs (EDORA, 2010b) although the lower 'connexity' of these areas tends to result in influxes of lower-skilled workers, unless the amenity value of destination areas is such that they attract skilled migrants. Agriculture's importance to the social, economic and environmental life of the countryside is such that it is given specific treatment in the next chapter.

The emergence of regional rural economies

Whilst there is strong evidence of economic diversification in Britain, there is great variation in the drivers and outcomes of regional rural economies across the country. There are areas where manufacturing or intensive agriculture have shaped an export-led model of growth, while elsewhere, particularly in areas of high landscape value and easier access to urban centres, a different demographic has developed due to these particular 'assets'. Numerous scholars have attempted to capture and project the nature of the ongoing 'differentiated' economic shifts, exploring an economy that spans a contextual and political redefinition of what the rural is, as well as the economic and social diversification occurring in rural areas (Marsden, 2006: 4). The diversity in the economies of the countryside, described in this chapter, connects to conflicts and tensions that are explored in later chapters. These conflicts – for example, between farming and tourism (Chapter 4), or between tourism and 'local' place communities (Chapter 10) – have influenced the trajectory of economic development; and despite attempts to achieve comprises and consensus, particularly at a community level and through the empowerment of local actors, these conflicts continue to challenge conventional forms of land-use planning.

In their progressive account of the local, national and global forces shaping economic and social change in the British countryside, Marsden and colleagues (1993) focus on the 'systems of relationships' that define rural localities. They single out the consequences of the consumption shift and different representations of property rights, increasingly as non-productive assets, as key to the mechanics of rural change. Reflecting on the situation in the early 1990s, Marsden and colleagues argued that the transfer of property into new, hitherto urban, hands is the key factor producing a spatial diversification of rural Britain since the demise of the hegemony of agriculture in the rural economy. With the changing structure of rural communities (Chapter 5) and economies (Chapter 1 and above), new interests (or new 'elites') gained power in rural land management and planning. This meant that localities began to differ in the way land-use was regulated, with external economic processes gaining prominence. Incomers tended to give priority to the maintenance and protection of aspects of the rural idyll that attracted them to an area in the first place, such as tranquillity, nature values and their particular understanding of rural community life. A new defence of the countryside began, which reinforced economic shifts in some areas, creating preserved and contested versions of the consumption countryside. In addition to these social and economic parameters, Marsden and colleagues (1993) singled out political parameters reflecting ideals of representation and participation as crucial to shaping change. Depending on the socio-material context, the main actors could mobilise around either production interests or, for example, environmental issues (Marsden *et al.*, 1993: 186; Marsden, 1998). In short, variations in the relative role and power of agricultural, residential and other commercial interests were seen to have led to different outcomes in terms of the type and extent of economic development that is experienced in different parts of the countryside. Box 3.2 contains a typology of four differentiated 'ruralities' that emerged in the 1990s' countryside, and which continue to shape rural economies.

Lowe and Ward (2007, 2009) provide an alternative, and more fine-tuned and recent, typology of regional rural economies in England and Wales, based on the key variables listed in Box 3.1. This comprises seven area types, each representing specific and tangible social and economic characteristics and the various ways that these currently manifest themselves. While the more recent statistical information from DEFRA suggests that rural economies have suffered to some extent from the economic downturn since 2007, Lowe and Ward's typology nevertheless provides a visual aid to understanding how place-specific assets shape an area's ability to benefit from emerging opportunities. Box 3.3 contains descriptions of the seven types of rural economies, identifying how proximity to, or distance from, urban cores may impact on economic wellbeing. It also highlights how particular demographics associate with different levels and types of economic activity. The mapping of these types was presented in Chapter 1, Figure 1.4.

Part 2b: The role of planning in the emergence of regionalised and diverse rural economies

Some aspects of rural economic regionalisation in Britain connect to the changing structures of economic governance – and particularly devolution of responsibility – that has occurred over the last couple of decades (or, more specifically, the 17 years of planning reform and institutional change depicted in Table 2.4 in the last chapter). The decentralisation of decision-making to

Box 3.2 A differentiated countryside (Marsden *et al.*, 1993)

The preserved countryside: These areas, situated in the English lowlands or the attractive and, crucially, accessible uplands tend to house a large proportion of new inhabitants mostly urban middle-class commuters. This new – and often highly educated and vocal fraction of the community – is able to influence the planning system in favour of amenity conservation and thus limit both agricultural intensification and diversification into unwanted mass tourism activity or other environmentally harmful development. The preserved countryside consists of areas where the reconstitution of rurality is often highly contested. While the new inhabitants express a demand for leisure services and residential property, they also aim to protect their positional goods against what they perceive as harmful agricultural diversification in the form of intensive tourism or farming activity.

The contested countryside: Formed in remoter locations and thus housing fewer urban incomers, the contested countryside still depends on farming as the main economic activity. In the absence of diversification opportunities provided by demand from new incomers or urban tourists, agriculture tends to follow a productivist logic. Farmers remain in an influential position in decision-making about local issues and are able to promote local economic priorities in the face of, as yet, weak consumption interests. In these areas the conflicts between new incomers and existing landowners are at their strongest.

The paternalistic countryside: In areas of large private estates and big farms, existing landowners retain a hegemonic role in the rural land-development process. But because of the need for economic diversification in the light of falling agricultural incomes, large landowners will often have sold assets such as farm buildings and estate housing. Conservation measures (attracting grant support) and affordable housing provide significant options for diversification to the often more socially conscious landlords that aim to remain in the area, maintaining a long-term management view of their properties.

The clientelist countryside: In remote areas unaffected by counter-urbanisation, agriculture remains the main economic interest, and state (or CAP) support for agriculture and rural development plays a major role in maintaining the local economy. Environmentally damaging development projects and types of economic diversification that would be rejected in other areas often encounter little opposition in the face of a strong need to sustain the existing, and create new, local livelihoods. These are also areas where the disappearance of whole industries such as coal mining has led to high levels of unemployment and severe polarisation of incomes. Employment concerns and the welfare of the rural community dominate often highly corporatist local development.

locally-accountable bodies, and particularly to cross-sector partnerships, has been justified on the grounds of responding to local circumstances: working with local assets and capabilities within the wider structural context. Two

Box 3.3 Typology of rural England

Dynamic Commuter Areas: Socially and economically dynamic and affluent, these areas have relatively high population densities and young to middle-age, high-income professional classes predominate. Economically, they are closely connected to adjoining urban areas with high levels of out-commuting. They can be considered as a form of wealthy outer suburb. A concentration of these areas is found in the South East of England.

Settled Commuter Areas: Similar to the above but less economically vibrant, these areas are located on the edges of the provincial conurbations such as Birmingham, Bristol, Manchester, Newcastle and Sheffield. The fortunes of these areas are closely linked to economic trajectories of their related city regions, again due to high levels of out-commuting.

Dynamic Rural Areas: These slightly lower density rural areas have fast growing economies and an increasing population. Characterised by high concentrations of professional and knowledge workers they experience lower levels of out-commuting and are less connected economically to urban conurbations than the above. The presence of major institutions such as universities or research centres tends to be a key feature of these areas, which are to be found mainly south of a line between Avon and the Wash.

Deep Rural Areas: These areas perhaps best reflect traditional notions of the 'countryside'. Agriculture, particularly livestock farming, and tourism form an important aspect of the local economy. Population and income levels are below rural averages. Economically, and in terms of net migration, they may appear to be in steady state, but recent events (such as foot and mouth disease in 2001) have revealed the fragility of their situation. Commuting to urban areas is limited (due to physical remoteness or poor infrastructure) as is the ability to attract younger in-migrants and entrepreneurial activity. These areas are found, for example, in Northumberland and mid-Devon.

Retirement Retreat Areas: These areas form popular retirement destinations and are found particularly along the coast, most notably in Southern England. Retirement-related services including leisure, social care and health are significant sources of jobs. Although income levels tend to be below average, the relatively high population density of these areas means that service provision is economically viable. With the ageing population fuelling demand, the economies of these areas are relatively vibrant.

Peripheral Amenity Areas: These rural areas present some of the most significant economic and social challenges in the countryside. Located in marginal zones often in coastal locations such as the west coast of Cumbria, the economic structure is dominated by agriculture, tourism and retirement-related services. However, they are secondary tourist or retirement destinations due to their isolation, or poorer quality environment, sometimes associated with past industrial or mining activities. Income levels are well below the national average amongst both existing and incoming populations, attracted by low house prices. Generation of new economic activity tends to depend on public intervention.

Transient Rural Areas: These areas are situated relatively close to urban centres and, as sources of local employment are not substantial, they experience significant levels of out-commuting. However, unlike the Dynamic and Settled Commuter Area categories, commuting does not imply high incomes. Although the majority of the population is economically active, average income levels remain very low. These areas are not attractive for entrepreneurial activity. The East Riding of Yorkshire and South Norfolk serve as examples.

major themes are considered in this final section. The first is the major trends in economic governance that are observable over this period. The section begins by looking at regional arrangements from 1998 onwards and ends by highlighting the future role of LEPs (see Chapter 2) as a localist economic governance arrangement, which seems likely to lead to the further differentiation of rural economies in the future. The second is the role of land-use planning in economic development. Statutory planning is often presented as a brake on development, but its impact differs depending on prevalent socio-economic and political parameters (Marsden *et al.*, 1993).

From regionalism...

Economic governance, promoting and regulating development and setting the focus and direction of investment, has its roots in the 1940 Barlow Report, with its concern for the distribution of industry. Later, this concern was manifest in the promotion of decentralisation and in the development of a regional agenda. In 1999, following the Regional Development Agencies Act of the previous year, the then Labour government established Regional Development Agencies (RDAs) in England whose role it was to develop regional economic policy tailored to the needs and assets of different parts of the country. The Agencies (one for each standard region; so nine in total) were also given substantial budgets and therefore the power to invest in key infrastructure and in specific sectors. The RDAs assumed responsibility for the economic performances of their respective regions. They produced Regional Economic Strategies (RES) that were nevertheless criticised for a lack of regional specificity (Ward *et al.*, 2003) as they often seemed to underplay how particular local assets provided anchors for specific types of development effort and investment. The strategies were thought by some to be too generic, emphasising standard development approaches – including agglomeration around hubs and export-driven growth – without linking these approaches to local conditions or getting down to a level that would enable them to build on particular area-based strengths. The Assemblies appeared to be working with standard development tool-kits and there was little evidence of locally-sensitive strategy.

Through this regional approach, government sought to enhance economic opportunities in rural England, and more specifically to '[...] reduce the gap in productivity between the least well performing quartile of rural areas and the English Median' (DEFRA, 2008: 94). This was one of Labour's Public Service Agreement (PSA) targets. These were developed for all areas of public policy intervention and government hoped to achieve its rural development PSA by channelling additional resource allocations through the RDAs. One way this was to be achieved was through the designation of a supported 'rural pathfinder area' in each of the regions outside of London. These were to be test-beds for innovation in economic development and service delivery. It was also hoped that the designation of pathfinder would also address concerns that too many initiatives were being targeted at urban areas as part of an 'urban renaissance' agenda, which was clearly Labour's priority in the early 2000s.

The pathfinder experiments, however, did little to counterbalance the focus on city-regional and urban agendas. The argument that regional economic governance had too crude a focus (and was excessively 'top down') seemed to be winning more support as the decade went on. Government began to look for ways to drill down below the level of the region, though it retained its emphasis on strategic planning and began focusing on the sub-regional agenda, now through the promotion of Multi Area Agreements (MAA) between diverse stakeholders and business groups (Morphet, 2010). This was part of the partnership-based governance shift noted in Chapter 2. MAAs were viewed as a tool for promoting joint-working and collaboration on 'strategic' issues of relevance to the effective functioning of city-regions, such as economic development, housing and transport. Nationally, sixteen MAAs were signed. Some focused on city-regions; some on small provincial cities and their rural hinterland; and others were more rural but containing key service centres. The MAAs were concerned with 'functional economic areas', which is why they tended to delimit urban hubs and their rural spokes: rural economic functions were invariably seen as subservient to urban ones, with the promotion of physical urban–rural connectivity viewed as the only means of growing rural economies. In policy terms, the spatial privileging of urban areas, at the expense of rural ones, was readily apparent from policy rhetoric – laced with the ideas of 'urban competitiveness', 'renaissance' and 'peripherality' – and the favouring of urban institutions and organisations in regional governance arrangements (Pemberton and Shaw, 2012).

Despite the urban inclinations of the RDAs, there were times when the Development Agencies played a major role in rural economic development. An outbreak of 'foot and mouth' disease in England in 2001 led to the setting up of a 'Rural Task Force'. This highlighted the significant connections between farming and tourism (see also Chapter 4), and observed that the foot and mouth crisis was likely to have a devastating long-term effect not only on farmers but on the broader economic well-being of rural areas. In that context, it was vital that RDAs became 'the catalyst for overseeing and targeting support for all sections of the rural economy' (Haskins, 2001: 9). Various support programmes were developed across England and for a while the perceived problems of top-down economic governance were forgotten.

...to localism

But it was still the case that many rural areas felt ignored by the Regional Development Agencies. The urban focus suggested significant faith would be placed on 'economic trickle-down', which would 'trickle down' only when middle-class households decided to counter-urbanise or spend money on day-trips to nearby countryside. Pressure was building for a more local focus. The Coalition Government has responded to this pressure by withdrawing Public Service Agreements and Multi Area Agreements and by dismantling all regional planning and economic governance infrastructure. In its place, it has invited local authorities and business groups to form Local Enterprise Partnerships (LEPs: see Chapter 2). The 39 LEPs that have formed (more than four times the number of RDAs, suggesting a more fine-grained, localised approach) now cover all local authorities in England and have attempted to align themselves with a more 'natural' economic geography, covering places that are functionally linked or display some shared economic reality.

The creation of LEPs signals a different approach to economic governance: one that combines the emphasis on partnerships that developed under Labour with a rejection of regionalism, which has become a key mantra of the Coalition. But perhaps more significant is that the LEPs are business rather than public sector led. The idea is that local entrepreneurialism rather than planning should lead economic regeneration, perhaps even weaning some regions off their reliance on public funding. LEPs have been tasked to bring forward growth plans for their areas and they will be responsible for distrib-uting the now smaller pots of public money earmarked for regeneration. This will include the competitively allocated Regional Growth Funds and Growing Places Funds, as well as European Regional Development Funds (ERDF: see Chapter 2). The idea is that these should be used to pay for interventions that address key infrastructure barriers and build the local capacities needed to exploit endogenous growth opportunities. For example, funds could be directed at training in support of local craft industries. Those industries might already have a market amongst tourists. But broadband infrastructure improvements, and the building of a distribution facility, might allow those industries to connect to wider markets, increasing turnover and earnings, and generating additional wealth to be re-invested in the local economy.

Such a strategy is known as 'smart specialisation', focusing on a small number of key activities with a view to making a significant positive difference within a local area. It is far too early to judge whether the LEPs are delivering sub-regional economic transformations, but there are already concerns over the institutional capacity of the partnerships. Each one has been allocated just £235,000 to cover its running expenses over a four-year period (Shutt *et al.*, 2012) which seems far too little to achieve the goals that government has set. Although the partners (the businesses themselves and local government) already have their own resources, there are now doubts over the capacity of LEPs to deliver against the new economic governance agenda. But the capacity issue is a complex one: different areas have different institutional capacities (not just financial resources, but pools of knowledge and skills), different levels of European funding, and some contain powerful business groups who are quite ready to lead these agendas and promote their own ends, which may or may not align with broader socio-economic goals.

The emphasis on smart, endogenous growth seems to suggest that economic governance will extend its reach and concern itself with a wider array of situations, finding and supporting new strengths within rural economies. However, there seems to be an ever-present danger that these emergent agendas – some of which link to social entrepreneurial action and micro-businesses that have grown from part-time entrepreneurialism – will be drowned out by urban growth paradigms. In some quarters, the LEPs model is too piecemeal and fragmented, lacking capacity and a clear institutional structure. There is a preference for stronger economic governance, based upon extended city-regions and mayoral political systems (Heseltine and Leahy, 2011). There is fear that such a model would re-privilege urban cores at the expense of rural interests (Pemberton and Shaw, 2012; Henderson and Heley, 2014). That said, it may be possible to integrate a more localised approach to promoting rural economies with a more strategic focus across city regions, emphasising the benefits of agglomeration and export-driven growth alongside investment in micro-businesses and environmental services (see Box 3.4).

Evidently, the changing frameworks of economic governance and planning have a profound impact on rural areas, not least by increasing or decreasing the visibility of rural concerns and channelling money to endogenous as opposed to exogenous development strategies. Another concern is the role of land-use planning in helping or hindering local growth. This provides the final focus of this chapter.

Land-use planning, neighbourhood planning and new economies

The role of conventional land-use planning has been perceived as limiting, if not completely stifling, the diversification and development of rural economies. Statutory planning is viewed as serving a consumption agenda in rural areas, which emphasises conservation and constraint over the promotion of new economic activities (see Chapter 2). There is a long-standing view, especially within the business community, that the way planning decisions are reached is too prescriptive and inflexible, unnecessarily bureaucratic, and far

Box 3.4 Integrating the rural into Local Enterprise Partnership agendas: The case of the Liverpool City Region (LCR)

In 1994, what was then known as Merseyside (now rebadged the Liverpool City Region) was designated a European Objective one area, the first in the UK, because its GDP was less than 75 per cent of the European average. At that time, there was an integrated budget incorporating money from European Regional Development Funds (for capital projects), European Social Funds (for developing human capital), European Agricultural Guarantee Fund (for rural development) and a small Financial Instrument for Fisheries Guidance.

From 1994 until 2000 little use was made of the EAGF allocation but in the second funding round from 2000–2007 a specific rural programme was established on Merseyside called the Integrated Countryside and Environment Programme (ICEP) which ensured that there was a focal point for rural enterprise within a city region where urban regeneration and urban social exclusion agendas predominated. By 2007, rural development on Merseyside was focused on a rural LEADER programme with a strong emphasis on economic development and rural growth with support focused on helping land management enterprises improve productivity and diversify their activities and helping businesses (including tourism) located in the rural parts of Merseyside to develop.

Moving forward into the European funding period 2014–2020 it is likely that there will be another smaller LEADER programme continuing to support rural businesses and hopefully contributing to the wider growth agendas of the Liverpool City Region. Outside of the main urban centres and towns within the Liverpool City Region some 90,000 people live, with 7,500 businesses generating over £30 million annually to the LCR economy.

The developing LEP strategies have acknowledged the importance of the rural economy in providing an important context for growth (this is not a new idea; see for example Gallent and Shaw, 2007) and for the development of micro-enterprises (60 per cent of LCR rural businesses employ 1–4 people). In the context of the LCR, LEP priorities for their rural areas are focusing on the Blue-Green economy, notably low carbon developments and green infrastructure, the visitor economy, and supporting new business within urban and rural locations.

too slow (Carmona *et al.*, 2003). This is despite explicit support in government policy, over a number of years, for the proactive delivery of new economic opportunities in the countryside. In 1999, Government's Performance and Innovation Unit noted, in a review of the rural economy, that the '[...] planning system can impose a heavy burden of compliance costs [...] and may have significant effects on competition and innovation (economic dynamism)' (PIU, 1999: 138). Echoing this concern, a comment recorded in the 2006 Report of the Rural Advocate for England expresses what appears to be the common view amongst small businesses:

I want to expand my business, but I can't. The planning system won't let me. If I expand I'll have to move to an urban area – and then what happens to the people who work for me here?

(CRC, 2006a: 11)

A general presumption in favour of development conflicts, in rural areas, with: a presumption against development in open countryside, outside settlement envelopes; with the prioritisation of agricultural development written into the 1947 legislation; and finally with the view imported into many rural areas from the 1960s onwards that development should happen only in urban areas. That latter view is often defended in district and parish councils where middle-class incomers hold sway. However, the recent NPPF has reiterated that planning should not be a brake on growth and recovery (Box 3.5). Indeed, these most recent instructions to planning authorities state that development proposals should be approved without delay when they accord with plan policy, and where the development plan is absent, or silent or relevant policies are out-of-date, permission should be granted unless there are adverse impacts that would significantly and demonstrably outweigh benefits, or the application is in contravention of any laws (including EU Directives).

Box 3.5 Positive national support for developing the rural economy from the National Planning Policy Framework (DCLG, 2012)

3. Supporting a prosperous rural economy

Para 28. Planning policies should support economic growth in rural areas in order to create jobs and prosperity by taking a positive approach to sustainable new development. To promote a strong rural economy, local and neighbourhood plans should:

- support the sustainable growth and expansion of all types of business and enterprise in rural areas, both through conversion of existing buildings and well-designed new buildings;
- promote the development and diversification of agricultural and other land-based rural businesses;
- support sustainable rural tourism and leisure developments that benefit businesses in rural areas, communities and visitors, and which respect the character of the countryside. This should include supporting the provision and expansion of tourist and visitor facilities in appropriate locations where identified needs are not met by existing facilities in rural service centres; and
- promote the retention and development of local services and community facilities in villages, such as local shops, meeting places, sports venues, cultural buildings, public houses and places of worship.

So, *prima facie*, there is strong support through the planning system for the development of rural economies. But in practice, many rural communities (and hence politicians) are sensitive to material changes in the use of buildings or new developments that will impact on landscape amenity or the 'visual character' of small towns and villages. Therefore, a great many planning applications are contested even where the viability of businesses, or the capacity of businesses to expand and provide local employment, hangs in the balance. That said, there has been a recent drive by government to improve both the speed at which planning decisions are made and their quality (measured in terms of either the proportion of positive approvals and/or the extent to which local authority decisions are successfully upheld at appeals). Within the Growth and Infrastructure Act 2013, there is provision for applicants to apply directly to the Secretary of State for determination of a planning application if a local authority is deemed to be under-performing, against time benchmarks for the determination of applications, or a significant number of a local authority's planning decisions have been successfully challenged (DCLG, 2013). The legislation suggests that planning is about to become more supportive of development than it was when the Performance and Innovation Unit reported in 1999. Planning Inspectors will be judging decisions against Paragraph 28 of the NPPF and if local planning committees are more restrictive in their decisions than government says they should be then it seems likely that decisions on rural development will be overturned. And if that happens regularly within a local authority, applications will be fast-tracked by PINS (see Chapter 2) and overseen by the Secretary of State. These changed rules, backed up by the payment of costs to appellants who have found themselves on the wrong end of a poor planning decision (i.e. contrary to the spirit of the NPPF and possibly denying a local business the opportunity to expand in a reasonable way), could result in a more supportive approach to economic development.

However, it was noted in Chapter 2 that the planning system works on the basis of local discretion. It is not compliance-based. Just because a planning application seems reasonable does not automatically mean that it will be approved. Decision-makers will weigh the merits of the application against its amenity impact, or the precedent it might set. Although the NPPF and the Growth and Infrastructure Act seem to stack the cards in favour of those seeking to grow rural businesses, it should be remembered that local planning practice has often constrained business growth, defending the property and amenity-based interests of households whose economic interests are unrelated to that growth – either because they are retired or derive incomes from urban employment – and who view the countryside as a retreat from, rather than a location for, new economic activity.

The extent to which the planning system inhibits development is highly variable, seeming to do so regularly in the preserved or contested countryside, but less so in the clientelist countryside. The advent of Neighbourhood Planning (again, see Chapters 1 and 2) seems to offer opportunities for a more flexible approach to business growth. Within the framework of Neighbourhood Development Plans, local people (within parish councils) can decide whether businesses need to expand for the good of a community. They can then permit that expansion by producing a Neighbourhood Development Order (NDO) that modifies the General Permitted Development Order. This could have the effect, for example, of allowing the conversion of agricultural buildings to small business use without the need to seek planning permission from the local authority. Such conversions would become 'permitted development'. But this would only happen if members of the community can agree on the need for such development. Within more conservative communities – of the type that oppose new economic activities within planning committees – the NDO strategy is likely to be no more effective than taking applications directly to committee. However, it is quite possible that decision-making in the context of 'localism' (where the threat of a committee-delivered outcome is removed) will be built on a community-based understanding of the need for economic diversity and jobs. Objectors who found it easy to provide written representations against development to planning committees may find it more difficult to oppose changes which have the broad agreement of neighbours and are supported within community-produced orders and plans. This, however, is largely conjecture and it will be several years before the full benefits and disbenefits of a neighbourhood approach to land-use planning are known.

Concluding remarks

The needs of rural economies are often overshadowed by dominant urban agendas. And although there has been considerable convergence in the structures of rural and urban economies in Britain, they remain distinct enough to warrant very different support strategies. Diversification away from agriculture is often service-sector-led, but there are also a large number of business start-ups in rural areas. Many of these are micro-enterprises that key into the consumption-based drivers of rural economies, especially tourism. Others are larger creative industries, which are foot-loose but exploit the connexity of some rural areas to urban companies and global markets. The distinct features of rural economies – determined by variable assets and local capabilities, and framed by broader political, regulatory and economic structures – point to a need for economic governance approaches with a distinctive rural flavour. These have not always been forthcoming. Regional approaches tended to privilege urban agendas, seeing rural areas as subservient or merely the beneficiaries of economic trickle-down. Growing criticisms of over-generic approaches to economic governance and planning have resulted in a localisation of support structures, paralleling a broader transition away from exogenous growth models to endogenous ones which promote smart specialisation. There may be some opportunities for LEPs to push this agenda, though the capacity and political longevity of these partnerships is currently in doubt. However, on the back of all of this there are signs that land-use planning may become less of a barrier to economic growth in rural areas than it once was. This is largely due to the push being given by government as it tries to balance its localisation of planning with an emphasis on growth promotion.

But one final conclusion, followed up in the next chapter, is that although the major economic narrative in Britain over the last 50 years is the transition from production to consumption – largely a transition away from farming as the linchpin of Britain's rural economy – farming retains huge socio-economic and environmental significance, not only as part of a wider agri-food sector, but because of associated on-farm diversification, its role in supplying many of the products on which the consumption economy relies, and its continuing role in making the countryside.

Summary

- Place-based soft and hard assets are becoming increasingly central in determining the fortunes of rural economies.
- The rural economy is now just as diverse as the urban economy in terms of economic structure, although it is characterised by a very large number of very small businesses.
- The farming economy contributes significantly to the context for other rural activities, although in employment terms it is becoming less significant.
- There is perhaps a lack of awareness amongst policy makers concerning the potential of small towns and rural areas to make significant contributions to growth, largely because traditional economic narratives emphasise the importance of core cities and urban economies that enjoy the potential benefit of agglomeration.
- Past economic governance arrangements seem not to have prioritised the needs of rural areas, adopting top-down perspectives and approaches to promoting growth.
- Planning policy is becoming increasingly supportive of rural growth (taking a positive approach to business expansion), but the implementation of policy is often contested locally as economic interests clash with those promoting other aspects of local amenity or village/countryside character.

Key readings

- Gray, D. (2014) Economic approaches to the rural, in Bosworth, G. and Somerville, P. (eds) *Interpreting Rurality: Multidisciplinary approaches*, Routledge: London, pp. 32–53.

(A concise guide to economic theorisations and their applicability to rural economies in specific locations.)

- Lowe, P. and Ward, N. (2009) England's rural futures: A socio-geographical approach to scenario analysis, in *Regional Studies*, 43(10), 1319–1332. (Explores the way that the economies of rural areas may develop over time, as a result of population change, rural–urban linkages or relative isolation. Also explores the notion of regional 'ruralities' and their possible future trajectories on the basis of robust empirical data.)
- Marsden, T., Murdoch, J., Lowe, P., Munton, R. and Flynn, A. (1993) *Constructing the Countryside*, UCL Press: London. (Provides an excellent overview of different types of rural area, their economic and political/governance characteristics. Also introduces many of the concepts central to current rural geography thinking.)

Key websites

- DEFRA resources on the rural economy can be found at www.gov.uk/government/collections/rural-economy
- European Council for Villages and Small Towns www.ecovast.org/english/about_e.htm

Note

1   The DEFRA Rural–Urban classification defines areas as rural if they fall outside of settlements with more than 10,000 residents. Census Output Areas (each with an average of 309 inhabitants) are all assigned to either one of four urban types (Major Conurbation; Minor Conurbation; City and Town; City and Town in a Sparse Setting) or one of six rural types (Town and Fringe; Town and Fringe in a Sparse Setting; Village; Village in a Sparse Setting; Hamlets and Isolated Dwellings; Hamlets and Isolated Dwellings in a Sparse Setting) (GSS, 2013).

4 Agriculture and beyond

Beyond the consumption versus production binary

The last three chapters have used the production-to-consumption narrative as indicative of the trajectory of economic change in the countryside, with social and landscape consequences that are considered in later chapters. This production versus consumption 'binary' has been widely used to depict the changing forces and opportunities facing rural economies in recent decades. The discourse suggests that agriculture is a sector in decline and, as seen in the previous chapter, there is some truth to this; at least in a purely economic sense. Yet, food security (that is, access to sufficient and nourishing food) is now re-emerging as a global priority in the context of rapid population growth and climate change. Recent volatility in global food prices has had a profound impact on many countries; and the impacts of this volatility have also been felt in the West. But while these important developments have led some to argue that a new phase of rural productivism is beginning, it is also increasingly evident – particularly in developed countries – that rural economies are diverse and hold significant potential over and beyond agriculture. The multifunctional potential of rural landscapes and the expansion of service-based industries have significantly changed the character of rural economies over past decades (see Chapter 3) leading some commentators to draw attention to the ways in which conflicting policy and developmental trajectories are a source of considerable tension in the countryside. At a global scale, some suggest that we must look beyond the production/consumption binary to understand the evolutions that rural areas are undergoing. As discussed in Chapters 2 and 3, globalisation – particularly when manifest in the notion of connexity – is a significant change force that presents a plurality of opportunities and outcomes contingent on place-specific assets. Another significant issue to note is that much of the consumption shift discourse is expressed in the vocabulary of neo-liberal capitalist accumulation, which highlights specific developments and opportunities over others as desirable and arguably narrows the scope of development options identified for rural economies.

This chapter will focus on the notion of sustainable rural development and explore the conflicting pressures and policy agendas that rural areas face in this context. It will explore the reasons for agriculture's economic decline and introduce a broader notion of the land-based economy that engages the idea of multiple functions of rural landscapes. A deeper insight into tourism and ecosystem services provision will help the reader gain a more nuanced understanding of some of the options available for diverse rural economies in the British context today.

Part 1: A sustainable land-based economy – key thinking

The 'dis-embeddedness' of agriculture and its implications for the rural economy

The notion of commodity networks is useful for understanding the developments underpinning agriculture's declining value to the rural economy (Hughes and Reimer 2004). The commodity network approach highlights the significant global interdependencies in the agri-food market, thereby

Figure 4.1

The disembeddedness of agriculture: Inside a greenouse growing watermelons for export in Almeria, Spain

highlighting the drivers that influence food production and pricing, as well as food consumption decisions. Food self-sufficiency (expressed as the percentage of consumed food that has been produced domestically) is the result of geographic, climatic and technological factors and perhaps most importantly, reflective of food preferences that are becoming locally more diverse and globally more homogenous (FAO, 2013). While an estimated three quarters of the world's agricultural holdings are smallholdings (farms of less than 1 hectares in size) the really significant decisions governing land-use and the availability of food are made by large-scale producers, investors and large retail companies whose decisions are motivated by consumer preferences (Lyson, 2006; Global Agriculture, 2014; HLPE, 2012). Box 4.1 explains the role of commodity speculation in food price formation. This network of interdependencies operates in a context of often complex government legislation, intergovernmental agreements and manufacturer and retailer standards regarding the quality and quantity of produce. While manufacturers purchase commodities at market prices, large retailers hold huge power over prices of manufactured foods as well as meat, fish, poultry and fruit and vegetables. The number of large retailers that dominate the

Box 4.1   Commodity speculation and food prices

Food substances (so-called soft commodities) such as wheat, rice, sugar and oilseed, are traded in international commodity markets that are governed by forces outside the influence of any single production unit, nation state or indeed retailer. The commodities market is based on a system of predicting stock availability and price buying and selling, mainly by large investors, financial institutions such as banks and building societies or pension funds, in order to gain profit.

For example, a price is forecast for a ton of wheat for a given future point in time and an investor buys this 'future stock' for the forecast price. This price will of course be speculative rather than based on certain knowledge of the following year's harvest and market conditions. The investor may want to make a profit by selling the 'future ton' of wheat and will only do so if the forecasted price rises. This may happen in response to changes in other market conditions such as the price of oil, for example, or speculative buying of wheat stocks by other investors, which will send an artificial signal of high demand and low availability.

Compounding this speculative price formation, most trade is now undertaken by so called High Frequency (HF) or algorithmic trading, where the whole buying and selling operation is undertaken by super-fast computers that monitor minuscule movements in interest rates, prices and other market conditions and execute deals often with minimal human involvement (McGowan, 2010). Add to this volatility the fluctuations in availability of crops due to weather and disease impacts and it is evident that food commodities are becoming increasingly unstable market goods (see Figure 4.2).

agri-food retail sector is dwarfed by the number of producers that they buy from, and therefore the former have immense influence on prices and significantly mediate signals from consumer demand (Lyson, 2006). It is necessary to consider these interdependencies and underlying institutions in order to really understand how and why agriculture, in any given context, is performing as it is.

According to Van der Ploeg (2006), rather than regionally embedded regimes, dependent on local environmental and socio-economic circumstances, farming is now to a large extent characterised by different socio-technical regimes shaped by schemes of prescription, control and sanctioning by state and increasingly, private agents. This has led to a homogenisation of agriculture and has been an ongoing process since the first half of the twentieth century and, some claim, much earlier. National and international pressures such as speculative price-setting (see Box 4.1) and subsidised technological investments have caused a 'dis-embedding' of agriculture from its local and regional social, environmental and economic context. The crucial change here is that in the new 'socio-technical regimes' that have replaced these (Van der Ploeg, 2006) it is the logic of 'un-embedded' finance capital that influences agricultural production, rather than any of the aspects of embeddedness (Lyson, 2006). While large supermarket chains encourage economies of scale in farm production in order to be able to offer the lowest consumer prices, critics point out that these price wars may not work for the benefit of the consumers in the long term if they simultaneously decrease farm incomes and corrode the basis of agricultural production, particularly in developed countries where input costs are high (Rayner, 2014). Moreover, due to the complexity of food commodity networks, consumer price rises rarely translate into farm gate prices. Much like the networked enterprises discussed in Chapter 3, large-scale intensive agriculture often brings few economic benefits to the immediate local economy because of its dependence on long supply chains and economies of scale (which add to the pressure to minimise input costs). It does not offer attractive employment opportunities or indeed invest much of its value added in the local economy (see the 'rent-seeking [rural] economy' in Marini and Mooney 2006: 96–97).

These developments in agriculture have accelerated in recent decades but are rooted in government policies dating back to the post-war era. The enlargement of markets and emerging technological innovations, production subsidies and other policy decisions and more recently, large-scale retail enterprises that sieve through consumer demands, have fundamentally changed what is being produced, how and in what quantities. While agricultural prices have characteristically been subject to fluctuations (Figure 4.2), in 2008 the impact of a range of emerging forces on the access to food was illustrated via a price peak that had global implications on people's ability to access food (FAO, 2013; HLPE, 2012).

Food price rises are particularly problematic for poorer countries where the price impacts are more significant at household level but recent experience of

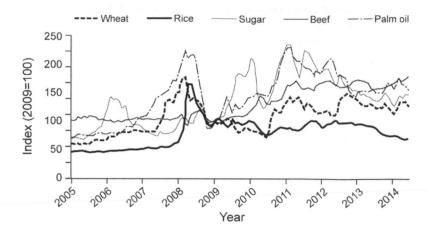

Figure 4.2

World agricultural commodity prices to June 2014
Source: DEFRA, 2014b: 32: 'Food Statistics Pocketbook' derived from UNCTAD:
http://unctadstat.unctad.org/wds/TableViewer/tableView.aspx?ReportId=28768

food banks in the UK have demonstrated that so-called developed countries are not immune to food crises. The role of price speculation in the 2006–2008 global food price crisis, resulting in a highest ever price peak in 2008, is hotly debated. However, its commonly cited drivers include: decreasing and low stocks; a combination of significant regional droughts in some locations and hot and humid weather in others; a simultaneous rise in crude oil prices; and, more controversially, export restrictions by major exporters, wanting to protect domestic supply in the face of decreasing stocks (OECD, 2011). As discussed below, bio-fuel production also played a role. Many suggest that in the long run, these developments together with the challenges stemming from climate change and population growth will seriously compromise global food security (Box 4.2).

Models of sustainability in diverse rural economies: the notion of the multifunctional rural eco-economy

The above discussion explains how global trade mechanisms that encourage price speculation and the emerging and potential impacts of climate change and population growth are colluding to make agriculture an increasingly precarious livelihood – and one increasingly dominated by big businesses that do not necessarily have food sovereignty or the strengthening of rural economies as corporate goals. However, this is not the whole picture. Many

Box 4.2 Food security and food sovereignty (HLPE, 2012; Food Ethics Council, 2006)

The HLPE 2012 report on food security and climate change outlines availability, access, utilisation and stability as the four central pillars of food security. 'Availability' of food is conditioned by the amount of food produced, which to an extent depends on trade and price signals as well as farming practices and access to land. The ability to produce beyond demand and to store food is decisive for maintaining stocks in preparation for fluctuations in production. 'Access' to food, on the other hand, depends on income levels and food prices as well as availability. 'Utilisation' refers to the ability of consumers (who have access) to select and prepare food that fulfils their nutritional needs and suggests that food and health education is relevant to food security. The final pillar of food security, 'stability', can be hampered by climate-related factors influencing availability, as well as political instability hampering access. To a large extent, stability depends on factors that are beyond the control of local governments.

Food security is by no means a goal that has been universally achieved, when a significant percentage of the world's population remains undernourished. Moreover, since the warming of the global climate system is now unequivocal, its impacts on food production are likely to further compound the already considerable challenges posed by the prominent trends of urbanisation and population growth that are projected to continue through to 2050. While the impact of climate warming will vary around the globe, it is certain to increase the vulnerability of populations everywhere to shocks that might compromise food security (see Chapter 9).

The four pillars of food security reflect the complexity of the challenge of achieving food security in the context of the present day global agri-food sector. Indeed, in the global context, some claim that the discourse should centre on 'food sovereignty', which highlights the role of trade agreements and access to land for maintaining food security, drawing attention to the locus of power in agricultural production and trade. The Food Ethics Council (2006: 29) defines food sovereignty as

> [...] the right of people to define their own agricultural and food policies. It is also the right of peasants to produce food and the right of consumers to decide what they want to eat and how and who will produce their food.

writers assert that, particularly in areas of considerable nature and landscape amenity value, farming in the 'developed West' is resisting the dis-embedding forces described above, and is now deriving increased benefits from a wider range 'non-market' goods which multi-functional agriculture produces (Wilson 2008). Lyson (2006: 293) draws attention to an alternative paradigm centred on 'local problem solving and civic engagement'. In some places agriculture is being re-embedded in the local rural economy via farm and non-farm diversification and the development of a multi-functional rural eco-economy (e.g. Kitchen and Marsden 2009).

The notion of multi-functionality highlights the opportunities provided by the so-called consumption shift (Marsden, 1999), drawing attention to the function of the rural as a space for leisure activities or as having specific characteristics making it a desirable place to live (see Chapter 1). The consumption shift is manifest in the declining importance of land-based production in the rural economy and in the functional switch from primary production to the maintenance and nurturing of 'quality', not only of goods directly produced but also of the countryside as a form of capital itself (Garrod *et al.*, 2006). 'Countryside capital' consists of the economic potential inherent in the visual, functional and cultural aspects of the rural idyll (see Chapter 1). This commodification of the rural idyll has been expressed in a number of ways. As noted in Chapter 3, recreation and tourism have been gaining economic momentum since the mid-twentieth century (Briedenhann and Wickens, 2004). While tourism introduces new demands on rural areas, particularly in relation to the protection of landscape and nature, it also provides local jobs and income for a broad range of local businesses contributing to the maintenance of skills and potentially of local traditions that make specific places attractive to visitors. Briedenhann and Wickens (2004) conclude from research in South Africa that the clustering of attractions and small-scale enterprises in remote locations can generate significant synergies between job creation, the local economy and also environmental conservation. The consumption shift has also been felt in agricultural production itself and marked by heightened concern over the impact of farming on the environment and the 'quality' (embeddedness) of 'local produce'. In the developed West, these concerns have triggered a process of diversification into niche products and different farming styles such as organic methods. The consumption shift has inspired local agricultural landowners to exploit the as-yet-untapped value in the increasingly multifaceted public perception of the rural idyll and engage its commercial potential accordingly (Crouch, 2006; Short, 2006). Non-agricultural diversification of farm economies has taken the form of both land development for residential property as well as for tourism and recreation services (Lobley *et al.*, 2005). As discussed in the previous chapter, an array of new small and medium-sized businesses have appeared in rural areas, as companies and people have relocated away from urban areas in search of a perceived better quality of life associated with the rural idyll (CRC, 2006b).

The rural eco-economy and ecosystem services and goods

This 'commoditisation' of the rural resource base is likely to benefit from the emerging framing of environmental values and resources as 'Ecosystem Services and Goods' (ESG; MEA, 2005) which is receiving increasing attention in nature conservation policy as well as development literature (Schomers and Matzdorf, 2013). These crucial functions, otherwise known as 'the benefits that people derive from ecosystems' (MEA, 2005: V) underpin all life on the planet. In policy discourse, food and fuel production, which are 'provisioning ecosystem services' (Dunn, 2011), are increasingly accompanied by the management of 'regulating services' (e.g. climate regulation via the storage of carbon and the mediation of temperature via permanent plant cover) and 'supporting services' (e.g. the nitrogen cycle; soil formation) and, not least, 'cultural ecosystem services' (the enjoyment that humans gain from visiting beautiful landscapes but also the spiritual and historic meaning attached to nature that underpin much of the rural idyll, for instance). The role of forests and farm land as 'carbon sinks' – playing a role in the mitigation of the carbon dioxide emissions that contribute to climate change – or the capacity of farm land to store water and mitigate flooding, if managed appropriately, are increasingly framed as potential income sources for landowners. Public funds are starting to be directed at the maintenance of these services in the form of payments for ecosystem services (PES) (Dunn, 2011). A frequently cited example of a private sector PES is the case of the Vittel-Evian mineral water company paying local farmers to change land-use in order to preserve the water purification capacity of the local ecosystem, crucial for the natural springs that act as the source of Vittel-Evian mineral water (e.g. Branca *et al.*, 2009). If anything, this emerging 'ecosystems approach' underscores the critical importance of rural areas, their role in addressing global challenges, and the potential that exists for rural economic diversification.

Recognising 'radical ruralities'

There is plenty of evidence, however, of significant obstacles to tapping into the resources of rural areas in the ways described above. There is often too great an emphasis, in policy, on economic efficiency and growth, which leads to an under-valuation of the potential of rural areas to deliver other services and development in alternate ways. Halfacree (2010) highlights how, in addition to countryside commodities such as 'viewed landscapes' and 'recreational services', rural areas are also consumed in other ways: notably through 'practised representations' including hobby farming and lifestyle entrepreneurialism. The latter refers to the ways in which people engage in alternative lifestyles, or downshifting, or try to escape from the conventional lifestyles of the modern world. It is argued that simple monetarisation downplays the significance of practices that address rural marginalisation and enhance quality of life: in fact, such practices can reconnect rural places to wider socio-economic networks and counter the isolation of some communities (Gallent, 2014). There is much going on in the countryside that does not register on the radar of the 'rural economy', either because it takes place in the so-called grey or voluntary sector (and delivers social goods – see Chapter 5) or because it generates value (for individuals and communities) that is not best expressed in economic terms and which cannot be readily commoditised (Lyson, 2006). Some have identified similar antipathy to the notion of 'farming as a way of life', often held by farmers themselves, in policies and initiatives that promote the neo-liberal or productivist agenda, seeking to measure agricultural progress, because of its uneasy fit with the neo-liberal logic of economic efficiency and productivity gains (Murdoch, 1996; Murdoch and Ward, 1997, cited in Halfacree, 2007). Broadly, there is a discourse that seeks to extend the production/consumption binary into new areas of non-monetarised activity, but which is nevertheless manifest in the production and consumption of a wide array of social goods.

Moreover, in the global development literature, development models reliant on connecting local economies to global markets are portrayed as a threat to the diversity of livelihoods and options for deriving agency from local natural resources (e.g. Escobar 2006). Escobar suggests that the engagement of subsistence livelihoods into global markets by the commodi-tisation of the resources that they rely on discriminates against traditional and alternative ecological and cultural order and practices and underpins many of the natural resource conflicts facing developers and conservationists around the globe. Many writers make reference to the narrow models of rural development that spring from the neo-liberal agenda alongside the implicit influence of the (uneven) relationships forged between the global South and Northern developed powers (for a summary, see Woods, 2011) on food security and rural wellbeing. Bio-fuel production serves as a good example. *Prima facie*, it presents a lucrative development opportunity, increasing land values and rents and bringing large-scale investments by international companies (HLPE, 2013). But as with any cash-crop production, bio-fuel opportunities tend to benefit those with existing market access and who are able to exploit economies of scale. This means that, particularly in a developing country context, it has the potential to destabilise the food security of local communities, stripping small holders, often women, of access to land and thereby accentuating existing inequalities in small-holder communities (Rossi and Lambrou, 2008).

The notion of the 'relational rural', conceiving the countryside as 'hybrid' and comprising networked relations of agency – binding together humans, ecological systems and non-material resources – may deliver a vantage point on rural change which assigns equal value to its economic, social and natural components (Woods, 2011). Kitchen and Marsden (2009) propose a model of sustainable rural development underpinned by a broadening of the rural resource base and supported by a mobilisation of commodifiable resources through a range of entrepreneurial actions. This model recognises that the practised representations noted above – including hobby farming or lifestyle entrepreneurialism that may connect to local traditions and practices – will add value to places, often increasing social capital and the potential for further agency (Halfacree, 2010), and are likely to involve production processes that are embedded within the local area, with shorter supply chains that mean 'commodifiable aspects of the rural idyll' will directly benefit local people. These new, and at times radical, conceptualisations of land-based commodities – as material and also experiential, place-specific and embedded

– could also provide some direction to land-use planning approaches or economic governance frameworks, as they seek to understand and respond to rural needs and help deliver benefits that are contained within and by rural communities.

Balancing food security and the sustainability of rural economies

As Potter and Tilzey (2005) have shown at a European scale, there are arguably two competing agendas shaping agriculture globally (see also Gibbs *et al.*, 2008). The first presents technological development and large-scale efficient production as the solution to the increasing demands for food by an ever-growing global population with its increasingly homogeneous tastes. This agenda is driven by actors such as the World Trade Organisation (WTO), which see the free movement of goods within the market as inevitably leading to greater efficiency, maximising wellbeing and overall food security. The second, counter agenda, often driven by smaller global actors such as Non-Governmental Organisations (NGOs) and individual governments, emphasises the negative implications this neo-liberal approach has for smaller and marginal actors such as family and subsistence farmers, the global poor and those who appreciate and value the non-market goods produced by family farming, particularly in the developed West (see above). Advocates of the latter agenda point to the inherent inequality and instability that flows from the neo-liberal model, and favour an approach to agriculture that values shorter commodity chains, smaller-scale production and greater devolution of decision-making power to the 'operative' level of producers and consumers. It was noted above that in the developed world, the notion of 'multi-functional agriculture' delivering a range of non-market goods and services (linked to landscape and biodiversity) is representative of this second agenda (Wilson, 2008). However, based on case studies of agricultural transitions in the UK, Australia and the US, Tilzey and Potter (2008: 57) suggest that:

> [...] initiatives to engender sustainability and post-productivism are subordinate to, or defined by, the dominant regime of accumulation in all three polities. The implication is that stronger models of sustainability are either marginalised, or transmuted, into various weaker models as, and if, they become embodied in state policy to constitute elements of modes of social regulation, a process designed to secure the 'relational sustainability' of capitalism(s).

It is difficult to say which model is better poised to provide global food security in the long run and it seems unlikely that one or the other agenda will ever completely dominate. Crucially, whichever model of development takes precedence, it will need to meet the requirement of a 60 per cent increase in the production of staple foods to meet the projected increase in demand over the coming decades to 2050 (FAO, 2013). Bearing in mind that production capacity is only one of the four pillars of food security alongside access, ability to utilise and stability, neither model offers a universal panacea (HLPE, 2012).

Part 2: The land-based economy in Britain: a broadening notion of rural development?

The remainder of this chapter focuses on the British situation and examines how some of the broad thinking set out above is manifest, or not, in a wide-ranging notion of rural development in this country. Inevitably, the trajectory of land-based rural development in Britain (and the UK as a whole) is framed by European Union membership and the rules and requirements that determine access to a range of funding streams. So whilst the focus is on Britain, Europe provides a crucial framing context.

Agricultural and rural policies and the rural economy

Food shortages in the aftermath of the Second World War, continuing into the late 1940s, provided the critical context for the emergence of the European Community's 'Common Agricultural Policy' (CAP). The five founding principles of that Policy are listed in Box 4.3, updated to reflect later policy and fiscal regimes.

Box 4.3   The five founding principles of operation of the CAP (Winter, 1996)

1   Free circulation of agricultural commodities within the Community;
2   The guarantee of single minimum prices for each agricultural commodity within the Community (the intervention or floor price);
3   Preference given to products from within the boundaries of the Community (import taxes for products from other countries);
4   Support given to exports to compensate for the difference of the Community floor price and the world price of the product in question (restitution payments for products exported outside the Community);
5   'Financial solidarity amongst Member States', which meant the formation of the European Agricultural Guarantee and Guidance Fund (EAGGF), financed from the income from levies on imports of agricultural products into the Community and of direct contributions from each Member State, based on each country's annual VAT yield.

The CAP has been reformed several times, most recently in 2013, and nowadays price-support for agricultural commodities no longer forms the mainstay of the subsidies, but instead, arable and livestock farmers receive a 'Single Farm Payment' (SFP), which is calculated not according to how much a farm produces, but according to farm size or historical production levels. The policy jargon singles out two 'pillars' of the present day CAP (see also Chapter 2). The first pillar is that of the SFP, and the second pillar is formed by funds directed at rural development. Although farmers are amongst the recipients of rural development funds, these monies are used to support a much wider and more diverse range of activities that have broader economic benefits for rural communities and support 'multi-functional' agriculture, which produces both food and environmental goods (Van Huylenbroeck and Durand, 2003; Wilson 2008). These changes to the policy mirror the consumption shift discussed above and can also be viewed as attempts at re-embedding agriculture in its social and particularly environmental context.

The first pillar of the CAP today comprises six policy instruments. Cross-compliance mechanisms mean that all payments to farmers are conditional on compliance with standards for environmental conservation and protection, animal welfare and food safety, as well as a set of standards defining good agricultural and environmental practice, mainly dealing with the quality of agricultural soil. A breach of these cross-compliance standards will result in the farmer having to return a part or all of the financial support received under any instrument of the CAP. In addition to the basic direct payment (the SFP) a mandatory green payment constituting 30 per cent of the national farm subsidy budget is made (European Commission, 2013). This latter instrument is designed to compensate farmers for the non-market goods (see Part 1 above) that they have to provide due to existing regulatory require-ments. This will cover the maintenance of permanent grass land 'ecological focus areas' and crop diversification. A voluntary Redistributive Payments Scheme can be established by Member States, facilitating the access of smaller farms to CAP funds and a separate Small Farm Payments scheme specifically designed to facilitate the access of small farms to CAP subsidies by simplifying the bureaucratic demands. Member States can contribute up to 15 per cent of the first pillar payments to support products that are particularly vulnerable. These 'coupled' payments are made on the basis of production targets and are therefore the only remaining directly market-distorting payments included in the CAP.

For a long time, the CAP was criticised for driving over-production, leading to product 'dumping' on international markets and related undercutting of market prices. However, the CAP is now almost completely 'decoupled'. This means that the level of support that a farmer receives is not dependent on crop or livestock output, as was the case with price support, therefore minimising price distortion. Instead, recent reforms have brought the issue of 'quality' of production to the foreground, in determining how agricultural subsidies will be allocated. Moreover, the reforms have cut the amount of subsidies that farmers receive so that, whereas in 1992 the CAP constituted 61 per cent of the EU budget, it now comprises 37.5 per cent (European Commission, 2013). In 2010, the European Commission proposed the following overarching objectives for both pillars of the CAP: viable food production; sustainable management of natural resources and climate action; and balanced territorial development. These arguably reflect the global concerns articulated in the HLPE (2012) report on food security and climate change discussed in the first part of this chapter. Accordingly, the newly-reformed second pillar of the CAP also includes a 'risk management tool-kit' comprising insurance schemes and access to income-stabilising measures (European Commission, 2013).

As a further attempt to address environmental problems associated with the dis-embedded productivist farming model, a third of Member States' pillar-two rural-development budgets will have to be dedicated to voluntary payments for farmers in Areas of Nature Constraints (ANC). These pillar-two payments are delivered in the form of agri-environmental schemes and offer locally tailored payments for farmers operating in areas with specific environmental values such as NATURA 2000 sites (see Chapter 10). According to the European Commission (EC) briefing paper on the future of the Rural Development Policy, its main aim is to deliver 'smart', sustainable and inclusive growth in the European countryside by supporting innovation and skills development, increasing resource efficiency and unlocking local potential for economic diversification supporting and going beyond agricultural restructuring (see Chapter 3, and see also Box 4.4 for specific aims). As with the first pillar, the allocation of spending to Member States is based on their historical share of the rural development budget. The measures in pillar

Box 4.4   The six overall aims of the EU Rural Development Policy 2014–2020 (European Commission, 2013)

1 Fostering knowledge transfer and innovation in agriculture, forestry, and rural areas;
2 Enhancing farm viability and competitiveness of all types of agriculture in all regions and promoting innovative farm technologies and sustainable management of forests;
3 Promoting food chain organisation, including the processing and marketing of agricultural products, animal welfare and risk management in agriculture;
4 Restoring, preserving and enhancing ecosystems related to agriculture and forestry;
5 Promoting resource efficiency and supporting the shift towards a low-carbon and climate-resilient economy in the agriculture, food and forestry sectors; and
6 Promoting social inclusion, poverty reduction and economic development in rural areas.

two are co-funded from national and European sources. In addition to the ANC payments, rural development funds are directed at business development in the form of grants and investment aid that aim to support new businesses with short supply chains, thereby retaining economic knock-on effects within a local area, as far as possible.

There are three main ways that European funding is being used to promote the non-agricultural development of the rural economy.

First, there is the Rural Growth Pilot Networks Fund. This was created as part of the previous Rural Development Programme (RDP 2007–2013). Five pilot projects have been established in Cumbria, the Heart of the South West (covering Devon and Somerset), the North East (covering areas of County Durham, Gateshead and Northumberland), Swindon and Wiltshire and in Warwickshire. By the end of March 2014 these pilot areas had between them received £2.4 million of DEFRA funding and an additional £4.2 million of European match funding. Pilot projects aim to overcome the known barriers to rural economic growth, including a shortage of premises, slow internet connections and the challenges related to the dispersed nature of rural business communities (DEFRA, 2014c; see also Chapters 3 and 7). The generic model for these rural growth pilots is based on the creation of various business hubs (often the market towns) that will provide premises for both existing and start-up businesses, often with access to local bespoke business support. In the South West, for example, several hubs have been created or are in the process of being created (see Box 4.5).

Second, there is the Rural Growth Scheme. This is a £177 million package allocated to the Local Enterprise Partnerships (see Chapters 2 and 3) to be spent on rural priorities, including business support, supporting micro-businesses, promoting rural tourism and investing in broadband and renewable energy (although other schemes are more directly addressing the broadband issue). The money comes from pillar-two CAP funds so is not 'new' money. Each local allocation is based on the number of rural residents DEFRA defines as living within the administrative boundaries of the LEP. As there are 39 LEPs in England and the budget covers a six-year period, it could be argued that the resources have been spread extremely thinly. However, depending on targeting, the funding can make a huge difference to small projects.

Third, £3.7 million has been allocated to England's LEADER+ Programme during the period 2014 to 2020 period, although new projects will not be developed until 2015. This is a continuation of previous LEADER+ activity. First introduced as part of the CAP European fund stream in 1991, the focus of this programme has been on the endogenous growth potential of rural regions (see also Chapters 1 and 3). The original objectives of LEADER were 'rural innovation' but not solely in an economic sense. Rather, there was (and continues to be) a significant emphasis on broader social, environmental and cultural activities and provision: i.e. a broader rural development agenda that is pursued most vigorously in areas judged to be more peripheral to the European core (European Commission, 2003).

Box 4.5 Rural hubs in the South West Rural Growth Pilot (Heart of the South West, 2014)

In the South West Rural Growth Hub, the plan is to develop approximately 38 acres of serviced employment land including a selection of high-quality, flexible business premises to support all types of enterprise within the rural areas of Devon and Somerset.

In Devon four hubs are envisaged:

- **Dartington Webbers Way** is a proposed £700,000 redevelopment of an industrial park and is part of the continued development of the Dartington Estate. The project will initially create 500 square metres of modern enterprise units and a further 3,000 square metres thereafter.
- **Holsworthy AgriBusiness Centre** is a £6 million investment to construct a new livestock market and agri-business centre. It has replaced the previous facilities in Holsworthy town centre, and also brought forward 18 acres of new serviced employment land.
- **Opportunity Oakhampton:** Devon County Council, with support through the Rural Growth Network, is leading on the development of this site, to provide a 20-acre serviced business, office and manufacturing park. It will complement the existing Oakhampton Business Centre, already up and running on the site.
- **Bicton EaRTH Centre/Home Farm Enterprise Centre:** The former Home Farm farmhouse, to the rear of the innovative Environmental and Renewable Technologies Hub (or EaRTH centre), is being converted into an enterprise and incubation space to complement the EaRTH renewables training centre. It will be supported by a digital apprenticeship scheme for ten people.

The Somerset hubs include:

- **Highbridge Enterprise Centre:** A £1.2 million new-build enterprise centre, consisting of two buildings and industrial- and office-based businesses. Ten start-up businesses will be incubated here, receiving access to superfast broadband as well as business mentoring.
- **Somerset Work Hubs:** Building on the successful network of Devon Work Hubs, the concept is being rolled out to Somerset, to provide up to three centres, providing flexible low-cost office space aimed at micro-enterprises and mobile workers.
- **Upgrade of Dulverton and Minehead Enterprise Centres:** the enterprise centres in Dulverton and Minehead will be refurbished to ensure their facilities are able to meet the needs of local businesses.

The eco-economy in Britain: farm diversification, leisure, tourism and radical ruralities

Lowe and Ward's (2007) typology of rural areas, described in Chapter 3, shows how specific rural regions in England and Wales have developed demographically and economically distinct features. The manufacturing sector remains a leading employer and source of GVA in the countryside (DEFRA 2014a). According to the Department of Business Innovation and Skills (DBIS, 2010b), the leading manufacturing sectors in the UK are now the food and drink, aerospace, pharmaceuticals, electronics and automotive sectors. New emerging technologies such as low-carbon, industrial bio-technology, nanotechnology, digital and advanced materials such as composites are also significant and illustrate the increasing emphasis on highly-skilled manufacturing in the UK, as in other developed countries. While a significant proportion of the rural workforce is employed within manufacturing, there are fewer actual businesses in this sector, suggesting that rural manufacturing companies tend to be large, or at least larger than typical rural employers. This point was made in the last Chapter. The sector has also suffered disproportionately from the recession with a high number of redun-dancies in the years following 2008. The impact of these redundancies will be felt in those areas with greatest reliance on this sector, especially in the dynamic or settled commuter areas or transient rural areas all of which, in Lowe and Ward's typology, are located close to urban centres.

However, the manufacturing sector does not extend its reach into all parts of the countryside. Elsewhere, there is greater reliance on small and medium-sized businesses, including micro-enterprises. Unlike manufacturing, these have tended to exploit the consumption shift and signal significant diversity in Britain's rural economies. Building on the overview of this diversity provided in the last chapter, we now consider three activities that express the increasingly multi-functional nature of rural economies: agricultural diversifi-cation, tourism-related business and renewable energy production underpinned by social enterprise.

Agricultural diversification

Rural areas in Britain tend to be host to a significant level of farm diversifica-tion, which Lange *et al.* (2013: 136) define as '[...] the extension of on-farm business activities into other sectors to enhance the continuation of agricultural activity, to broaden the income base, and to enable steady farm development'. Numerous examples of such diversification are set out in Box 4.6.

Tourism-related services such as accommodation provision are common survival strategies for smaller European farms but these tend to be less able to benefit from pillar-two CAP payments for a broader range of diversification activities (Lange *et al.*, 2013). Landscape features, in particular, are decisive for farmers' ability and willingness to diversify into recreation and tourism-related businesses. So too is proximity to major urban areas. Urban fringe farms often concentrate on food production, but may also try to connect more directly with customers in towns and cities through a range of direct supply initiatives including 'box schemes' (often a doorstep service to urban households, or local rural ones, who wish to place a regular order for farm

Box 4.6   Agricultural diversification (DEFRA, 2014d)

Adding value to agricultural products: On Merseyside, the current tenants of Reeds Farm took on a 30-year tenancy from Knowsley Estates in 2006 and turned the farm back into a predominantly agricultural enterprise growing potatoes on high-quality grade one agricultural land. Not only do they grow potatoes but they also brand, sort, wash and store potatoes. Recently they have secured a contract to supply potatoes to a major retailer, which provides a regular income and requires a constant supply throughout the year. They now work with a further 18 local farmers and suppliers within the business to ensure the contract can be delivered and through direct marketing of a finished product, all farmers receive a better return for their product. The investment in the potato processing facilities was supported by a LEADER grant worth £120,000 and has created four new jobs.

Tourism and Local Heritage: In the Lake District (particularly in the upland areas) farming for food production on its own is at best a marginal activity and farmers cannot survive without agricultural subsidies and many have diversified their business portfolio. The National Trust, which is a major landowner in the area, is keen to work with their tenant farmers to ensure that farm business is viable as a means of ensuring that the cultural landscape can be maintained. Hence they are often supportive of appropriate forms of farm diversification amongst their tenants. Yew Tree Farm in Borrowdale is an example of a farm that now generates more income from its off-farm activities than farming per se and provides local employment opportunities. They offer high-quality Bed and Breakfast in the farm house. They also offer farm experience tours for groups of more than ten. In relation to continued production, they butcher their own meat and sell it directly to their customers, both restaurants and individuals and they have a tea room selling locally produced cakes and gifts.

Office Development in Rural Areas: Hope Draughting Ltd recently set up a 3D quality modelling service to engineers, design consultants, installers and manufacturers, utilising the latest draughting, visualisation and point cloud software. They located themselves on the family farm by converting a redundant agricultural building. LEADER funding helped to pay for the construction of the building and the fitting of ground source heat pumps for both heating and cooling the building. Originally it was intended to create 3 full-time jobs but by 2014 this had risen to 13 and the business was planning to further expand its premises.

Socio-economic and Heritage Benefits: Thornton Manor was the family home of Lord William Lever, founder of what today has become Unilever. The manor itself is a Grade II listed building and the gardens and lake are on the Register of Historic Parks and Gardens maintained by English Heritage. The estate received £64,000 of LEADER funding to help create one of the largest outdoor marquees in the country capable of accommodating up to a 1,000 guests at a wedding. Seven additional full-time jobs have been directly created, and trade with local suppliers has increased significantly adding to the multiplier effect. Some of the profits will be used to restore the lake and public access to the grounds will be maintained. Hence the intervention brings both socio-economic and heritage benefits.

produce via the internet; requiring some investment in marketing, management and distribution) or farmers' markets (Lange *et al.*, 2013). But despite this variability, it has been reported that more than half of all farm businesses in the UK have now diversified in some way, and have seen an average income gain of £10,400 per year, per farm (DEFRA, 2014d). It was noted in Chapter 3 that land-use planning has often been seen as an impediment to farm-based diversification (as well as other business start-ups). Recent planning policy, however, seems to suggest a more relaxed approach to land-use changes, where those changes support the economic viability of businesses and bring broader community benefits. The 2012 National Planning Policy Framework (see Box 3.5) is unequivocal in its support of a 'strong rural economy' and calls upon both local plans and neighbourhood (development) plans to allow the expansion of all sorts of business and enterprise in rural areas by permitting the expansion of existing buildings and the construction of 'well-designed' new ones. There are examples of this happening in the cases cited in Box 4.6 and it now seems likely, given the devolution of some planning responsibilities to Parish Councils, that greater support will be given to farm diversification and there will be fewer examples in the future of planning blocking new enterprise.

Figure 4.3

Multifunctional agriculture – a stile providing access to recreational opportunities on farmland in the Chilterns

However, much will depend on the nature of that enterprise and the attitudes within local communities. There may well be broad support for some diversification into tourism-related activities or clear 'creative industries' (represented by Hope Draughting Ltd in Box 4.6), but there may be continuing resistance to diversification into the processing and direct distribution of conventional farm produce. Middle-class households have a reputation for opposing large new farm buildings, erected to house new on-site abattoirs, processing facilities and direct distribution centres: all of which are perceived to be noisy, potentially smelly, and counter to the expected serenity of the countryside.

Tourism and the consumption shift

Rural areas have become increasingly significant as sites for the consumption of various lifestyles, for example in the form of health farms, spas, restorative breaks and sports, and so-called 'extreme' sports (Lowe and Ward, 2007). Garrod *et al.* (2006) suggest that rising demand for tourism and recreation services in the countryside signals the need for both planners and rural entre-preneurs to regard the countryside as an economic asset – rich in 'countryside capital' – that should be managed and maintained in order to derive economic benefits for local areas. Kitchen and Marsden (2009) cite as an example of a tourism enterprise *Twr y Felin* in Pembrokeshire (or TYF for short), which engages with broadening and mobilising the rural resource base in new ways by offering extreme sports holidays on the Welsh coast. It provides corporate events, targeted at companies who want to develop their employees' management and business skills in an outdoor environment. This type of business commoditises not only local skills, in business and sports coaching, but also utilises the added value that the spectacular coastline provides for extreme sports such as climbing and kayaking. Inspired by the local coastline, the company staff invented a new form of sport and exploration, 'coasteering' that they are marketing as a very particular tourism experience:

> We call it *Pure Ocean Play* – surrounding yourself with clean salt water then exploring the wonderful Atlantic coast in the most playful way that you could imagine. Coasteering with TYF is a mischievous combination that'll see you walk, scramble, climb, swim and cliff jump your way around the best of Pembrokeshire's rocky coastline in the company of friends, family or like-minded adventurers.
>
> TYF are recognised around the world as the pioneers of this epic sport and took the first guided groups onto the cliffs nearly 30 years ago. We're pleased to see that, as the sport has been spread by the tide to other oceans and coastlines, the essence of fun, respect, discovery and adventure still abound.
>
> (www.tyf.com/adventure/coasteering/)

However, not all tourism manages to deliver on the potential of connexity and deeper and broader engagement of local resources. Tourism is often associated with low-paid and seasonal employment and thus mostly seen as a weak driver of economic stability and growth. While its influence on the rural economy remains limited to specific areas (see again Lowe and Ward's typology in Chapter 3), Haven-Tang and Jones (2014) contend that rural tourism opportunities often attract so called 'lifestyle entrepreneurs' (of the type who founded TYF) whose business goals are congruent with their lifestyle goals rather than economic growth and efficiency (Hall and Rusher, 2005, cited in Haven-Tang and Jones, 2014). This, together with the small scale of many rural tourism enterprises, contributes to the clear limitations of the sector as an income source for all rural areas. Geographic distance contributes to a lack of cohesion and makes co-ordinated 'destination development' difficult in some areas (Hall and Rusher, 2005, cited in Haven-Tang and Jones, 2014). The Commission for Rural Communities (CRC, 2013) found that rural micro-businesses (of ten or fewer employees) tend to face similar types of problems due to high transport costs and, most problematic of all, poor broadband connections, which inhibit advertising. Recognising the significance of this, the 2010–2015 government committed to providing superfast broadband to at least 90 per cent of premises in the UK and to provide universal access to standard broadband with a speed of at least 2 Mbps (see Chapter 7).

Administrative costs and having to deal with bureaucracy were also found to be relatively more taxing for small rural tourism businesses. As with farm diversification, acquiring planning permission featured prominently as one of the major bureaucratic hurdles facing these businesses (CRC, 2013). However, the CRC's report predated changes to the land-use planning promoted by the Localism Act 2011 and the NPPF (DCLG, 2012). As noted above and in the last chapter, the unequivocal support for rural growth and enterprise, set out in the NPPF, offers some succour to smaller rural businesses, and this support may well be delivered through neighbourhood planning (and Neighbourhood Planning Orders, see Chapter 3) if not directly through local planning committees. And whilst the lion's share of public funding tends to be targeted at enterprise clusters in rural areas, especially those locating in service hubs or market towns (CRC, 2013), there is some evidence that many LEPs are offering tailored support for rural micro-businesses, providing advice on how to access European LEADER funds (CRC, 2012).

Social enterprise and renewable energy production

Some of the values reflected in Halfacree's (2010) analysis of rural consumption and his notion of the 'radical rural' are arguably manifest in the significant role that 'social enterprises' are now playing in rural development. Sepulveda (2009: 2) defines social enterprises as 'businesses that trade for social purposes'. Examples are cited in Chapter 7, particularly of enterprises that form to run villages stores. The relationship of these enterprises to the market economy is variable. Some resemble conventional businesses, but are constituted to plough profit back into local projects. Others operate only to cover costs and provide a specified service to their community. The latter type range from networks of enterprises that operate 'local currency' arrangements, attempting to eliminate entirely all the standard market and monetary orthodoxies, to housing co-operatives. They represent an alternative form of production and consumption in the countryside, delivering important 'social goods' and forming on a platform of human and social capital (Chapter 5). It can be seen as a more embedded form of life-style entrepreneurialism, with rural communities releasing themselves from the shackles of urban economic and political norms (Featherstone *et al.*, 2012). However, Teasdale *et al.* (2013) caution against too optimistic an interpretation of the role of social enterprise in local economies, despite some very positive estimates of the number and growth of these by various agencies in recent years. While the United Kingdom is lauded as having a 'world class institutional support system' for social enterprise, estimates of the number of social enterprise range from 62,000 in 2007 to 8,507 in 2009, depending on how stringently the notion of trading 'for social purposes' is defined (Teasdale *et al.*, 2013: 121). A great deal more is said on entrepreneurial action and its roots in community life in the next chapter, and examples of community action around specific business types – shops, post offices and pubs – are provided in Chapter 7. Despite the uncertain contribution of social enterprise to solving a variety of rural development and service dilemmas, the exploitation of social capital has become an important development paradigm (Lyson, 2006), which can be exemplified here using the case of renewable energy and the mitigation of climate change.

One way in which 'local networked agency' – drawing on a community's store of social as well as material capital – is challenging the dominant model of economic production, is through the local promotion of renewable energy. Kitchen and Marsden (2009) present the case of a community wind-turbine in Wales, run as a social enterprise and exemplifying how communities can engage with local resources in order to diversify their economic bases, enhance local skills, and support job creation. *Bro Dyfi Community Renewables* (BDCR) has two wind-turbines near Machynlleth in Mid Wales, which produce electricity for local distribution. The company's shareholders are members of the local community and also investors in green energy from outside the area, and the turbines are run by local experts (the idea of setting up a community turbine came from a local resident in 2000 who was the voluntary director of a scheme elsewhere and worked for a conventional renewable energy company). Its profits, when there are any, are returned to the community in the form of advice and assistance with household energy efficiency. Bro Dyfi Community Renewables presents itself as:

[...] an example of communities and environmentally conscious investors coming together to create and deliver local schemes that are a direct response to climate change, boost local regeneration and strengthen self-reliance.

(http://bdcr.org.uk/)

Kitchen and Marsden (2009) point out that by providing energy and an annual cash contribution from profits to *Ecodyfi*, a local regeneration organisation, the benefits of the project are retained locally by the community.

There have been recent attempts to encourage similar social enterprises. In 2010, the Labour government launched its 'Low Carbon Communities Challenge'. Funding was made available for projects similar to the Bro Dyfi one, in which there was a closed-circuit connecting renewable energy production and other community projects (DECC *et al.*, 2009). According to DECC (2014), renewable energy installations falling under its 'Feed in Tariff Scheme' (which enables small-scale [under 5MW per annum] producers of renewable electricity to sell surplus electricity to the National Grid) accounted for 12 per cent of all renewable energy infrastructure. Whilst not all renewables infrastructure is run by social enterprises, recent figures suggest that local energy production, including on a household scale (small wind-turbines, solar panels and green heating systems), is increasing. There is a huge potential for more schemes of the type currently being operated near Machynlleth, and the benefits are clear: retained resources for community projects and direct benefits for rural households. Furthermore, although commercial onshore wind-farms often endure an acrimonious journey through the planning process (Rule, 2015), being seen as visually obtrusive and incredibly harmful to the countryside, community-based schemes command much broader support, owing to their scale and the direct benefits they bring.

Conclusions

Rural economies have seen dramatic change since the Second World War, being subject to the production-to-consumption transition outlined in the previous two chapters and becoming increasingly dominated by service sector activities. But in spite of this transition, agriculture remains critically important, not only because of growing concerns over future food security, but because how land is cultivated has much broader environmental and social implications. It is clear from the discussion presented in this chapter that rural land and the rural landscape can produce a range of environmental services and goods, some of which deliver direct and retained benefits for communities. How is planning supporting the delivery of these benefits? Important adjustments to European funding programmes have facilitated a shift away from the over-production of food-stuffs to support for 'non-agricultural development'. This is one important message. Another is that communities themselves – led by lifestyle entrepreneurs – are playing a significant role in diversifying rural economies beyond their previous agricultural base, and they seem to be receiving increasing support from a more flexible approach to land-use planning. This is because the scale and impact of community-based projects tends to be far less than commercial ones; the latter needing to be large and profitable in order to attract conventional investment.

Summary

- The economic fortunes of agriculture are affected by supra-national and national support systems and by the power of large retailers, operating in centralised markets, to determine farm-gate prices.
- Critics suggest that the integration of rural economies into broader markets is highly problematic, causing a leakage of potential local benefits.
- In developed countries, the consumption shift has triggered a move beyond agriculture into a plethora of new farm-based businesses attempting to derive value from increasingly multi-functional rural landscapes, often trading in niche food products, tourism experiences and ecosystem services.
- In Britain, the social enterprise sector provides some examples of more self-reliant development models, which commoditise local assets and skills, retaining benefits within closed circuits.
- However, agriculture remains a significant land-use and the back-drop to much rural economic activity. In some areas, it is responding to global market pressures through intensification of activity. But elsewhere, it is at the centre of efforts to mobilise local assets, connect with local markets, shorten supply chains, and exploit the multi-functional potential of rural areas.

Key readings

- Halfacree, K. (2010) Reading rural consumption habits for difference: Bolt-holes, castles and life-rafts, in *Culture Unbound*, 2, 241–263.
  (Emerging critical thinking on rural development.)
- Haven-Tang, C. and Jones, E. (2014) Capitalising on rurality: Tourism micro-businesses in rural tourism destinations, in Bosworth, G. and Somerville, P. (eds) *Interpreting Rurality: Multidisciplinary Approaches*, Routledge: London, pp. 237–250.
  (A contemporary source on developments in rural tourism, citing international data and cases.)

- HLPE (2012) *Food Security and Climate Change: A report by the High Level Panel of Experts on Food Security and Nutrition,* The Committee on World Food Security: Rome.
  (Explores the potential impacts of climate change and population growth on the agri-food sector globally.)
- Woods, M. (2011) *Rural,* Routledge: London.
  (Key readings on representations of the countryside and on rural development.)

Key websites

- DEFRA resources on the rural economy can be found at www.gov.uk/government/collections/rural-economy
- European Council for Villages and Small Towns www.ecovast.org/english/about_e.htm

Part 3

People and communities

5 Rural society and communities

....

A peculiar social context?

In the opening chapter of this book, it was suggested that one of the 'realities' of rural life is that weaker or 'thin markets' for private enterprise, and for public services, generate a strong sense of self-reliance. Where neither private enterprise nor public services are viable (or struggle to address local needs) communities look inwards for solutions to the everyday challenges they face. But modern social entrepreneurialism is rooted in a particular rural 'way of life' that is itself built on a communitarian tradition that is centuries old and grew from common (land-based) occupation, shared values and religion, and relative isolation. Rural communities have long been considered distinct from their urban counterparts, but more recently this notion of social distinctiveness has fused with a particular community governance agenda – and with a 'greater likelihood of community control over services and community action born of a communitarian spirit' – that is today a key component of thinking on the countryside and the nature and governance of planning in rural areas. Practically, it means that 'community action and planning' is often an important part of rural planning (see Chapter 1).

This chapter aims to provide a context for the three chapters that follow, dealing with the accessibility and opportunity challenges that arise in rural areas, the social entrepreneurialism that can take root in response to those challenges, and the ways in which communities may become involved in the specific area of housing delivery. It is again divided into two parts. The first part begins by re-examining more established writings on rural society and the nature of social relations in the countryside, before considering whether there are unique rural conditions that incubate a particular kind of society and – more specifically – a richness in social capital and a desire to engage in social entrepreneurial action that may connect with established government and seek to deliver various outcomes through (state) supported self-help. The second part then focuses specifically on Britain, unravelling key changes in rural society and communities, defining these concepts in the British countryside and tracking onward links to liberalism, neo-liberalism and communitarianism. On that same conceptual platform – introduced in Chapter 1 – the chapter then explores ideas of social capital and action, active citizenship, and the changing governance of rural areas, focusing ultimately on the new 'framing' of community action and planning within a broader policy emphasis on 'localism'.

Part 1: Society and community – key thinking

The 'traditional' rural community

The idea of the closely-knit rural community has international resonance. Cities are places of intermittent and incidental social encounter; the small towns and villages of countless countries around the world are considered, by contrast, intimate and as places of regular and sustained social exchange. Neighbours are genuinely neighbourly and people look out for one another. There is an abundance of 'familial trust', with rural intimacy contrasted with what Tönnies (1887) called a 'being together of strangers' in many urban locations. Television and other forms of mass media have tended to perpetuate this idea of rural familiarity, be it in small towns in the US, archetypal English villages, or the Australian outback. Community spirit is often seen as part of the rural idyll outlined in Chapter 1, generating and sustaining the 'traditional' rural community in which common bonds – built of land-based tradition (e.g. farming, forestry or fishing), shared values and religion – are strong and endure over time. Time is an important ingredient: whereas cities are places of transient encounter, interactions in villages are maintained: people truly 'dwell' and in doing so create a place for themselves in the world (Heidegger, 1971) and generate an agreed 'stable culture' (King, 2004: 23) which provides a strong foundation for collective action.

Yet, this image of rural society – populated by like-minded, neighbourly people, all drawing a living from the land – is, as MacGregor (1976: 524) points out, '[...] much nearer to the jolly village green on the pantomime stage than reality' and it also gives little indication of what is meant by the word 'community'. Today, the idea of groups of individuals possessing certain 'capitals' is useful. Unconnected to others, a single person embodies 'human capital': their potential to get something done; fix an engine, plough a field, or mend a roof. But through connection to others, a group of people come to possess 'social capital': the 'resource potential of relationships' draws the person who can fix an engine to the person who owns a broken bus; the bus runs again and becomes available to others in the group who need to make a journey. Communities are built on connections, or relationships that transform an aggregation of individuals into a *social* group with pooled skills and capabilities. There is perhaps something inherently instrumental in modern thinking on 'community', which is rooted in the idea that people come together for a reason (to achieve something) and not merely because of residential proximity or propinquity. Such instrumental ideas often assign greater strength to virtual over place (socio-spatial) communities, arguing that those who come together online have a reason and rationale for doing so that may be more potent than those who just happen to find themselves living on the same street. Bell and Newby (1971) argued more than 40 years ago that residential proximity in a rural area does not guarantee any reason or desire to interact. More recently, Delanty (2003) has been critical of social-spatial notions of community, arguing that they imply, but rarely test, claims of interaction and social cohesion. In contrast, communities that are 'identity-driven' (centred on sexuality or belief), politically motivated (convened to fight a perceived injustice), or technological (centred on chat rooms) all have clear rationale. They display the outward characteristics of community, and nothing is merely assumed.

Arguably, the 'traditional' view of community is one of a static co-existence of individuals, which contrasts with a more modern, dynamic and instrumental view – now bound up with the idea of *social capital* and a propensity towards self-help, introduced below. However, there has been an evolution in thinking rather than any fundamental social shift (that shift has occurred, delivering a more complex social world, but it has not changed the basic nature of human sociability). Communities today often need to do more, want to do more, and can do more. But these *actions* remain grounded in *interactions* that generate some level of common understanding, a degree of trust and a pooling of resources that facilitates the action. Even the earliest contributions to the community discourse accepted that social exchange and its consequences – a sharing of values, opinions and beliefs – delivers what Tönnies (1887) called a 'unity of will', and acts as a crucial bonding agent, transforming an aggregation of individuals into a social group (Blumer, 1948: 546) able to act with common purpose.

One of the difficulties, however, with the earlier literature on community, starting with the work of Tönnies, is that it sets the discourse on a path towards assumed social cohesion and unity. The 'dialectic' community proposed by Tönnies ends up a happy one, sharing common beliefs, and able to muster support for actions that enhance cohesion and wellbeing. It provides an intellectual buttress for the picture painted above of the 'traditional' rural community. More recently, Ruth Panelli (2006) has noted that although everyday interactions may generate occasional agreement and bring people together, those same interactions may also catalyse division between competing interests. Communities may be divided, or fragmented into interest groups or cliques that co-exist in the same space. Discourses of community should embrace the potential for conflict. But if the unity of will is lost, or never really existed, is there any value in the community concept? Panelli (2006: 68) suggests that today's socio-spatial groups are more-often-than-not heterogeneous, displaying a mix of needs, values and beliefs. Tönnies was not wrong, but was dealing with very different social realities: nineteenth-century realities of strict social stratification, common religion, and subordination of a rural peasantry by powerful landed interests. Much has changed in the last hundred years. The economic shifts outlined in Chapters 2 and 3 have had a transformative effect on rural society, resulting in conflicts centred on the control of resources, disputes around the nature and extent of various rural challenges, and disagreement over how resources should be managed and needs addressed. Social complexity has generated a new governance challenge, which was alluded to in the first chapter, and also a challenge for future planning. This is returned to in the British context at the end of this chapter.

Social and community transformations

It was shown in Chapter 3 that the rural economies of the nineteenth century were ripped apart by industrialisation. Urbanisation was the immediate outcome, followed in the later twentieth century by agricultural modernisation (see Chapter 4) and then by spatially-selective counter-urbanisation. These processes had a fundamental impact on the economic basis of rural society, with the population no longer divided between those controlling land and those subservient to landed interests. It was noted above that Tönnies' distinction between rural and urban communities – the former viewed as homogenous and closed and the latter as heterogeneous and open – was critical to the notion of the 'traditional' community, described by a number of authors (including Martin, 1962 and Newby, 1979) and transported into popular culture. But whatever truth lay in these past generalisations, the economic transformation of both rural and urban areas, and consequent counter-urbanisation and amenity-driven rural migration, has rendered them largely meaningless today.

The countrysides of many advanced industrial nations are much more complex places, in which clear and obvious communities, expressing unambiguous and unified values, are far more difficult to identify. Fifty years ago, the English sociologist Ray Pahl argued that the terms rural and urban had lost all meaning 'in a sociological sense' (Pahl, 1965): in the space of seventy years it had become impossible to differentiate between rural and urban communities, and attempts to draw out simple universal features tended to suppress 'ontological difference' (Young, 1990: 339). More recently, Panelli and Welch (2005) have tried to untangle the complex web that is community, concluding that it is a 'social construct to be variously and continuously negotiated' (Panelli and Welch, 2005: 1589). Researchers or others concerned with the notion of community must deal with a complex mosaic of values underpinned by patterns of gender, social class, ethnicity, disability or sexuality (Little and Austin, 1996; Cloke and Little, 1997; Delanty, 2003; Neal and Agyeman, 2006). Whilst it is still agreed that community is a product of social interaction, these values may provoke conflicts that lead to different patterns of inclusion and exclusion. Transformed rural communities are rarely unified and almost never homogenous, except in extremely rare cases where isolation has shielded them from fundamental economic shifts. In most cases, they express the economic and social changes that have rippled across Western society, brought to the countryside by population movement, displacement and gentrification. In the British context, we will look later in this chapter at how planning needs to negotiate this complexity in two important ways: first, in addressing the diverse needs of rural places; and second, in drawing diverse communities into the process of planning itself.

Community spirit and communitarianism

In summary, communities comprise networks of social transaction. They may be more or less cohesive, and often contain antagonistic elements. What does this mean for the 'way of life' or 'community spirit' that is sometimes claimed as an important feature of rural areas? Despite the changes that have affected the countryside, a number of simple differences in function and form between urban and rural places remain. Policy makers frequently try to delimit urban communities, attempting to map the boundaries of neighbour-hoods. If they delve more deeply into the personal and group interactions that underpin urban sociability, they find that all boundaries are porous: social networks extend beyond mapped neighbourhoods. Work-based, family-based, church-based, university-based, or leisure-based relationships are often more significant than those shared by near neighbours. And these relation-ships are relatively easy to maintain: the neighbourhoods of the city are contiguous, often joined by reasonably good public transport. Even when this is not the case, these same relationships can be sustained online or in the work-place. The same is also true in villages or small towns, but the form of rural settlement provides a very different context for sociability. There is likely to be less choice in 'interactional infrastructure' (Kilpatrick et al., 2014) in a village: a single pub, one primary school, a post office or shop. This may mean that the likelihood of a chance encounter, between the same people using the same facilities, is greater. There are also likely to be fewer public transport opportunities, perhaps encouraging residents to stay put and make the most of what they have locally, including the opportunity to socialise with neighbours. And lastly, although the 'community' may not be socially bounded – cars and computers provide the means to sustain external rela-tionships – rural places may have much clearer physical boundaries than urban ones. They may comprise small islands of development in a patchwork of green fields, connected to the nearest larger settlement by a single road and a twice daily bus. Or they could even be actual islands, off the coast of Scotland or Finland. This stronger and clearer physical boundary may well foster a heightened sense of identity, of belonging or place attachment. In some cases, it may counter the implosive effect of the introduced social fractures noted above, giving a rural community enhanced cohesion and a greater sense of itself as a social entity; but which can nevertheless break down when resource conflicts loom.

There are perhaps also a number of instrumental reasons why rural communities may *appear* more cohesive. First, when confronted with external challenges, their clearer boundaries can be more easily defended. Their sense of self as physically discrete may produce a feeling of being socially bounded, although the defence of that boundary – against new housing or other types of development – may expose differences in need and opinion within a community. Second – and perhaps more importantly – different groups within a village of say 500 residents may come to believe that, in the face of private sector disinterest or fiscal austerity (the 'thin markets' noted above), the challenges it faces are on a scale that 'the community' can deal with. There may be a belief in some rural places in the capacity of local people to do more for themselves, because they are confronted with clearly bounded issues affecting a fixed number of residents. Again, in urban areas, porous boundaries and a sense of self that is often confused by being part of a bigger contiguous whole – alongside patterns of transient residence and greater private sector investment/public intervention – may reduce *apparent* cohesiveness.

Much depends of course on the nature of challenges faced and the characteristics of residents. Many urban communities appear extremely cohesive, take forward incredible collective projects, and feel a strong sense of identity, sometimes born of common religion, ethnic grouping, shared poverty or simply a feeling that they have been neglected or badly treated by insensitive public policy or top-down decisions. Injustice often brings communities together (Ciaffi and Gallent, 2014) even where there is uncertainty or ambiguity surrounding the exact boundaries of those communities. However, there remains a sense that 'communitarianism' (as defined below) has frequently gained an easier foothold in rural areas and, as suggested in Chapter 1, that community-based action has become a critical feature of rural planning.

Normative 'statutory planning' (see Chapter 2), wherever it happens, tends to pursue some broad construction of the 'public good'. It sets itself fixed goals, using words like (economic) 'prosperity', (social) 'wellbeing' or 'equity', (environmental) 'resilience' – all badged under the heading of 'sustainability' – as abstractions of what it seeks to achieve, and as justifications for a range of very different interventions. Not having new housing, or having lots of it, may be the claimed path to greater sustainability. Such abstractions, and justifications for different planning policies, are professionally produced, often by experts – trained in universities around the world – who may have limited contact with affected communities. Because of the complexity of needs, and the range of opinions, in those communities, these types of planning outcome are likely to receive a mixed response and are frequently contested. More generally, where the planning system seems closed, top-down and highly bureaucratic, its defence of the 'public good' as justification for all manner of outcomes, may appear to challenge the right of communities to define and pursue their own goals. Planning comes to be seen as an imposition, insensitive to community ambition and priority. It may even become a source of perceived injustice, and a trigger for protest, conflict and social mobilisation. Whilst that may not be an entirely bad thing, at the root of planning-based conflict is the insistence that communities should accept expert constructions of the public good and take contributory actions – i.e. accept new housing, roads or other forms of development because experts have professed that these are necessary and in the 'national interest'.

In recent decades, there has been frequent rejection of the contributory actions demanded by public planning. Sage (2012: 267) argues that such top-down planning sits in opposition to a liberal worldview which '[...] holds that society consists, or should strive to consist, of an association of free individuals, detached from imposed duties and obligations and free to form their own aspirations and interpretations of the good life'. It should not be the role of the state to define the common good but to defend individual rights. Neo-liberalism in the 1980s extended this thinking to the emancipation of markets from state interference, giving rise to a new politics of anti-planning and greater faith in the capacity of market efficiencies to deliver an unspecified public good through economic trickle-down (Jessop, 2002: 470). It might be argued that more 'forceful' planning is grounded in a very different view of society from the liberal/neo-liberal agendas. It attempts to shape and steer the whole, whilst the (neo) liberal worldview emphasises the value of individuality and views notions of society and community as potentially oppressive. Britain and the US in the 1980s found themselves in the grip of neo-liberal thinking, which had a profound effect on all forms of state regulation, including planning.

But between more forceful, top-down, public planning and neo liberalism sits a middle view that emphasises the importance of community:

> The starting point for communitarian theory is the basic tenet that the existence of strong community life – expressed as a state of affairs in which individuals belong to and participate in a wider group (or groups) of common interests and shared goals – is of inherent value in human society.
>
> (Sage, 2012: 267)

Community is viewed as a producer of 'social goods' that arise from the interactions and reciprocities of 'community life', described earlier in this chapter. The fixed bus, able to transport neighbours on vital shopping trips, might be seen as such a social good; as might the soft support of general neighbourliness or the setting up of a village shop, run by volunteers. Communitarianism (a word that suggests the centrality of community life, but which is rarely uttered) focuses on what local people, coming together, can achieve for themselves. It may also suggest the relegation of an imposed public good behind the priorities and goals set within a community. The actions that communities take, and the plans they make, can therefore be framed within communitarian theory. But beyond this theory, there is a problematic reality. It was shown above that communities are often fractured by different private interests rather than led by a single communitarian ideal. Community agendas may be hijacked by vocal individuals who act to protect, for example, private property values and are not interested in the delivery of broader social goods. Likewise, communities cannot cut themselves off from wider society; they will retain obligations (those frequently rejected 'contributory actions') and will often need external support, in the form of capital or expertise, in order to achieve some of their goals. A negative view is perhaps that '[...] communitarian planning – with its role in defining and extracting social goods – might be conceived as an ideal that is corrupted, on the one hand, by self-interest, and on the other, by state manipulation' (Gallent and Ciaffi, 2014). A more positive view is that community action needs an element of framing, to minimise private corruption and to ensure that what happens in a village or neighbourhood connects and contributes to wider social goals. The question of framing community action, and the production of social goods, is a governance question, which is returned to in the British context.

Active citizens and social capital

The idea that community is a producer of social goods has become a very important one in social science. These goods arise from interaction, as noted several times above, but are more specifically products of the 'resource potential of social relationships' (Agnitsch *et al.*, 2006: 36) or the 'social capital' that accumulates between active citizens. Thinking on social capital is complementary to communitarian theory: it offers an explanation as to how social goods arise from 'community life'. Robert Chaskin and colleagues (2001) explain that innate human capital – the skills, knowledge and resources that all individuals possess, including the ability to fix a bus – is turned into social capital through the building of relationships and, consequently, social *capacity* is increased through a coming together of '[...] interactional resources and social capital existing within a given community that can be leveraged to solve collective problems and improve or maintain the wellbeing of that community' (Chaskin *et al.*, 2001: 7). The term 'social capital' was first used at the beginning of the twentieth century, but was popularised through the work, in Italy and the US, of Robert Putnam in the 1990s. It is rooted, in part, in the classical analyses of community and community life, though its modern use and revival owes more to Habermas (1984: 286) and his distinction between individualistic actions designed to achieve personal goals and 'communicative action', grounded in dialogue and subsequent agreement around collective remedial or preventative actions. Hence Woolcock (1998) claims that 'reciprocity and trust' are vital to community life and key to the production of social capital. This implies a need for regular and direct social contact, often face-to-face, between individuals who acknowledge their inter-dependence. Context is important, with rural locations having the right opportunities for neighbours to get acquainted, and the right scale for subsequent social enterprise and entrepreneurialism – for the reasons set out above (see also Moseley, 2000).

However, it was noted above that social boundaries are porous, even though the scale and location of rural communities may generate some sense of social enclosure. This is an important point in relation to social capital. Everyday interaction is crucial: the 'bonding' of near neighbours sustains the socio-spatial community, and is the basis for building social capital. Putnam (2000) contends that bonding between friends and neighbours generates the trust and reciprocity that Woolcock (1998) identifies as vital, delivering solidarity within a community (Putnam, 2000: 22). But the porosity of social boundaries – in any community – creates opportunities to 'bridge' to extra local resources. Communities can do a lot for themselves, but they cannot do everything. How much a closed community can achieve is limited by the extent of local skills and knowledge (Agnitsch et al., 2006) and also faces the risk of turning into a clique and becoming prejudiced against outside interests and ideas (Agnitsch et al., 2006: 40; see also Rydin, 2014). Fortunately, such closure is rare, and rural communities are able to bridge to outside resources in a number of ways. Acknowledging the complexity of modern communities, Putnam (2000: 23) draws a distinction between bonding and bridging social capital, calling the former the means by which communities 'get by', and the latter the means by which they 'get ahead'.

The linked discourses of community, communitarianism and social capital provide an important basis for thinking about the community element of rural planning. Entrepreneurial action is not unique to rural areas, and urban needs are as complex as those found in the countryside. However, there is a strong tradition of community involvement in the planning and delivery of rural services, and this has become an important area of government support during recent decades. Furthermore, a wider socio-economic mix in many rural locations means that these areas now possess a broader array of knowledge and skills, which is mirrored by a broader range of behaviours and tastes, resulting in competing views on what a future countryside should look like. Social change in the countryside has brought new needs, new opportunities and a potential for conflict. And yet, 'community' has become a focus for governments looking to renew local democracy and deliver planning and other services through local action (Gallent, 2013).

Part 2: From spontaneous social entrepreneurialism to 'localism'

There has been a gradual shift in Britain from policy pursuing rural modernisation (see Chapter 1, 3 and 4) to targeted support for local enterprise and endogenous development that arises from community processes. But even during those periods when communities were largely ignored by higher-level planning, social entrepreneurialism was alive and well in the countryside. The desire of local people to do more for themselves was clear in the 1970s when these issues were first studied and voluntary networks, including the Rural Community Councils in England, started to offer structured support for community projects. But by the 2000s, government had begun to embrace these projects and started to assimilate them into broader efforts to deliver rural services and planning more attuned to local needs and ambitions. In this part of the Chapter, we examine how transformations in the nature of community in rural Britain, followed by a politicisation of community as an instrument of intervention, led eventually to a jump from spontaneous social entrepreneurialism to the localism (that is, involvement in statutory planning at a neighbourhood or community scale) that was introduced in Chapter 2.

Rural communities in Britain

Britain's 'traditional rural communities' – those that, socially at least, seemed to adhere to the pastoral view of village life depicted in popular media – were built on a platform of common occupation, which resulted in shared values and a common pastoral culture. Traditional communities were *land-based*, with social status determined by ownership of land or subservience to landed interests. Britain's modern rural communities tend to be *mixed*, being a product of more recent population movements and a consequent social re-patterning of the countryside. Those movements and consequences are examined in the next section.

E.W. Martin's *The Book of the Village* (1962) paints a picture of village life from the Peasants Revolt of 1381 through to the late 1940s. In it, Martin charts many changes over this 600 year period – affecting farming, religion, local politics, rural industry and education – and concludes that '[...] every type of person in the village – the squire and the parson, the farmer and the labourer and craftsmen – has been affected by the new circumstances' (Martin, 1962: 127). However, the sub-text is that the social structure changed very little during this period. Those affected by the 'new circumstances' in the 1940s were the same groups living under the manorial system of the medieval period, all joined together by a landed/landless relationship. The 'squires' were the major landowners, employing tenant farmers, who in turn employed labourers and engaged the services of craftsmen. The spiritual needs of this land-based community were overseen by the local parson. This 'squirarchy' remained intact in the early years of the twentieth century, but between 1900 and 1950 it was subject to some adjustment because rising taxes and death duties after 1914 meant that property and land was often sold and the estates broken up during the inter-war period (Woods, 2005b: 31–32). Tenant farmers sometimes benefited from this process, buying former estate land cheaply and becoming owner-occupiers: mechanisation made their farms more profitable and the acquisition of wealth sometimes elevated their status close to that of the old squires (Martin, 1962: 129). The position of labourers also changed during this period. Mechanisation put paid to some rural jobs: tens of thousands of labourers left the land in the 1940s and 1950s, but those who remained also achieved higher status, being able to operate complex equipment and being treated, and paid, as highly skilled labour. Craftsmen fared less well during this period, 'suffering most from the competition of the factory' (Martin, 1962: 130). The boots worn by farmers or chairs sat on by squires could now be sourced more cheaply from elsewhere. Despite acknowledging that rural areas had not 'stood still', Martin's picture of the English village of the 1940s was one in which the parson still remained a leader, where land was still the commodity at the heart of the social structure, and where the pressures of in-migration remained largely unknown.

But this situation changed rapidly after the Second World War. During the hundred years before the War, the relative share of the national population living in rural areas fell from 50 per cent in 1851 to 20 per cent in 1939. But thereafter, it remained at about that level: roughly a fifth of people in Britain live in 'the countryside': in small towns, villages and in dispersed clusters or single dwellings. A hundred years of depopulation, in absolute and percentage terms, was followed by stability, both in the numbers of people living in rural areas and in the rural population share. But why did the population stabilise, and how was an apparently stable population subject to fundamental social change? The answer is simple: the population was *replaced*, like for like in terms of numbers, but 'newcomers' had very different characteristics from the replaced residents. Numerical stability (and some rises in population, especially in near-urban *and* some remoter rural places) and social shift was a result of counter-urbanisation. Such counter-urbanisation, or an 'inversion' of the 'traditionally positive relationship between net migration and settlement size' (Buller *et al.*, 2003: 8) was, and remains, a potent force, underpinning the social reconfiguration of rural Britain. During the post-war period of 'modernisation' and reduced need for farm labour, it countered the drift from the land and brought new consumers of rural space to the countryside, including second home buyers, commuters, down-shifters and retiring households. More recently, it has added about 250,000 residents to Britain's rural population in the last decade.

Champion (2000) suggests that the movement of people from urban to more rural areas is driven by a 'trilogy' of factors, categorised under the headings 'flight', 'quest' and 'overflow'. 'Flight' suggests a one-way movement away from cities by people – including those retiring – pushed by negative urban drivers. 'Quest' is the process that leads some people to move away from urban areas in search of alternative lifestyles (e.g. those retiring or 'down shifters') or particular residential qualities on which they place a personal premium. The final process in Champion's trilogy is that of 'overflow': here, a lack of room in cities – and high property prices – results in a drift away from urban areas. This third process also sits squarely with the concept of a 'cascade', with population gravitating firstly to satellite towns (in mid-life) and eventually to locations further afield (on retirement). The draw of rural areas has been examined in a number of research projects. In the post-war period – and more particularly, since the 1960s – the perceived qualities of rural living (Fielding, 1982) – together with a broader idealisation of the countryside as an idyll (see Chapter 1) and some of the anti-urban 'flight' motivations noted by Champion – have magnified the demand for rural property (Mitchell, 2004). In England, in particular, the accumulation of personal wealth in 'escalator' areas – a product of rising disposable incomes and equity growth in property – has encouraged some households to export their wealth to 'importing' regions and to join the steady flow of migrants to nearby rural areas, or venture further afield in search of investment opportunities. Darling (2005) has drawn attention to the significant economic returns on refurbishment and extension of rural dwellings and agricultural buildings, with similar returns being noted by researchers in the 1990s, encouraging many British households to export their wealth to rural France (Buller and Hoggart, 1994). Commentators have been able to apply the ideas of Ruth Glass (1964) and the notion of a 'rent gap' to the analysis of social change in rural areas, and to the exchange of people and of lifestyles that has been witnessed over an extended period. The upshot has been a very significant social transformation: the replacement of Martin's land-based communities with mixed communities displaying a range of needs, aspirations, interests and tensions.

The post-war transformation

The transition away from the type of *rural* community described by Martin has been the subject of a large volume of sociological analysis (see Pahl, 1965, 1975; Bell and Newby, 1971; Clout, 1972; and Newby, 1979). The movement of people in Britain resulted, according to Clout (1972: 50), in the '[...] traditional social structure [being] transformed to approximate with conditions [...] in urban Britain'. Urban communities, as Tönnies observed at the end of the nineteenth century, were more mixed and more transient. They had also become fractured along class and income lines, and tended to be ever-more outward-looking and less concerned with socio-spatial fixity (Hall, 1996; Ward, 2004). In the 1960s, the English sociologist Ray Pahl noted that villages were frequently 'restricted in outlook and in their involvement outside the village area' (Pahl, 1965). This outlook contrasted with that of the newcomers arriving at that time, whose '[...] spheres of association and contacts is wider than the village community which is seen largely as a dormitory and a place to spend the weekend' (Pahl, 1965:163). Pahl's work focused directly on the tide of social change washing over near-urban rural communities in the 1960s and into the 1970s. Counter-urbanisation was bringing newcomers to the countryside who tended to be middle class and who had no link with the land-based economy. This marked the end of what Newby (1979: 156) called the rural 'occupational' community.

By the end of the 1970s, Newby was able to claim that '[...] the lowland English village as an occupational community [had] virtually disappeared, except in a very few remote areas' (Newby, 1979: 164). 'Occupational', or Martin's 'land-based', communities had been dependent 'upon agriculture for a living' which 'conferred a sense of order on village life' (Newby, 1979: 157). But new transport infrastructure, especially the opening up of a network of motorways in the 1960s, made villages more accessible and ended their isolation. The in-migration that followed had a profound impact:

> There are now few villages without their complement of newcomers who work in towns. These new 'immigrants' have brought with them an urban, middle-class life-style which is largely alien to the remaining local agricultural population. Unlike the agricultural workers in an occupational community, the newcomers do not make the village the focus of all their social activities. [They] maintain social contacts with friends elsewhere, and if necessary, make use of urban amenities while living in the countryside [...] The newcomer, moreover, does not enter the village as a lone individual [...] newcomers arrived in such large numbers – perhaps due to the building of a new housing estate by a local speculative builder – that the individual 'immigrant' found himself one of many others whose values, behaviour and life-styles were similarly based upon urban, middle-class patterns of sociability.
>
> (Newby, 1979: 165)

These population movements, driven by the factors noted in the last section, brought gentrification and frequent conflict between the 'rump of the old occupational community' (Newby, 1979: 166) and those importing their middle-class lifestyles. That conflict centred on a clash of cultures, and on divided opinion as to the future needs and development trajectories of many villages. It led, according to Newby, to the remnants of the old occupational communities closing themselves off and becoming 'encapsulated' within the mixed communities that have dominated the British countryside since the 1970s.

The community development, planning and governance challenges encountered today are rooted in the fundamental shifts described by Pahl and Newby. Modern communities now display the features described by Panelli (2006), being socially heterogeneous and 'encapsulating' many competing interests, producing a range of governance and planning issues, which are examined below. But there are also perhaps simpler tensions as groups distinguished by wealth and income compete for scarce resources. Right from the beginning of this transformation, newcomers have competed against members of the occupational community for housing, changing the nature of rural housing markets, pushing up prices and producing housing access difficulties in many areas, as a result of demand pressures and planning constraint (Satsangi et al., 2010). These access difficulties have been compounded by middle-class opposition to new development and anything that is deemed 'detrimental to the character of the village' (Newby, 1979: 167) or to the preservation of property values. There has been long-standing opposition to 'social' or 'affordable' housing in many locations, which continues to this day (Gallent and Robinson, 2011), and is most readily explained by the fact that affordable housing is not generally needed by many middle-class commuters, second-home buyers or retiring households. In relation to other resources – village shops, post offices, buses or pubs – the same lack of need generates further tensions. The 'middle class patterns of sociability' noted by Newby (1979: 166) may mean that the pub no longer retains the same appeal (or is transformed by introduced tastes and expectations) and middle-class income and wealth reduces reliance on buses or cheap convenience stores. Social change has altered the profile of rural needs and changed fundamentally the *expectations* of many communities. This has placed considerable stress on residents whose incomes and expectations differ from those of their relatively affluent neighbours.

The change in expectations of what the countryside should be, since the 1960s and 1970s, is crucial to understanding the tensions and conflicts that now characterise rural planning debate. The gentrifiers brought with them 'strong views on the desired social and aesthetic qualities of the English village', arguing that 'it must conform [...] to the prevailing urban view – picturesque, ancient and unchanging' (Newby, 1979: 167). The erection of modern farm buildings, new housing, or even quite modest changes to the physical fabric of many villages can now attract significant opposition as the middle-classes seek to substitute a working countryside with a 'rustic suburbia'. Arguably, this pattern of division has softened over the years and a great many established in-migrants now back the type of development, including the conversion of former agricultural buildings and new housing for local needs, which might support economic development and the preservation of community life (Gallent and Robinson, 2011, 2012b). However, the conflicts that arose during the post-war transformation of the countryside seemed to point to a simple division between 'locals' and 'newcomers'; between the traditional 'rump' and the middle-class gentrifiers. But it is important to note that the social structure of villages prior to the transformation was not flat; rather it displayed a steep hierarchy (see above). And the newcomers were not all the same; their motivations for heading to the countryside, and their personal wealth and income, varied greatly. The countryside experienced a collision of two very different social structures, creating a hybrid with urban traits and rural remnants. This was described by Pahl (1975) and generally holds true today.

The first group in Pahl's hybrid rural structure were (and remain) the *large property owners*, sometimes surviving squires, but more usually the successors to those farmers who had benefited from the break-up of the great estates at the beginning of the twentieth century. Today these form an agrarian elite (Woods, 2005b: 33), retaining power over land and often benefiting from the sale of that land for development. The second group were (and are) the *salaried immigrants* with good incomes derived from urban jobs and with enough capital to take advantage of the 'rent gap' between urban and rural property prices. These are the gentrifiers; the bulk of the middle-class migrants who introduce and sustain a rural version of urban sociability. The third are a sub-set of the second: *spiralists* according to Pahl who move themselves and their families to a village or small town as their professional careers progress and as they are forced to move home. They prefer near-urban locations where they believe fast integration will be easier. The fourth group are the *reluctant commuters*, with limited income and little capital. These are the overflow migrants (Champion, 2000: 14) who would rather have stayed in the city but, for reasons of cost, move to cheaper parts of the countryside. Whilst the salaried immigrants look to trade-up to larger rural properties, old and rustic or new and executive, the reluctant commuters take advantage of cheaper new-build opportunities in the edge of town or village estates (Newby, 1979: 15). They may not have the means to bear the costs of rural living, ultimately encountering financial difficulty as a result of high commuting expenses. The fifth group are the *retired households*. Many are wealthy after a lifetime working in higher-paid urban jobs and following the release of equity from urban homes. They are seduced by the lure of the rural idyll. However, retired households have 'varied financial backgrounds' (Pahl, 1975: 43) and the less well-off members of this group may have been attracted by a *combination* of perceived tranquillity *and* the expectation of lower costs. Like the reluctant commuters, they may encounter financial difficulty and may end up 'lonely, bored and isolated' away from life-long friends, especially as they become less mobile (Pahl, 1975: 44).

Apart from the large property owners, all of the above groups are in-migrants to the countryside. The first remnant group of the occupational community are now *former social housing occupants*. They are the successors to the agricultural workers once housed in tied accommodation or private lets. Their importance to the rural economy along with a recognition of their need and rights by local authorities led to their rehousing in council homes. These have since been purchased (through the right-to-buy – see Chapter 8) and this group has therefore benefited from rising property values. Many will have sold on and moved out of their former homes, now living in new private red-brick developments (alongside the reluctant commuters) built on the edges of many small towns and villages since the end of the 1960s. The next remnant group is the *rural poor* – a 'low wealth, low income' housing class according to Shucksmith (1990) – which often comprises private or tied tenants, or nowadays the occupants of new housing association homes, who find themselves increasingly marginalised within many mixed and gentrified rural communities. Pahl also noted the presence of *tradesmen and small business owners*, identifying them as a separate class. It seems likely that today this group is distributed across the new social structure. Those rooted in the occu-pational community may find themselves better or worse off, in the red-brick homes or in social housing. And they may have been joined by next-generation newcomers who have chosen to set up small businesses, perhaps responding to new rural appetites (for restaurants or gastro-pubs) or the rediscovery of rural crafts, for which there is a strengthening market in the 'consumption countryside' (see Chapters 2 and 4).

Mixed communities – with more varied needs, appetites and opinions – present planning, in all its forms, with potentially more complex problems. These problems and the planning responses are examined in Chapters 6, 7 and 8. However, the transformation of rural communities has been paralleled by a tendency towards community-based action, partly because of that trans-formation itself (bringing a different range of skills and behaviours to the countryside) and partly because governments have come to view 'community' as a potential instrument of service and policy delivery.

The politicisation of community

The political journey in Britain from the neo-liberalism of the 1980s to the localism of the 2010s (which appears to value and embrace many outcomes of 'community life') has been marked by ever-greater emphasis on 'community' and 'consensus' as instruments of governance. Margaret Thatcher (UK Prime Minister from 1979 until 1990) was a proponent of neither of these things, placing little value on the notion of community, and dismissing consensus as '[...] the process of abandoning all beliefs, principles, values, and policies in search of something in which no one believes, but to which no one objects' (Thatcher, 1993). Thatcher's strident views on community and consensus ran contrary to a prevailing wisdom that had taken root in the 1960s: society had transformed, becoming evermore pluralistic, and planning needed to re-scale to deal with the complexity of a society in which professional constructions of the public good were being regularly rejected (Rittel and Webber, 1973). Thatcher's departure from government coincided with renewed interest in creating opportunities for local choice in the systems of government, and ultimately in what has been called a 'governance shift' (Jessop, 2003) away from the exercise of executive power towards various attempts to involve communities in decision-making and -taking. During the 1990s, the Joseph Rowntree Foundation funded a number of research projects that examined local leadership (1994a) and the governance gap (1994b). By the time 'New Labour' came to power in 1997, the community agenda seemed to have unstoppable momentum. 'Community' became the buzzword of all major initiatives, notably the 'Sustainable Communities' plan of 2003 (OPDM, 2003) – a broad strategy to promote new housing in southern England and market renewal in the north – and of a number of Local Government Acts that aimed to promote community and business involvement in local decision-making. The Labour Government set up 'Strategic Local Partnerships' (LSPs) in 1999 and, in the following year, handed these partnerships a responsibility to produce 'community strategies' (see Chapter 2). These were to be 'strategic visions' for a place, in which the 'community's vision' is fully integrated and 'at the heart of creating sustainable communities' (DCLG, 2006a: 101). More broadly, Labour promoted the idea of 'community leadership', viewing this as central to all local planning and to a broader 'modernisation' of the local state.

But a central concern during the 2000s was how well community strategies reflected local needs and ambitions. The LSPs had a strategic focus and the community input was delivered by representative bodies: residents' networks or associations of community councils. The influence over policy was indirect (Gallent and Robinson, 2012b). This contrasted with the direct action and projects being undertaken in many neighbourhoods and villages: drawing up plans, running local services, or taking charge of key assets. When the Conservative–Liberal Democrat Coalition came to power in 2010, its agenda seemed to be to turn 'community' into more of a direct delivery vehicle. Since then, there has arguably been more direct involvement of place-communities in the formulation of planning policy and greater support (in terms of cash and framework flexibility) for local projects. What Labour began in earnest, the Coalition has continued and extended under a banner of 'localism'. Community continues to be crucial to the roll-out of government policy. But the extent to which this is the complex, fragmented reality of community described in this chapter has to be doubted. The politicised community, whose virtues are regularly extolled by politicians, has a 'unity of will': it wants to take charge; it reaches consensus on a vision of the future; and it acts in unison. But it is often the case that when communities become 'delivery vehicles' (or even when action is spontaneous) that delivery is taken forward by a narrow section of people: those with time on their hands, with the right skills, who are educated and politically aware. This is not the case in all instances. Some groups work hard to involve as many people as they can; to extend and 'democratise' community-based planning. Others, however, represent only a narrow set of interests and may pursue very specific, private agendas. The danger of 'private corruption', noted earlier in this chapter, may be mitigated through the *framing* of community action and also through external help to ensure that a wide cross-section of people are able to participate.

Community action framed: localism and planning

A parallel journey that has taken place in Britain over the last 40 years has been from relatively spontaneous social entrepreneurialism to more 'framed' community action, as governments have sought to harness the energy and legitimacy of that action. This journey started in the countryside. Owen and colleagues (2007: 49) note the existence of 'very-local community-based planning' from the 1970s onwards, represented by the work of rural Parish Councils and confined to the drawing up of occasional 'parish appraisals'. Evidence gathering by Parish Councils in England (and equivalents elsewhere) was initially intended to provide a basis for actions internal to the communities. But it quickly became apparent that local politicians were often referring in council to the views of their constituents, prompting some local groups to draw up more aspirational Parish Plans and Village Design Statements. They were frequently disappointed, however, by the lack of general take-up of their ideas and wishes.

Generally, local planning teams found it difficult to take the wishes of parishioners into formal plans as many of the issues raised were *ultra vires*: outside the remit of land-use planning. However, the first Labour government published a Rural White Paper (DETR and MAFF, 2000) setting out how communities could 'play a much bigger part in their own affairs and shaping their future development' (DETR and MAFF, 2000: 145). The White Paper went on to suggest that community-based plans should identify how new development might be fitted into villages, what design standards development should meet, and which 'valued local features' should be preserved through future planning decisions (DETR and MAFF, 2000: Para. 12.3.1). The aim was to encourage Parish councils to focus directly on issues within the remit of planning, with the White Paper signalling government's view that 'framed' community action is of potentially greater value to the planning system. Four years later, a 'Vital Villages' initiative was launched which, amongst other things, sought a more consistent approach to the production of Parish Plans, offering advice and funding to parishes considering drawing up their own plan (Countryside Agency, 2004: 6). However, it is notable that the initiative focused not only on the content of plans but also on the role of parish planning in providing an outlet for local energies and 'invigorating community spirit' (Countryside Agency, 2004: 9).

In the same year, the government brought forward its Planning and Compulsory Purchase Act, setting out key structural changes to the planning system (see Chapter 1 on 'spatial planning' and Chapter 2). In that context, parish plan preparation was seen as an aid to local engagement, part of planning's local evidence base and a means of delivering 'distinctiveness' in policy design (by connecting to the diversity of needs in particular areas – see the earlier discussion of social complexity). Similarly, a focus on parish planning was also viewed as one way of localising the 'community strategy' agenda emerging from local government reform. The overarching aim was to bring a degree of professionalism to community-based planning and hence make it more 'useable'. Very little headway was made, however, in this direction. There is no evidence to support the view that Parish plans became more professional during the 2000s or that they were more regularly absorbed into local plans. For the most part, it seems to have been business as usual in many rural areas with groups of highly-motivated parishioners continuing to draw up aspirational plans that were seldom taken account of by either planning officers or committees (Gallent and Robinson, 2012b).

Either the formal planning system had failed to harness the added-value of community-based plans in a systematic way, or those plans offered little to the system that was worth harnessing. The Coalition Government, elected in 2010, appeared to take the former view and came to power on the back of a 'localism' agenda that included a new approach to community-based planning: one that would be framed by normative (planning) expectation and integrated into local plans. In the run-up to the Election, the Conservatives pledged to strengthen the part played by communities in local statutory planning. The Coalition's Localism Act 2011 introduced the Neighbourhood Development Plans (NDPs) and Orders (NDOs) described in Chapter 2, stipulating a process for their production and requiring that they be 'made part of' local plans once they have been approved by referendum and providing they respect 'the overall national presumption in favour of sustainable development, as well as other local strategic priorities such as the positioning of transport links and meeting housing need' (DBIS, 2010a: 24). It is likely in many rural areas that old Parish Plans will be substituted with NDPs and that a great many communities will comply with the new rules in the hope that their compliant plans will have a greater future influence on planning policy and decisions. But for other communities, the prospect of having to participate in the planning of development that they are fiercely opposed to may not be an appealing one. It seems likely, therefore, that the new system will not be universally embraced.

And what of the spontaneous action that springs from community life? Has this been replaced by a new alignment with formal planning? Some communities will probably continue to follow their own paths, setting local priorities and engaging in voluntary action that is independent from official processes. But others will grasp the opportunities for supported community action provided by the new framework. The NDOs allow Parish Councils to extend permitted development rights (Chapter 2), potentially allowing them to take forward community projects (for example, converting old farm buildings to community use: see Chapter 8) without recourse to local planning, under the rubric of a 'community right to build' (see DCLG, 2013). There are signs here that government is willing to devolve far more responsibility to communities, the big hope being that these communities will pursue genuine social goods rather than private interest – contributing to a broadening of rural planning in response to a diverse set of needs.

Summary

- Understanding the nature of rural communities – their complexity and their capacity for communitarian action – is critical to understanding the new diversity of challenges that face rural areas.
- Communities can be understood as networks of individuals engaging in a range of social interactions, with those interactions transforming the innate human capital of individuals into collective social capital.
- Rural communities are thought to have a particular propensity towards communitarianism and social entrepreneurialism, rooted initially in their high degree of social homogeneity (displayed by occupational

communities) and then in the import of new skills and knowledge during the period of post-war counter-urbanisation (and the transition to mixed communities).

- The politicisation of community, from the 1990s, as a policy delivery tool seemed to suggest a move from spontaneous to framed community action. But the reality today is one of more regularly supported (but often still spontaneous) action as government tries to harness the energy of communities to solve complex local challenges.
- Community action is an important aspect of rural planning, and a vital response to the increasing diversity of rural needs.

Key readings

- Gallent, N. and Robinson, S. (2012) *Neighbourhood Planning: Communities, Networks and Governance*, Policy Press: London.
  (Recent research on the transition from parish plans to neighbourhood plans facing ten communities in Kent.)
- Newby, H. (1979) *Green and Pleasant Land? Social Change in Rural England*, Hutchinson: London.
  (A timeless classic, dealing with the post-war transformation of near urban areas faced with counter-urbanisation.)
- Pahl, R. (1975) *Whose City? And Further Essays on Urban Society*, Penguin: London.
  (Another timeless classic dealing with the social structure of rural communities in Chapters 1 and 2.)
- Rydin, Y. (2014) Communities, networks and social capital, in Gallent, N. and Ciaffi, D. (eds) *Community Action and Planning: Contexts, Drivers and Outcomes*, Policy Press: Bristol.
  (An accessible overview of community and social capital.)

Key websites

- Action with Communities in Rural England: www.acre.org.uk/.
  (The umbrella body for the 38 Rural Community Councils operating in England.)
- Neighbourhood Planning Portal: www.planningportal.gov.uk/inyourarea/neighbourhood.
  (Government portal outlines processes and opportunities for neighbourhood planning arising from the Localism Act 2011.)

6 Accessibility, services and opportunity

•••

A complex domain

The last chapter drew attention to the complex local challenges faced by modern rural communities and the stronger sense of 'community' that may flow from relative isolation and the inability of the public sector to meet all service needs, leaving local people in some instances to fend for themselves. This chapter considers the role played by voluntary, community-based, action in responding to transport needs in the countryside, but its broader aim is to examine the fundamental rural challenge of accessibility, conceived as the degree to which personal mobility and transport availability renders key services 'get-at-able' (Moseley, 1979) and the extent to which those living in the countryside are able to participate in desired activities. Accessibility is an important component of the service challenge in rural areas; as important perhaps as the social transformations outlined in the last chapter, which have altered the needs profile and created new patterns of disadvantage. The combination of social change and spatial accessibility provides the key driver of the rural service challenge, generating distinct geographies of opportunity.

Conventional notions of accessibility emphasise physical access (Hutton, 2013: 90) or the transport options available for reaching key services – schools, shops and so on – at regular and predictable intervals. However, the field has recently taken a 'mobility turn' (Sheller and Urry, 2006) marked by an acknowledgement that the social objectives of travel extend beyond the school run or visiting a doctor, as important as these things might be. Mobility allows individuals to participate in social networks, and 'new mobilities' (Sheller and Urry, 2006) – including access to telecommunications infrastructure comprising fixed-connection internet or 3G services – make many modern services 'get-at-able'. There are a great many barriers confronting both physical transport delivery and good internet access. Dispersed populations, especially in remoter rural areas, create the thin markets noted in Chapter 5. Changes in the socio-demographic structure of some areas, leaving an ageing population in the wake of outward migration by younger people, may create heightened demand for transport which is nevertheless dispersed and difficult to service. Car-owning incomers (or seasonal residents) may be able to meet their own needs, but in doing so reduce demand for public provision – which thereafter lacks the viability it needs to offer good-quality services for older or poorer residents. This cocktail of market challenges, from a more dispersed, economically mixed, and seasonally variable population (swollen by summer visitors who disappear in the winter months) has resulted in generally poor-quality public transport services across European peripheral regions and contributed to:

- **Social and economic marginalisation** – relative to ever-improving connectivity within and between Europe's core regions.
- **Car dependency** – reflected in higher rates of car ownership particularly in deeper rural areas, creating cost and affordability issues for households and sustainability problems more broadly.
- **A heightened perception of remoteness** – beyond the actual reality, but serving to discourage investment and tourism.

With reduced access to services, employment, friends and family, and other opportunities, the challenge of exclusion can quickly arise; the deprivation that follows is both *economic* (and compounded by the costs of owning, running and constantly using a car) and *social*, locking people out of social networks and denying them access to soft social infrastructure. The challenge of mobility – and of car dependency – enlarges the carbon footprint, with rural areas eventually viewed as inherently unsustainable (see Taylor, 2008; and see also Chapter 2)

Towards the end of this chapter, we turn to look at community responses to the problems of rural transport. But we begin here, in Part 1, by unpacking in greater detail the accessibility challenges that now characterise rural areas, and the changing role of the state in addressing them. Part 2 of this chapter then examines the narrative of response in Britain before considering local initiatives.

Part 1: Accessibility: transport supply, demand and the role of the state

Access

People and communities in the countryside have basic social needs, which were once met within their village or a nearby market town: the original setting for Newby's 'occupational community' with its neat 'sense of order' (Newby, 1979: 157). A corresponding sense of order prevailed on settlement geography too, with lower-order hamlets clustered around villages which were typically within a day's walk from a market town where higher-order, non-essential goods and services could be accessed. In lowland agrarian areas, the village was the source of all essential services, accessed via dense networks of footpaths and other 'rights of way' from surrounding settlements and farmsteads.

Industrialisation during the nineteenth century, in Britain and elsewhere, began to alter these spatial relationships. Railways, in particular, suddenly connected many market towns and larger villages to major cities, blurring the boundary between town and country. The result was both suburbanisation beyond previous urban limits – the creation of 'metrolands' – and the eventual transformation of some satellite towns into urban dormitories. A spatial unravelling of life–work relationships ensued; some existing rural residents could now access jobs and opportunities in the cities; and some former urban residents could access cheaper housing and different lifestyles in the countryside. The old hierarchy of nearby rural settlements, walkable from one another, became less important as new spatial relationships – reliant on the railways – came to the fore. By the second half of the twentieth century, rural hinterlands beyond the market towns and key service villages had been opened up by car-ownership, creating the mixed communities described in Chapter 5 and producing a rural society less tied to local goods and services.

These changed spatial relationships, precipitated by industrialisation and underpinned by social change, are the essential driver of today's access challenge: they thinned the market for essential village services, diminished their viability and eventually resulted in the closure of shops, post offices, pubs, schools, chapels and so forth (see Chapter 7). And a gradual rationalisation of public transport services, underpinned by the same driver and examined later in this chapter, has meant greater dependence on private cars to access everyday essential services – accentuating the disadvantage felt by the rural poor and giving rise to the concept of 'transport poverty'. According to Britain's Department for Transport (DfT, 2006) this form of poverty exists where:

> Inadequacies in transport provision [...] may create *barriers* limiting certain individuals and groups from fully participating in the normal range of activities [...] this concern focuses attention on the link between transport provision and activity *participation* and the role of *accessibility*.

Transport poverty in turn results in transport(-based) exclusion; that is, exclusion from a range of socio-economic activities, from employment to activities that are part of the reproduction of community. This happens for reasons that are *spatial* (the location of the activity is inaccessible without private means), *temporal* (the location cannot be reached at the right time using available public transport), *financial* (the costs of reaching it are prohibitive for a particular individual) or *personal* (a disability, for example, makes it impossible for an individual to utilise the available public transport) (DfT, 2006). The temporal dimension of transport exclusion is neatly illustrated in the following example from Sloman (2003):

A CAB (Citizens Advice Bureau) in Cornwall reported that a nurse working shifts sought advice about the problems she experienced in getting to work, following the introduction of a new bus timetable. She now had to get up earlier to walk 20 minutes to the nearest bus stop to her home, where there was no shelter. The client could not get home by public transport if she worked into the evening, as the last bus home left at 7:30 pm.

(Sloman, 2003: 4)

This single example points to wider problems in rural transport provision. According to the Countryside Agency in England:

> Weak rural bus services are commonplace. Where services operate, they are often inadequate, even at peak periods. Rural residents also have a longer walk to bus and rail pick-ups, reflecting the differences in public transport proximity and interchange.
> (Countryside Agency, 2001: 63–64, quoted in Sloman, 2003)

Finding a solution that addresses the interplay between dispersion, distance and demographics, and delivers adequate levels of access to, and participation in, public transport is a major challenge in all countries. In the sections that follow, we explore the complex nature of rural travel demand, the difficulties of public sector provision, and integration between urban and rural systems.

Travel demand

Population dispersion makes the cost-effective provision of rural transport services extremely difficult, although the degree of difficulty will vary from one place to another depending on the level of settlement distribution, population density at key hubs and the demographic structure, which will skew the spatial patterning of need. All of these features determine the demand base, for either market-based or public-sector provision. However, there is a degree of predictability in travel demand and behaviour: at key density thresholds, modal switches occur and the amount of travel declines. In Britain, it has been found that car-based travel lessens in areas with densities above 15 persons per hectare (pph) as travel alternatives such as regular buses become viable and available; at 30 pph, trains tend to displace cars as alternative modes for longer journeys; and at 50 pph, overall travel demand sharply declines because that level of density supports the provision of multiple forms of alternative transport and a multitude of nearby services. However, these density thresholds tend only to be achieved (and sustained across wider areas) in larger towns and in cities. The typical rural household travels twice the distance each year when compared to someone living in Inner London or Glasgow, and most journeys will be by car (Headicar, 2009). However, the figures in Table 6.1 show that the overall number of trips, and

Table 6.1 Summary of annual trips, travel time, trip length and distance travelled (DEFRA, 2014a)

	Trips per person		Travelling time (hours) per person		Trip length (miles) per person		Distance travelled (miles) per person	
	2002–2006	2008–2012	2002–2006	2008–2012	2002–2006	2008–2012	2002–2006	2008–2012
All urban	1,034	959	383	366	6.4	6.4	6,644	6,158
All rural town and fringe	1,061	1,002	387	378	8.3	8.7	8,777	8,763
All rural villages, hamlets and isolated dwellings	1,085	990	424	394	9.8	10.2	10,680	10,057
England	**1,041**	**966**	**386**	**369**	**6.9**	**7.0**	**7,141**	**6,725**

actual time spent travelling, is similar for rural and urban residents. Interestingly, the same data also suggest that hours and distance travelled in England fell in all areas between 2002–2006 and 2008–2012. Similar declines have been measured in other Western nations, leading some commentators to conclude that countries have reached and passed the point of 'peak travel' (Goodwin, 2012). However, trip lengths among rural residents appear to be long and lengthening. In the Republic of Ireland, 40 per cent of the population resides in rural areas (double the UK level) and car users drive, on average, 24,000 km per year (Government of Ireland, 2010). This is 8,000 km more than their UK counterparts and has been attributed to counter-urbanisation to the countryside (where property is more affordable) and back-commuting to urban jobs (Wenzell, 2009).

The relationship between settlement size, population density and travel is a global one (Newman and Kenworthy, 1999), explaining high levels of rural car dependency around the world. However, car *use* in rural areas is a preference for some and a necessity for others: new residents have actively chosen a rural lifestyle that will include seclusion and car use; remnants of the old rural poor, described in Chapter 5, become car dependent as travel alternatives decline. That said, for many people living in the countryside, walking to a nearby shop or pub remains a possibility. The big difference between larger urban areas and rural areas, is that in the latter, residents are far less likely to use a bus or train and far more likely to be the driver of a private car (Figure 6.1). The result is a far higher CO_2 footprint for the typical rural resident and a strong possibility of transport poverty resulting in residential displacement, as lower income groups – unable to access essential services or employment – are forced to relocate to larger towns.

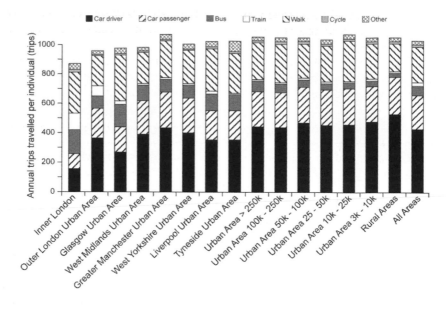

Figure 6.1

Annual trips per person by region and settlement size in Great Britain, 2002–2006
Source: Headicar, 2009

But relocation is not always viewed as indicative of a problem. Rather, it may be seen as part of the solution. Better accessibility may be achieved by creating – through public funding, voluntary action or private means – new transport opportunities or by reducing the need to travel. For urban areas, there is a substantial literature – and accompanying debate – on land-use mixing and compaction as means of shortening travel distances and maximising the take-up of green modes including walking and cycling. The 'city of short distances' is now a goal of much urban planning (Hall, 2013). Mixing uses in a compact, walkable 'key settlement' (or market town) where there is a concentration of new housing, retail and leisure has been an aim of rural settlement planning for several decades (Cloke, 1979; see also Chapters 2 and 3). The underlying rationale has been to 'make' a viable market for both key services and public transport by moving towards critical population density thresholds. It is a planned attempt to counter the dispersion and thinning of population after the Second World War (produced by a combination of counter-urbanisation and flight from land-based economic activities) and return to the type of pre-industrial settlement hierarchy described earlier in this chapter and outlined also by Walter Christaller in his Central Place Theory (Christaller, 1933). This approach has been employed in Britain since the late 1960s, with policies set out in Structure Plans (see

Chapter 2) for the enlargement of 'key settlements', 'key service villages' or 'market towns' and the cutting of services in smaller villages or hamlets (Sillince, 1986).

Whilst this policy has produced a network of rural holding points, it has starved lower-tier settlements of their services, causing the closure of many small primary schools, post offices, pubs and bus services. Car-owning incomers often choose to locate to these more secluded (and increasingly exclusive) locations, which are free from the 'threat' of development, pricing less affluent households out of the local housing market and shunting them further up this planned settlement hierarchy. Non-service centres have, in many locations, become increasingly gentrified (Gallent, 2009) and are viewed as places of housing policy failure. At the same time, they are populated by regular car users, reinforcing the view that small villages support unsustainable lifestyles and should not, therefore, receive any additional development. The gentrification of the English countryside, in particular, has resulted from a combination of market forces and restrictive land-use planning.

But in this same context of selective service investment, there are households who struggle with the costs of car ownership and 'muddle' their way through the rising fuel, repair and other associated costs until these become insurmountable:

A 62-year-old man in poor health sought advice from a CAB in Lincolnshire. He lived in a village with few amenities and poor public transport, so he felt obliged to run a car even though he could not really afford it. As a result, the client had gone overdrawn and had taken out loans to cover car-related costs.

(Sloman, 2003: 2)

Given all of these problems, the supply of rural public transport – or a viable and affordable alternative – would appear to be critically important, if more vulnerable and less affluent residents are not to be excluded from services and a raft of social and economic opportunities. At the same time, reduced *mobility* will have profound consequences for the rural labour market: distance and dispersion may deprive rural businesses and key services (remember the nurse in Cornwall) of the labour they need to prosper or even to operate. Agriculture remains an important activity in rural areas, joined in some instances by hospitality (tourism) and other service industries (see Chapter 3). Often, the accessibility that people need to different opportunities and activities is physical although in some instances, *new mobilities* may compensate for physical access difficulties. However, although new technologies have played a part in improving communications within, from and to rural areas – ensuring that residents are better connected – its impact on travel behaviour is difficult to measure. Internet access, broadband speed and mobile phone connectivity varies greatly between areas and across the UK, achieving standard service levels has become a key priority in rural development (DEFRA, 2014a). But even in well-connected areas, changes in travel behaviour are not easy to detect (Kenyon, 2010).

This is probably because a great many rural businesses still need workers on-site, and although goods and services can be bought and accessed online, the social need to travel remains largely unaffected (Mokhtarian and Solomon, 1999). In the US, Wellman (2001) has argued that the internet creates additional opportunities for social interaction, but it does not replace face-to-face contact, and may in fact be a catalyst for new interactions in the real world. It is doubtful whether people move to, or remain in, rural areas with the intention of interacting with their neighbours solely via their computer screens or smart phones. Opportunities to engage in more traditional forms of sociability is part of the rural idyll that many people buy into, meaning regular outings to the local pub or nearby village shop. These retain their social function within the community, and even longer journeys – to a market town or shopping centre – fulfils a purpose which is broader and deeper than the mere purchase of goods. Travel is experiential and an end in itself: '[...] for my part, I travel not to go anywhere, but to go. I travel for travel's sake. The great affair is to move' (Robert Louis Stevenson, 1878).

Public transport supply

Although good public transport may appear vital for the wellbeing of rural communities and their residents, its provision poses many challenges. Internationally, there is a spectrum of public transport models, from Britain's *laissez-faire* system of loose control over privately operated services (and subsidy support of particular routes and passenger groups) through to the Swiss model of full state-operated and funded provision. A closely related issue is how services are governed and delivered, whether by wholly rural authorities who must carry the entire burden of operating uneconomical routes, or by city-regions able to extend services from urban centres into rural hinterlands – and hence cross-subsidise different parts of an integrated system.

In many countries the role of the state has evolved from active transport provider to that of transport regulator in the context of a neo-liberal, market-led approach to public service provision. Yet the journey from one to the other has seldom been simple or linear. The UK, for example, had a long history of *privately* operated and maintained transport services prior to state intervention. This ranged from the centuries-old turnpike trusts, which built and maintained sections of roads in return for toll fees collected by parishes, to the railways that were later built on the back of venture capital from share-holders and corporations. State involvement in transport is a relatively recent phenomenon. The railways, built from the mid-nineteenth century onwards, were not nationalised until 1948. And interestingly, state control and funding did not guarantee the continued delivery of less economic routes. In fact, quite the opposite: a third of the rail network was scrapped in the wake of a 1963 report into the 'Reshaping of British Railways' (which gained subsequent notoriety and became known simply as the 'Beeching Report' after its principal author, Dr Richard Beeching). However, rationalisation of rail services from the 1960s onwards was not solely a British phenomenon, and nor was it purely a result of public-sector cost saving. Big population shifts and climbing rates of car ownership, combined with *either* nationalisation *or* transfer into the hands of larger private operators resulted in heightened concern for operating efficiency or profitability. The contraction of railway services happened across Europe and North America, although the speed of that contraction in Britain was unparalleled, with Beeching personally blamed in many rural areas for severing 'a vital lifeline [...] that can mean the difference between prosperity and poverty' (Loft, 2014: 2).

The reality of course is that Beeching was merely formulating a response to major structural pressures. Private car ownership meant fewer rail passengers using rural branch lines, which in turn affected the viability of freight services. The same pressures were evident across the whole of Europe, where many loss-making branch lines were closed from the 1950s onwards. Others were down-graded to single lines and the number of stations reduced. Money saved was then re-invested in road construction, in response to increased car ownership and the transfer of freight to road vehicles (Banister, 2002). Closure in the 1950s was followed by a rapid expansion of motorways, autobahn, freeways and autostrada in the 1960s. The 1963 report delivered cuts of approximately one third or 5,000 miles of track and over 2,000 stations (Loft, 2014). In Belgium, the Vicinal narrow gauge tramways, which once formed a dense network across the entire country, shrank from a peak of 4,811 km in 1948 to 977 km by 1960 (Laconte, 2013). Today only fragments of the original network survive; the largest being the 61 km *kusttram* stretching along the Flemish coastline from Knokke to De Panne.

Beeching's analysis of weaknesses in the rail network, and his response, continue to have a profound impact on rural transport patterns in Britain. He dismissed the development of light rail services (which could run on existing infrastructure, and which were being pioneered in Germany at that time) that might exploit branch line demand, arguing that '[...] it is not immediately apparent [...] why it is thought that rail buses would give a better standard of service than a road bus in most rural areas' (British Railways Board, 1963: 18). Beeching favoured 'inter-modality' or the idea that road-based transport would be used to reach a rail-head or 'parkway' station, from which the journey to a larger centre would be continued by train. This happened in a portion of cases, but in many others commuters, shoppers or day-trippers opted to continue their journey by private car, contributing to urban congestion and reducing ridership on the railways. This in turn led to ticket-price rises and further transfers to the roads.

Many German cities adopted an entirely different approach to rail-based urban-to-rural connections, using the redundant capacity on the regional heavy rail networks to provide light rail services able to run onto the urban tramway networks, providing interchange-free connections from villages to city centres. This idea was first adopted in Karlsruhe in the early 1960s where the Abtalbahn regional line was converted to light rail use. Other schemes were developed in the 1990s which comprised 'tram-trains' able to mix with heavy traffic on the national lines but also utilise lighter tram infrastructure. This model has been replicated across Europe: in the Netherlands, France, Spain, Portugal, and also in the UK where the concept is currently being tested in Sheffield. Such integrated transport solutions offer a range of advantages, the most obvious relating to financing and accessibility. Higher ridership in more densely populated areas will help finance those parts of the network covering more sparsely populated areas, ensuring broader and more even accessibility. However, this model works best where there is no jurisdictional division between rural and urban areas; where these are seen as an integrated whole.

Rural–urban transport integration

An important rationale of the Karlsruhe scheme was the idea that transport has an economic function, working for the benefit of a single urban and regional economy, comprising both the urban core and rural hinterlands. The programmes of public provision, underpinned by this rationale, are overseen by municipal transport organisations: umbrella bodies which retain control over the organisation of public transport, either by operating services themselves or by signing service-level agreements with licensed private operators. For example, they are able to require operators to guarantee connectivity between services, or even connections by taxi or minibus to outlying villages after onward bus services have ended for the day. One persistent problem, however, is that bus operators will be reluctant to enter into contracts for the least profitable routes, and even subsidy support may not compensate for the logistical difficulties of guaranteeing connectivity. This may, in some instances, provide a trigger for voluntary-sector activity which will aim to bridge the gaps in formal provision. This issue is returned to later in this chapter.

However, multi-modal municipal or regional umbrella bodies provide what is now a standard approach to transport organisation across Europe. In Britain, a patchwork of six regional transport authorities was established between 1968 and 1972: for Greater Manchester, Merseyside, Tyne and Wear, the West Midlands, South Yorkshire and West Yorkshire. These were all designed to mimic the London integrated transport authority – now 'Transport for London'. Outside of these city-regions, with their multi-modal co-ordination, public transport provision is led by local councils although a 2008 Local Transport Act made provisions for the creation of further 'Integrated Transport Authorities' at different scales.

Few transport services in the UK are now directly owned and operated by the local state, although local councils and transport authorities retain their licensing and co-ordinating roles. More broadly, the European Union has encouraged greater competition between transport operators by requiring the break-up of state monopolies, and through the active promotion of competitive, privately operated transport services (Directive 2007/58/EC; Office of Rail Regulation, 2014). But a generic and persistent problem with competitive, market-based approaches, which continually strive for greater efficiency (alongside increased profit) is that they seldom result in the integration of different transport modes to create 'seamless' journey chains. Rather, breaks in service (both temporal and spatial gaps) seem to be a common outcome where different operators manage different services. The result, as Krygsman *et al.* (2004) note, is that:

> [...] much of the effort associated with public transport trips is performed to simply reach the system and the final destination. Access and egress stages (together with wait and transfer times) are the weakest part of a multimodal public transport chain and their contribution to the total travel disutility is often substantial.

This is particularly true of rural to urban connections. According to Potter (2010), five elements are needed for the integration of public transport services:

1 *Locational Integration*: being able to easily change between transport modes;
2 *Timetabling Integration*: services at an interchange which connect in time;
3 *Ticketing Integration*: not needing to purchase a new ticket for each leg of a journey;
4 *Information Integration*: not needing to enquire at different places for each stage of a trip (or that different independent sources are connected and appear seamless to a user);
5 *Service Design Integration*: that the legal, administrative and governance structures permit and encourage integration.

In Switzerland, different levels and modes of public transport are integrated in a predictable way through the application of the *taktfahrplan* or 'clock-face' principle, which was introduced nationally in 1982 to deliver a consistent 'cascade' that requires minimal prior planning by the passenger. Departures at each hub correspond with different positions around the clock-face, according to the overall distance that the service will travel (see Table 6.2). As all major cities have hub timings at 00 and 30 minutes past the hour and basic 'interval families' of 30 minutes for inter-city trains, 15 minutes for regional rail and bus services, and finally 7.5 minutes for local tramway and bus services, service times at smaller stops are similarly predictable.

The Swiss clock-face principle is made possible by close co-operation

Table 6.2 Service intervals for hierarchy of different travel modes in Switzerland

Frequency (minutes)	120	60	30	15	7.5
International services	•	(•)			
Inter-city services		(•)	•		
Regional/s-bahn services		(•)	•	•	
Regional bus services			•	•	•
Local bus/tramway services				•	•
Hub times (minutes past the hour)	**00**	**00 or 30**	**00/30 or 14/45**	**00/15/ 30/45**	**No hub**

between Switzerland's 51 state-owned transport operators. Similarly, the Swiss model (and its German counterpart) is built on the territorial governance arrangements that bind larger cities to their surrounding rural regions, often giving rural areas metropolitan service levels. In contrast, Britain retains a very strong distinction between urban and rural, not just in terms of transport co-ordination, but in broader planning terms:

> [...] One marked difference [...] is that other European countries do not perceive quite the same urban/rural divide as we do in Britain. Transport in neighbouring rural and urban areas is planned as a coherent whole. The same body is responsible for both, and rural transport networks are strongly focused on the nearest main city or town. Fares and connections are integrated between bus and rail and across the entire city region, including rural areas. In contrast, transport integration in Britain tends to stop at the city boundary.
>
> (Sloman, 2003: 9)

These boundaries are not easily overcome as they divide the decision-making and budgetary responsibilities invested in local authorities and may be reinforced by political differences. In Britain, many neighbouring authorities have a history of working together effectively, but co-operation in areas like public transport (or strategic housing development, for example) is often dependent on good relations between local politicians and civil servants (the officers of the adjoining authorities) and such co-operation may be temporary or adversely affected by political change, especially if one authority (e.g. an urban one) believes that it is subsidising the services of another (e.g. a rural one). Unlike other countries, the UK (and particularly England) lacks a stable tier of strategic governance that is able to guide or provide a framework for 'larger than local' co-operation. As noted in Chapter 2, the UK General Election in 2010 returned a Conservative–Liberal Democrat government to power which quickly set about dismantling the regional planning apparatus (in England) in order to localise decision-making and restore local democracy, which it was claimed had been undermined by regional assemblies and unaccountable development agencies (Gallent *et al.*, 2013). The upshot was that England lost its equivalents to the French provinces or German Länder. However, the loss has since been balanced by the formation of Local Enterprise Partnerships, and Local Transport Boards in 2013, which seem to offer potential opportunities for integrated transport planning.

But perhaps the biggest opportunity for realising better place-connectivity in many Western nations comes not from the vagaries of public planning and governance arrangements, but the recent surge in rail ridership. Half a century after the Beeching axe fell on Britain's branch-lines, railways across Europe are experiencing a renaissance. Higher demand for services is fuelling new investment, upgrade programmes and, in some cases, the reinstatement of lines with the continued development of lighter rail vehicles to provide cost-effective services on marginal rural routes. In other instances, disused railway alignments have been converted to guided bus-ways and cycle routes, examples of which will be explored in Part 2.

This has, in part, become possible because of the social value that is now attached to greater 'accessibility'. The Beeching Review utilised a simple cost-benefit analysis (CBA) to demonstrate that too much public money was being spent on rail services, relative to revenues. The same approach suggested that motorway building in the 1960s represented good value for (public) money (Banister, 2002). An evolved version of CBA – adding a 'social' prefix to give SCBA – was used to support London's Victoria Line project, revealing wider benefits for users (Foster and Beesley, 1963). Slowly, greater credence has been given to softer benefits and to a wider range of transport solutions. In France, Multi-Criteria Analysis (MCA) has been developed to look across the various environmental, economic and social impacts of different types of project. Sometimes, this has resulted in weaker connectivity into rural areas as the economic costs seem to outweigh the social benefits. However, today a range of Strategic Environmental Assessments (SEA), Environmental Impact Assessments (EIA) and Sustainability Assessments (SA) combined to give a broader view of how transport investment or closure will affect different places and populations. These, however, combine to produce an extremely complex decision-making environment: once all the data have been crunched, many investments are '[…] a matter of judgement alone' (Starkie, 1982).

Part 2: Accessibility in rural Britain

Public transport provision across Britain was, in the last section, presented as being highly fragmented – often being controlled by (or between) public bodies but operated for the most part by private companies in a highly competitive market. The outcome, according to Glasson and Marshall (2007:

192) is that '[...] the UK is consistently one of the worst planners and managers of transport systems in Western Europe'. The environment for public transport provision, in terms of governance arrangements and viability, is particularly challenging in deeper rural areas where the majority of households suffer either acute transport disadvantage or are car dependent. England, Scotland and Wales are broadly the same in this respect. But it is precisely these problems that make Britain a fascinating case study. This is because, first, many Western countries have shifted to a similar neo-liberal approach to public transport provision; and competition between providers is being encouraged across the European Union. This means that the British experience offers potentially useful lessons for other countries. Second, the perilous state of British provision – set against rising demand, latent capacity in old infrastructure, and the bigger role being currently handed to communities in addressing their own needs – suggests a fertile context for innovation. And last, despite (or perhaps because of) all of the challenges noted above, *walking* remains an important aspect of mobility and accessibility debates in the British countryside. Few rural regions of the world match lowland Britain's density of public footpaths, bridleways and byways, while recent 'right-to-roam' legislation (see Chapter 10) means that much of upland Britain is accessible to walkers. Walking accounts for a quarter of trips (see Figure 6.1) in all types of area. Although this proportion shrinks slightly in the countryside, walking is the means by which many visitors access pubs, tearooms and shops. The network of paths supports accessibility within communities, supports rural economies, and also contributes to health and wellbeing.

More generally, however, transport has been identified as a major concern within rural communities (Countryside Agency, 2001) because of its role in determining accessibility and because of spiralling costs. Recent figures (DEFRA, 2014a) show that rural households typically spend nearly 14 per cent of their disposable incomes on transport, compared to 11 per cent for their urban counterparts; largely because public transport is more expensive and trips by car are longer and more frequent. These costs mean that many households wish to reduce their car use. A study by DEFRA (2007) found that 58 per cent of rural households wanted to use their cars less, but were unable to do so because of the lack of transport alternatives. This compared to 39 per cent of households in cities. Rural car ownership is currently running at a rate of 1.5 cars per household compared to a national average of 1.15 (Lucas and Jones, 2009). Sixteen percent of rural households thought it would be easy to reduce their car use compared to 27 per cent living in larger towns and cities (DEFRA, 2007).

However, the transport difficulties and levels of public provision are not the same for all types of rural area. Rather they are a product of affluence versus poverty and settlement distribution and density (i.e. 'sparsity' – see Chapter 1). Sloman (2003) has produced the following typology of rural transport issues in Britain:

1 *Prosperous Commuter Belt* – characterised by heavy traffic, where reasonable transport services may exist, but tend to serve local employment centres while bus services tend just to run along main roads only;

2 *Dispersed Settlements* – clustered around market towns and often with a minimal skeleton bus service on market days, leaving households with enforced car ownership;

3 *Tourist Honey-pots* – which experience severe traffic problems in peak season, but with generally poor transport provision for local people, and where bus services may frequently be disrupted by road congestion;

4 *Remote Villages* – where long journey distances and high fare costs can pose a problem for low income residents, and where car ownership can be vital.

In general, it is the situation of more remote villages that is often presented as being typical of the 'rural transport question'; distance and cost combine to produce isolation and transport poverty. But the challenges are mixed and require different solutions in different places. Some might lend themselves to integrated urban/rural solutions of the types discussed in the first part of this chapter whilst others might only be redeemable through non-market, voluntary-led (but public-sector supported) initiatives. We begin now by looking at conventional public sector answers to this rural transport question before turning to community projects.

Rural public transport policy

Although rationalisation of Britain's 'conventional' rural transportation services has continued unabated from the 1960s to the present day, 'demand responsive' services – that often rely heavily on volunteers and the support of local community groups, but continue to attract public funding – have come to the fore, as state provision has dwindled. But the general retreat of the public sector from rural transport has not been linear; rather it has been episodic, accelerating and then slowing with each change of government. The Conservatives have tended to promote private sector involvement in what has hitherto been 'public transport', also favouring an agenda of 'private mobility' achieved, where possible, through lower taxes and investment in roads. Labour governments, on the other hand, have tended to consolidate state involvement in transport; not reversing Conservative policy, but slowing privatisation initiatives. The rate of road building during the 1960s and 1970s (2,300 km of motorway and 2,500 km of dual-carriageway; Banister, 2002) was largely unaffected by political shifts. But the deregulation and privatisation of public services by right-wing governments during the same period tended to be interrupted by more left-wing Labour governments. Hence, the 1964–1970 Labour Government implemented the recommendations of the Beeching Review but also introduced subsidy for some loss-making railway lines. However, the sweeping deregulation and privatisation of the New Right during the 1980s and 1990s was not reversed by the Centre-Left Labour (i.e. 'New' Labour) governments of the 2000s. Rather, subsidy was redirected from the suppliers of transport to its users, whilst further impetus was given to 'community leadership' and voluntary sector efforts (Table 6.3).

A significant shift, from the late 1990s onwards, was towards greener, more sustainable transport and improved integration between modes. The Transport Act 2000 handed local government the responsibility to produce Local Transport Plans (LTPs), typically for a five-year period but this was often extended as these Plans were aligned with Community Strategies (see Chapter 2) and with local planning frameworks. The LTPs needed to assess existing

Table 6.3 Key transport policy developments in England and Wales since 1951

Government	Policy agenda	Impact
Conservative (1951–1964)	• Rationalisation of public transport • Road construction	• Beeching Review recommended a reduction in the rail network by a quarter • Creation of the motorway network
Labour (1964–1970)	• Continuation of existing programmes	• Implementation of Beeching recommendations and continuation of road construction programme • National Bus Company formed in 1969 • Introduced subsidies for loss-making railway lines
Conservative (1970–1974)	• Public transport promoted/car use discouraged	• Motorway construction reconsidered • Prioritisation of bus services
Labour (1974–1979)	• Land-use and transport studies	• Road-building programme curtailed • Local Transport Policies and Programmes (LPPs) introduced
Conservative (1979–1990)	• Market-oriented reforms	• Motorway construction resumed • Railway de-regulation
Conservative (1990–1997)	• Privatisation	• Railway privatisation
Labour (1997–2010)	• Sustainable transport agenda • Social justice	• National bus pass scheme for retirees • Local Transport Plans (LTPs) introduced • Community transport schemes encouraged
Coalition (2010–2015)	• Austerity • 'Big Society'	• Rural bus services rationalised • Community-led transport developed • Modest infrastructure investment

levels of accessibility and set 'challenging but achievable' targets for addressing identified problems. Thereafter, they were expected to contain an action plan, which might include bids for grant funding from central government. The first wave of plans appeared between 2001 and 2006. They were viewed as experimental, corresponding with a period in which local authorities were finding their feet and exploring different options. More recent plans tend to contain tried-and-tested formula for planning, executing and governing integrated transport solutions and are now more closely aligned with local plans, becoming material considerations in decision-making.

The Coalition Government has signalled its intention to devolve funding for major local transport schemes to Local Transport Bodies from 2015, which will act as 'hubs' between local authorities, Local Enterprise Partnerships (LEPs) and other relevant organisations (DfT, 2011). These bodies will prioritise, review and oversee the delivery of new transport investment programmes in a move that suggests government is now persuaded of the merit of integrated, multi-modal approaches to transport planning, overseen by umbrella bodies. The LEPs are viewed by some as successors to the now defunct regional development agencies, providing a framework for larger-than-local planning. Whether this is genuinely the case or not, the need to look across jurisdictional local government boundaries, when thinking about transport, now seems to be widely accepted. The LEPs tend to cover functional economic areas; often comprising core cities (or major towns) and their hinterlands, which are joined by daily patterns of commuting, or which consist of contiguous markets for goods and services. They therefore appear (at least in some cases) to be sensible areal units for Local Transport Bodies, able to link urban and rural services.

Conventional transport services

However, at the present time (in early 2015) public spending has been curtailed as part of wider efforts to cut the UK's budget deficit. Combined with a cap on local council tax, designed to ease pressure on households, this has meant a reduction, in real terms, of money available for transport services.

The impact in many rural areas has been particularly significant, adding to existing accessibility difficulties. Rural bus services have been hit hardest. The Campaign for Better Transport (2013) have pointed to a 41 per cent budget cut in 2012–13 across all local authorities: in that same year, less than half of all residents in villages and hamlets could access an hourly bus service within a 13-minute walk from their home compared to 97 per cent of urban residents (DEFRA, 2014a). Doubts over the future of services (once they have been reduced, will they be increased or reinstated?) often lead to reduced ridership and a switch to transport alternatives, especially private cars (Sloman, 2003). This means that even relatively short-lived reductions in service levels may have a lasting psychological effect, permanently changing people's transport choices and habits. The social impact of this can be profound. Much is often made of the community role of pubs and shops as points of interaction (as 'interactional infrastructure'; Chapter 5), and the same is also true of buses:

> [...] it might sound silly but we miss the company because when we all got on the bus you all said 'good morning', 'good afternoon', and they were all people you went to school with because we're all pensioners [...] but now we don't get to see each other. You know, I mean it's very rare [that] you meet the people now.
>
> (Female, 65+, Northampton, quoted in Age UK, 2013: 6)

Perhaps partly for this reason, the support of bus services is often a focus for community action, enabling less mobile residents to access the shops and services they need in their daily lives. On the other hand, the fate of railways has been of more central concern and issues around service levels can rarely be dealt with through community initiative. The McNulty Review (into *Realising the Potential of Great Britain's Rail*) in 2011 shone a light on the outstanding challenges facing rail in general and regional rail in particular. As well as finding that British railways are roughly 40 per cent more expensive to run than other European networks, it also showed that the greatest ineffi-ciencies occur in the regional infrastructure (Table 6.4), where the public sector bears 61 per cent of the overall cost. McNulty's proposed solution to this was to devolve infrastructure and operations to regional authorities of some description, perhaps even to the Local Transport Bodies, who would then be able to move away from the single set of operating specifications and develop bespoke services on local branch lines. This type of differentiated approach could foster greater innovation and allow less expensive rail solutions in less intensively used parts of the network. He argued that:

Table 6.4   Cost of different parts of Britain's rail franchise (McNulty, 2011: 64)

	Passenger miles (billion)	Net cost to government (£million)	Net cost pence per passenger mile	Net cost to government as % of total cost
Long-distance franchises	9.4	693	7.3	25
London and South East franchises	15.7	760	4.8	19
Regional franchises	6.0	1,873	31.1	61

Experience elsewhere in Europe suggests that it is possible to define a more appropriate level of specification for both infrastructure and operations that can maintain existing standards of safety, but which can reduce the costs of supporting networks which are used less intensively.

(McNulty, 2011: 64)

Following on from the Review, a further consultation exercise identified '[...] services linking smaller towns and rural areas with larger towns' as providing greatest potential for devolution (DfT, 2011). Once devolved, a variety of infra-structure options would be available to increase the viability of urban–rural services, including the deployment of smaller and lighter rail vehicles that have been designed specifically to run on existing branch lines. Alternatively, it might make sense to 'downgrade' (or lighten) some infrastructure for use by a range of light rail vehicles.

Green and Hall (2009) have argued that such upgrades in the actual transit infrastructure need to be matched by a better overall experience of public transport, including convenient and high-quality access and egress points, and interchanges between services and modes, providing easy access, safety, shelter, timely information and, in the case of larger interchanges, facilities to make for a comfortable and productive waiting period. Rural facilities – likely to be located in market towns or other key service centres – need to have all of these attributes, combatting the uncertainty of public transport: including timetabling and transfer ambiguities. They argue that good interchanges reduce doubt through accessible design, timely information, and an environment that offers a positive overall experience of using buses and trains rather than sticking with the comfort and relative predictability of car journeys. There is, according to Green and Hall, a need for good practice in small interchange design (see Box 6.1).

Community transport

The term 'community transport' can denote one of two things: either transport solutions that arise from local, voluntary initiative, as a 'social good' (Chapter 5); or public-sector-led projects that fit the definition of 'community infrastructure'. These are very different things but they can overlap when local groups become involved in the management and operation of services that have been paid for from public funds, sometimes levered through the planning process. Groups within many communities, witnessing the gradual decline of rural transport services and the concomitant rise of transport poverty (often felt most acutely by their elderly neighbours or by young people), may seek to fill the organisational and funding space vacated by the state; a space that private enterprise is unable or unwilling to enter. Moseley (2000: 428) has identified five ways in which such groups may engage with transport issues:

Box 6.1 Good practice in small interchange design

A 2009 review of railway stations (Green and Hall, 2009) looked at examples from across Europe of good practice in the design of small interchanges; examples where there was good access and a positive passenger experience. It was shown that rural market towns often perform a 'gateway' function between larger urban centres and their wider (rural) regions and that these market towns, or key service centres, often have redundant transport infrastructure that can be upgraded as well as capacity adjacent to the hubs that can be used to develop commercial or community facilities. These smaller rural hubs need the following things:

Access
- Cycle Parking Space for up to 5 per cent of joining passengers;
- Car Parking Space for up to 15 per cent of joining passengers;
- Bus Information: Displayed in or near station entrance (where practical);
- Taxis: if no taxi rank, phone number(s) prominently displayed;
- Street Direction Signs: Station signed from main road(s) and pedestrian/cycle routes;
- Station Signs: Standard signing in 'Brunel alphabet and pictograms';
- Totem Pole Rail symbol and station name.

Facilities
- Staffing: Part-time presence with opening hours published for ticketing;
- Clock: Each platform with scheduled service;
- Seating: On each platform with a scheduled service minimum 12 seats;
- Staff accommodation: Smart and well cared for.

Environment
- Cleaning: Station regularly cleaned and graffiti free: litter bins (emptied at least daily);
- CCTV Security: Station platforms.

1 *Finance* – helping to raise money to support a new or existing local service, such as a community minibus. Moseley (2000: 429) also notes that it is often the case that '[...] locally generated funding is a requirement of bigger funding packages'; groups try to raise the money needed to access government or European funds, or they may pay for a feasibility study which, if it points to the workability of a local project, may result in external support;

2 *Management* – community volunteers may be engaged in the organising of transport, such as a local dial-a-ride scheme;

3 *Labour* – volunteers may help by driving community transport vehicles or help in the task of maintaining facilities, for example those supported by a community rail partnership (see below);

4   *Other services* – community-based and operated services may be instrumental in reducing the need to travel, such as a post office in a village or a community pub, supported by shareholders in a locality;
5   *Strategic involvement* – this may be a key contribution and can occur in a variety of ways including involvement in a parish council or business group that has a substantial involvement in assessing the demand/need for, and planning of, future transport. Such strategic involvement may now be framed, in England, within the Neighbourhood Planning process. At one level, this can mean that Parish Councils become more directly involved in the planning decisions and material considerations that impact on transport planning by local authorities. At another level, it may give them access to Community Infrastructure Levy (CIL) income (levied by local authorities on residential or commercial development), allowing them to fund additional community-based services.

One can imagine members of a Parish Council becoming actively involved in discussions around both public investments and local projects. Gallent and Robinson (2012a) have shown that these Councils tend to be hubs for a large number of interest groups in rural areas, some of which operate voluntary services aimed at supporting more vulnerable people: for example, purchasing and running community buses (see below). They may be associated with well-known charities such as the Lions or Rotary Clubs. Such voluntary action can be more or less visible. A community bus may have its own livery and its daily or weekly shopping runs will be widely advertised. But other voluntary actions will be more hidden and less structured: these might include occasional car-sharing, or taking an elderly neighbour shopping once a week. Such actions can be parcelled under the heading of soft support or 'social infrastructure': the social goods that arise as a product of community life. They are distinctly different from the 'community infrastructure' provided by local government, but often in partnership with user communities.

However, some projects labelled 'community' have very limited 'community' involvement. Government has tended to use the label to denote scale, or community representation on the partnership set up to operate a particular service. Examples exist of both rail- and road-based government-led community initiatives. In 2005, the Department for Transport published its 'Community Rail Development Strategy' (CRDS) setting out approaches to the management of lightly used lines to better fit local circumstances. Its aim was to find ways to raise revenue by reducing costs and establishing patterns of service provision aligned with local needs and objectives (ACORP, 2014). Seven pilot schemes were established but very quickly more than fifty rural routes were placed under the management of a local partnership. The partnerships were empowered to modify services (run by different private operators) to suit local users.

The CRDS attempted to move rail services in the same direction as road-based demand-responsive transport (DRT) which, because of cost and flexibility (it is not reliant on the same level of infrastructure provision and does not share fixed infrastructure with national services) may be more directly controlled by community organisations. In urban areas, DRT tends to comprise 'para-transit' services that are operated solely for the use of people with particular mobility or other special needs. Vehicles are specially adapted for that purpose and they will operate in tandem with regular bus services. But in some rural areas, para-transit may be the only alternative to private cars. It will comprise bookable, non-time-tabled services. The vehicles involved are usually mini-buses, but taxis and private-hire vehicles may be used by some voluntary operators. In some instances, the services are tailored to specific groups (like their urban counterparts); perhaps to older or vulnerable people or those with special needs. They are advertised as 'dial-a-ride' or 'ring-and-ride' services. But in other instances, para-transit is simply a means of extending network coverage to harder-to-reach areas, or at times of the day when conventional public transport stops. The more formal services will be run by private operators on behalf of a local authority, with an element of subsidy. Others will be managed and run by community groups, with comparable state support and finance.

These types of rural para-transit schemes perhaps draw inspiration from the 'post-bus' (postal delivery vehicles with fixed seating, allowing them to carry passengers as well as parcels) services run by the Royal Mail from 1967 onwards, to replace dwindling rural bus services. Until recently, the Royal Mail was running more than 200 post-bus services across the UK, including in remoter and less populated areas. However, these services now appear to be winding down as Royal Mail finds it increasingly difficult to support its network of post-buses owing to competition from other mail services and the internet. It is in these same areas that para-transit schemes, either privately or voluntarily run, tend to operate. They have a vital role to play in raising levels of rural accessibility and can certainly be viewed positively, as a means of aligning service provision to needs and, in some instances, as a social good that symbolises the spirit of self-help found in a great many rural communities. Framed by recent government rhetoric in England, community buses can be presented as the 'Big Society' in action; but these buses are rarely well-co-ordinated. They tend to be poorly integrated with other transport infrastructure and do not always provide the means to access services or other activities easily and at predictable times. In that sense, they may be seen as a poor substitute for fixed public services. That said, some authorities are now experimenting with 'hybrid' para-transit schemes that do operate to fixed timetables and have the potential to interface with other services (see Box 6.2).

Walking and cycling

It was noted at the beginning of this section that movement by foot remains important in rural areas, with opportunities to walk or ramble drawing many visitors to the countryside. Indeed, the Countryside and Rights-of-Way (CRoW)

'Cango' is a transport scheme operated on behalf of Hampshire County Council (2014) to serve rural communities across the county with a fleet of minibuses that operate along fixed routes and to a regular timetable as conventional bus services, but have wider 'roam zones' within which they are able to pick-up or drop-off passengers by prior arrangement. The service is available to the general public and although it is not pitched specifically for the mobility impaired, vehicles are designed for high levels of accessibility and staff are trained to assist those with particular needs to provide a comprehensive single service to cater for a diversity of needs.

Act 2000 (see Chapter 10) has opened up many new areas to walkers. Within villages, networks of rights of way, cutting through back fields and straddling streams, can enhance the experience of being in the countryside and can also shorten journeys for residents, allowing them to avoid main roads.

Cycling is a bigger challenge in the countryside, and often much more dangerous for its participants. There were 64 fatalities on rural roads in 2010 compared with 47 on urban roads, although the number of non-fatal encounters was far greater in cities where the bulk of cycling takes place (RoSPA, 2014). There are many reasons why cycling along lanes or busier rural roads can be dangerous; less road lighting, faster vehicular traffic and a lack of segregated cycle infrastructure. There have been some efforts to create the latter, with dedicated cycle routes installed along discussed railway lines (forming 'green-ways') and efforts to direct cyclists along quieter lanes. It has been recognised that creating safe cycling opportunities may significantly improve accessibility to a number of key services including employment centres, primary and secondary schools, further education colleges, GPs, hospitals, town centres and shops selling groceries (DEFRA, 2014a). But improved provision for cyclists will have the biggest impact on access to nearby services that can be found within the nearest village: primary schools, grocery shops and perhaps a GP. It will have less of an impact on accessibility to those services – town centres, secondary schools, hospitals, or employment – located further away in key service centres, accessible via a busy main road.

Concluding remarks

Questions of accessibility and mobility are central to all aspects of rural planning and service delivery. Connectivity to employment will have a huge impact on the way a rural economy works; connectivity to services will make it more or less difficult to live in a rural area; and connectivity between individuals – between friends and family – will impact not only on the quality of rural lives, but also on the capacities that communities have to do more for themselves. Transport opportunities and systems provide the essential networks connecting people to people, people to jobs, and people to all the things they need on a daily basis. But those networks struggle to function in situations of population dispersion and often require some sort of subsidy or 'in-kind', community-based, support. They are dependent on public or voluntary investment, or a combination of both.

The UK narrative, in relation to public investment, is one of rationalisation. This affected the railways and bus services from the 1960s onwards. It was underpinned initially by rising levels of car ownership, and was afterwards compounded by a focus on road building and the prioritisation of 'private mobility' (for those able to afford it) over public transport. It was only when gaps in private mobility – affecting the very young, the very old and those whose broader economic disadvantage became a form of 'transport poverty' – became obvious that the state became more centrally involved in alternative transport strategies, supporting a raft of initiatives that are designed to substitute for public provision without imposing the same cost burden on the local state. Hence there has been a focus on lighter rail options and on para-transit solutions that can draw on voluntary effort, minimising public cost. Voluntary investment has become a regular feature of efforts to resolve rural transport problems.

Many of these same problems affect rural regions around the world; population densities fall away and public services become less viable. However, Sloman et al. (2009: 9) have pointed out that other European countries appear to have responded far more effectively to rural transport dilemmas than the UK. Although they often share the UK's ideological attachment to market-based approaches to transport provision, the co-ordination of that provision tends to cut across urban/rural boundaries with dendritic whole-system approaches emphasising the connectivity of urban hubs to rural hinterlands. Planning and governance arrangements transcend local authority boundaries, and although integrated arrangements are now operating in many parts of the UK, there is a tendency to forget or ignore some more peripheral rural areas. Within those areas, disparities in life chances may be acute, with a great many households facing daily accessibility and mobility disadvantage and unable to 'get at' a great many of the services and activities everyone else takes for granted.

Summary

- Centred on the question of rural transport, there are many challenges facing communities and people in the countryside: the challenge of accessibility to services and activities; the challenge of personal mobility; and the challenge of substituting public provision with community-based alternatives.

- A market-based approach to transport provision since the 1960s has resulted in declining public services in many rural areas. The viability of public transport is a function of population density, and as densities decline, so the market for rail-services and buses becomes thinner. If rural markets are viewed as discrete entities, the case for public transport investment seems to weaken.
- Broadly, there are two possible (planning) responses to this situation. The first is an integrated response, accepting that rural areas are not in fact discrete but part of larger functional areas. The second is to substitute an element of public investment with voluntary effort.
- Integrated responses to public transport provision in rural areas acknowledge the interdependencies between urban and rural places, which are connected by flows of goods and services. Once this is acknowledged, governance (and funding) arrangements need to be put in place which take a whole-system view of transport provision, with public regulators working with private operators and voluntary efforts to reach all parts of the system, cross-subsidising those parts where demand is less with revenues from parts where demand is high.
- Independent of this approach, voluntary effort (by community groups) can play a vital role in delivering para-transit to communities where regular public transport investments, or private operations, are not viable. But such voluntary efforts are likely to be more effective where they are integrated with a whole-system approach of the type described above, ensuring connectivity across a wider network.
- Support for rural transport initiatives is likely to grow when it is shown that rural accessibility is crucial for the proper economic and social functioning of an area, in which urban/rural divisions need to be broken down.

Key readings

- DEFRA (2014a) *Statistical Digest of Rural England*, DEFRA: London.
  (This document, which will be updated in the future, is a great source of statistics on rural issues, from housing to services. These digests were previously produced by the CRC but are now being undertaken in-house by DEFRA.)
- Sloman, L. (2003) *Rural Transport Futures: Transport Solutions for a Thriving Countryside*, Transport Research Laboratory: Wokingham.
  (This is a good overview of rural transport challenges alongside a number of possible remedies.)

Key websites

- Department for Transport: www.gov.uk/government/organisations/department-for-transport.
(As well as explaining general responsibilities, details of particular projects are also provided and periodically updated.)
- Campaign to Protect Rural England: http://cpre.org.uk.
(CPRE is a general lobby group but has run particular initiatives around 'integrated' rural transport solutions in recent years.)
- Plunkett Foundation: www.plunkett.co.uk/whatwedo/RCT.cfm.
(Supports a range of community initiatives, including those focused on transport and mobility.)

7 Rural services, public provision and local action

Introduction

In the previous chapter 'accessibility' was conceived in terms of personal mobility; how people are able to gain access to the goods and services that are essential to their economic and social well-being. Accessibility affects different groups of rural residents, in different places, in a wide variety of ways, depending on their individual circumstances. Mobility plays a big part in determining accessibility or the degree to which services are 'get-at-able' (Moseley, 1979), but what has been happening to those services themselves over recent years? And how are different agents and communities working together, or separately, to ensure the countryside's increasingly complex and differentiated communities have access to the things they need in order to survive and prosper? Actions in response to current challenges need to be set against an understanding of the complex and long-standing trends of privatisation and declining service delivery. Because markets for rural services are 'thin' (as a result of lower population densities and difficulties around mobility), there is continual pressure to concentrate provision, which leads to rationalisation, reinforced by introduced market disciplines and basic cuts in public spending. Attention has been drawn to the decline, rationalisation and concentration of rural services for many years (CRC, 2006c). These processes have created new patterns of disadvantage as the ease and convenience of accessing services declines and certain groups – characterised as the rural poor in Chapter 5 – are effectively denied access to the everyday things they need to live. There are many examples that might be cited here, from post-offices to pubs and from general stores to schools, but one very simple example is that of cash-points. In the countryside, these are few and far between, often located in small shops where a fee will be applied for each transaction. In cities, they are commonplace and offer 'free cash withdrawals'. It is more difficult, and costlier, even to get access to your own money in many rural areas.

This chapter is concerned with these challenges and with public and local responses. It was noted in the opening chapter of this book that the role of the state with regards to service provision has been changing over a number of years. Across numerous areas (including transport, examined in the last chapter, and housing, examined in the next), it has retreated from its traditional provider role and now focuses on commissioning and supporting the delivery of services by others. This change has been accelerated since the Coalition government came to power in 2010, although the outsourcing of basic services – such as waste collection, public transport and housing responsibilities – is a trend that has set in over several decades. This is partly a result of the compulsory competitive tendering (CCT) of some former public sector services, or big shifts in housing and transport policy, which have created a space for private and voluntary sector provision as a means of bringing market efficiencies to areas that were once the preserve of the public sector.

But changes in the pattern of public-service provision today are driven as much by the rhetoric of local choice and control as any further attempt to save money. Innovation and competition is still viewed as a way of delivering value for money for service users and taxpayers (Cabinet Office, 2011), but there is increasing concern for the way in which these services are tailored to local circumstances and needs. The planning and delivery of services are being localised. The sectorial shifts witnessed over a 50-year period are now accompanied by spatial shifts, with an increasing focus on local delivery by a range of community and voluntary actors.

The British experience is not unique, but state withdrawal from service delivery has been faster and more pervasive than elsewhere, with especially profound consequences for rural areas. Neo-liberal approaches to service provision have tended to accentuate spatial disadvantage. In some places, less state intervention creates a space for stronger communities (with a greater store of social capital; see Chapter 5) to address many of their own needs. But communities that lack the capacity to do more for themselves may find that services disappear and are not replaced, by either private provision or voluntary action. Important issues of spatial equity and fairness arise:

> There are important issues of fairness for people in rural areas accessing individual services, including ensuring that there is sufficient diversity to give people meaningful choice and targeting funding to address the higher cost of providing services in remote areas.
>
> (Cabinet Office, 2011: 18)

The basic government response has been to channel additional resources to rural areas through revenue support grants, factoring sparsity and the higher costs of providing services to rural populations into its grant formula allocations. Because the rural areas of Scotland and Wales are more sparsely populated than the lowland areas of England, the costs of providing services, per capita, is higher. This is part of the reason why public expenditure in those countries, and allocations through the Barnett Formula (which became an issue of public debate during the 2014 Scottish Independence Referendum), is greater than in England (Gallent, 2000). However, many campaigning groups believe that not enough grant support is provided to the poorest and most peripheral rural areas. The Rural Services Network and the Rural Fair Share Campaign are vocal advocates of higher rural spending, which would better reflect the transport and distribution costs (and hence time costs) of doing things like collecting waste from dispersed, lower density, rural households. Overall, economies of scale are less evident and less easy to exploit in the countryside making everything more expensive to provide (Hindle *et al.*, 2004; see also Chapter 4). Moreover, there may be fewer providers of services (Office of Fair Trading, 2012), reducing competition and raising prices.

In the remainder of this chapter, we explore access to goods and services across rural communities, drawing attention to the 'discriminatory' nature of increased competition and the local responses of communities to address their own needs. The challenges of public, community-based and private provision are examined, but particular attention is given to the social infrastructure that emerges from communities themselves. The current Coalition Government has promoted the idea of a 'Big Society' doing more for itself in the context of localism, with key community actors coming together to deliver key services – including the para-transit looked at in the last chapter – through social enterprise. There is some tradition of this in the countryside but it now appears to enjoy greater government support, being celebrated as the restoration of the 'natural bonds that [once] existed between people – of duty and responsibility' in place of the 'synthetic bonds of the state' (Cameron, 2009). But as well as charting the current direction of travel towards community-based action, the chapter also considers the ways in which new technologies are changing – and may continue to change – the way services are provided and accessed. Fifteen years ago, Malcolm Moseley observed that:

> The future of village services in a rapidly changing context is seen to lie with a combination of local innovation, collaboration between service providers, partnership between the statutory, private and voluntary/communitiy sectors, [and] the judicious introduction of ICT and outright subsidies.
>
> (Moseley, 2000: 415)

Nothing much has changed since this statement, which continues to neatly summarise the rural service challenge.

Part 1a: National government's response

Acknowledging the rural challenge

Government has, for many years, at least acknowledged the problem and scale of this challenge. The 2000 Rural White Paper *Our Countryside: The Future*

– *A Fair Deal For Rural England* (DETR and MAFF, 2000) set out the case for improving accessibility to high-quality services, ensuring greater choice and establishing patterns of provision that would be responsive to the needs of more vulnerable social groups. These aspirations found their way into DEFRA's 2004 Rural Strategy, to be delivered through a 'rural service standard' setting out acceptable levels of provision and access. Separate standards were issued for education (including pre-school provision), post offices (and access to online banking), transport, libraries, health care (including social care), emergency services and job-seeking support.

In 2011, the Coalition Government published its White Paper on *Open Public Services* (Cabinet Office, 2011) setting out five key principles for future service provision:

1   Give people more choice and direct control over the services they use, either individually or via directly elected local representatives.
2   Ensure that delivery is organised at the lowest level possible, which may include providing additional powers and opportunities to neighbour-hoods through localism and through a community-based right to deliver those services (see below), in instances where they think they can do so more effectively.
3   Wherever possible, open up public services to a range of different service providers, whether public, private, voluntary, or community-based.
4   Put fairer access to public services at the heart of reform, focusing on the needs of disadvantaged households, ensuring in particular that their experience of accessing services is improved.
5   Make sure that those delivering services are accountable to users.

The broad thrust of the White Paper was one of local choice and action, whether that action is to be taken by private or community actors. The particular challenges of rural service provision received some lip-service within the White Paper, with government promising that its 2012 Rural Statement would tackle issues of

> [...] fairness for people in rural areas accessing individual services, including ensuring that there is sufficient diversity to give people meaningful choice and targeting funding to address the higher cost of providing services in remote areas with less well established infrastructure.
> (Cabinet Office, 2011: 18)

That Statement (DEFRA, 2012: 10) pledged: an additional £1.34 billion of funding to support the Post Office Network, promising no closures and to look at alternative provision where traditional counter services were not viable; further funding for community-based rural transport; innovation in the delivery of library services; and a particular focus on health and social care, encouraging social enterprises to compete for public service contracts.

Service rationalisation and key settlements

But support for innovation and new delivery models, largely community-based, needs to be set in the context of sustained pressure to rationalise service provision and concentrate it in larger market towns and key settlements. This has been the principal government response for several decades, based on a 'hub and spoke' model of creating service centres for surrounding rural hinterlands. As noted already, that approach is rooted in Structure Planning (Sillince, 1986) and the designation of settlement hierarchies (Cloke, 1979; Cloke and Shaw, 1983), ranging from market towns (which are thereafter viewed as appropriate places for all sorts of development) to small hamlets (which are viewed as appropriate places for very little). Despite the end of Structure Planning a decade ago, many of these old settlement hierarchies were saved within local development plans and continue to shape the way in which services are planned (Taylor, 2008). The approach holds that it is only in larger towns and villages that critical population thresholds can be achieved, through a concentration of housing development, to support key primary and secondary services. Obviously, this approach is grounded in what might be considered a natural state of affairs: market towns, throughout the history of British rural settlement, have performed a vital service function for outlying villages and hamlets. Planning has merely supported the continuation of this function since the 1960s, and continues to do so today (see Box 7.1 and Figure 7.1).

But despite the apparent logic of the approach, it has faced recent criticism. Taylor (2008) has argued, in his rural economy and affordable housing review, that if used insensitively the approach can push smaller settlements into a 'sustainability trap', denying them the development that would deliver against a broader mix of needs. Lower-tier villages become increasingly attractive to affluent incomers, buoyed by planning's promise that no further development will be permitted and therefore that rural vistas will not be spoilt and property prices will be kept high through housing scarcity. Such communities may become increasingly gentrified, lacking affordable housing, transport or a local shop. They become the preserve of wealthier multiple car-owning households seeking seclusion and exclusion. And ironically, the denial of development and its concentration in key settlements may undermine sustainability rather than promote it. These lower-tier settlements become:

1 *Less economically sustainable* because 'local people' are unable to live locally leading to labour market shortages. And because families move out (owing to rising living costs) demand for local services declines, resulting in further closures;

2 *Less socially sustainability* and more exclusive. Poorer sections of society (the young and elderly, in particular) are forced out through housing pressure, which in turn hits the viability of rural services, which will

Box 7.1 Eden District Council's settlement hierarchy

In April 2010 Eden District Council, a rural local authority in the north west of England, formally adopted its Core Strategy. This document sets out the strategic vision for the area and identifies the key policies that the local authority (and others) should use when making individual decisions about development in the district for the next 15 years. The overarching vision statement seeks:

> To develop, maintain and improve a vibrant Eden economy and to provide affordable housing, supporting active and inclusive sustainable communities, building on natural assets, protecting and enhancing Eden's unique environment and heritage.

(Eden District Council, 2010: 19)

New development would be focused on a settlement hierarchy as follows:

1 The **Key Settlement** of Penrith will be the regional centre and sustain development of that of a large town. It is to this location that the bulk of new development is likely to be attracted and can be accommodated;
2 Three **Key Service Centres** of Alston, Appleby and Kirkby Stephen can accommodate moderate new development suitable to the scale of the towns, which will help to retain sub-regional service centre functions;
3 Forty Six **Local Service Centres** have been identified, which might grow to help retain local services;
4 **Smaller villages, hamlets and the open countryside**, where development will be restricted unless there is demonstrable local need.

accelerate the loss of vulnerable groups (low income, elderly and immobile), which in turn may adversely affect the social networks on which vulnerable people rely;

3 *Less environmentally sustainability* largely due to 'reverse commuting' as rural employees are forced to live elsewhere (usually in the key settlements) and commute back to their place of work. Good-quality public transport becomes essential, but because it is rarely available when they need it, they become car-dependent and face spiralling transport costs.

(Taylor, 2008)

Taylor has called for a more sensitive and permissive approach to the planning of rural development and services. He was largely concerned with affordable housing, but any loosening of planning's stranglehold on lower-tier settlements would extend to service provision, ensuring that people have what they need close at hand, if that meant contributing to more sustainable patterns of travel and living. Moving away from a rigid key settlement

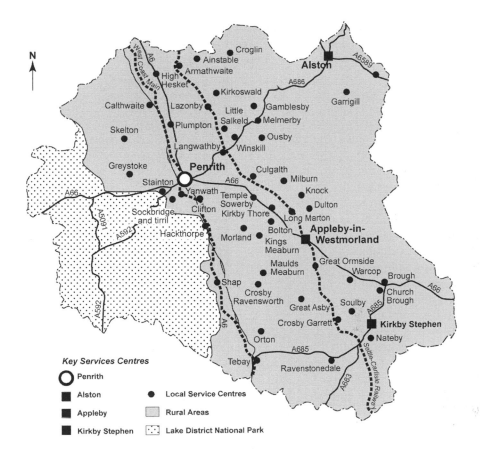

Figure 7.1

Map of Eden District Council's settlement hierarchy

approach would be one response to Hoggart and Henderson's (2005) claim that too little has been done, thus far, to address the social reconfiguration of rural areas since the Second World War.

Taylor's views were very much reflected in the Upper Eden Neighbourhood Plan (2013) which promoted a looser approach to settlement planning than that set out in Eden District's Core Strategy. The authors of the Neighbourhood Plan sought a more flexible approach to development in villages and hamlets (Sturzaker and Shaw, 2015), allowing small-scale development in support of local needs and community wellbeing (see Chapter 8). This burgeoning debate on rural sustainability, and the polemic against existing statutory planning orthodoxies, which fail to adapt generic ideas of sustainability to specific rural situations, seems to be gaining some traction. Nearly 20 years ago, Morphet (1998: 10) argued for a matching of 'sustainability objectives with the rural way of life' grounded in an 'understanding that rural areas differ in their needs and priorities'.

Rural proofing

Rural heterogeneity is perhaps most clearly acknowledged today through the 'rural proofing' of all areas of public policy. Such proofing aims to ensure that national policy approaches are appropriate for rural areas by requiring ex-ante evaluation of the likely consequences of policy on different types of rural place and various groups. It aims to ensure, as far as possible, that adverse impacts are avoided, or that measures are put in place to mitigate those impacts. Rural Proofing was pioneered by the Countryside Agency in England and then taken forward by its successor, the Commission for Rural Communities (CRC), after 2006. The CRC was abolished by the Coalition Government in 2013 and its remit subsumed into DEFRA. However, the Commission had been a powerful and persuasive advocate of proofing and left a legacy of concern over how national service priorities are interpreted and delivered in rural areas. It was particularly concerned with health care, education, leisure services and policing.

In relation to *health care*, it is a basic aim of the National Health Service (NHS) across the UK to ensure equal access to services, which need to be free at the point of delivery. An Institute of Rural Health was created in 1997 to examine how this aim might be met in different rural contexts and through the continuing cycles of NHS reform. The CRC incorporated the various findings of the Institute into its own proofing and recommended, in 2007, that NHS funding formulas should be revised to reflect: population dispersal and the costs of delivery; the particular demographic of many rural areas, resulting from retirement and in situ ageing; the greater need for mobile services/clinics; and the clear rationale for co-locating some health care provision with other public services. In relation to *education*, the Department for Children, Schools and Families (DCSF) set a target in 2005 of 90 per cent of 17-year-olds remaining in full-time education by 2015, up from 70 per cent at that time. However, within three years – and in conjunction with the CRC – the Department acknowledged that there were particular barriers to staying on at school or entering further education in rural locations, including: distance between home and school and the costs of travel; and the lack of large employers in these locations, requiring the qualifications that would encourage children from particular backgrounds to remain in education. It was concluded that greater investment in e-learning was needed to overcome the barrier of accessibility. Similar barriers were identified in relation to *leisure and sport*. 'Sport England' (a non-departmental public body) was established in 2009 with the aim of promoting greater participation in a range of sporting activities. It quickly identified, with CRC, that participation levels are especially low in rural areas, again because of issues of access and the cost to individuals of getting to appropriate facilities. The simple solution in that case was to introduce a rural programme, aimed at working with communities to deliver new infrastructure, with an initial budget of £10 million. And finally, in relation to *policing*, the National Policing Improvement Agency (another quango that was closed in 2013) worked not only with the CRC but also with the National Farmers' Union and the Farmers' Union of Wales to deal with the issue of police visibility in rural areas. This had become a major concern over a number of years. The closure of stations in many villages (and concentration of police services in key centres) had left many people concerned over potential response times and the ability of the service to deter crime, owing to its physical absence from many areas. The partner organisations published guidance for local forces on maintaining visibility in the countryside, conceding that the relative inaccessibility and remoteness of some places presented a particular challenge to policing, though much can be done to alleviate community concerns (CRC, 2009).

Building on this work, DEFRA published national rural proofing guidelines in 2013. Referring to these guidelines, the Treasury set out Government's commitment, in its *Green Book on Appraisal and Evaluation in Central Government* in the same year, to ensure '[…] that all its policies take account of specific rural circumstances' (HM Treasury, 2013: 9). Indeed, it is necessary for 'appraisers to assess whether proposals are likely to have a different impact in rural areas from elsewhere' (HM Treasury, 2013: 9). DEFRA's short two-page guidelines (DEFRA, 2013) cite a number of examples of how different government policies, for example relating to the regulation of domestic energy suppliers, have been revised and tailored to better serve rural households.

So the acknowledgement of particular rural challenges is expressed through discourses that question a one-size-fits-all approach and through policy adaptation. But still those challenges remain, not because public policy fails to address them but because they are deeply embedded in the nature of rural situations. It appears that they cannot be resolved through external intervention, but require a local response. External interventions tend to be top-down and motivated by a professional desire to deal with systemic problems; to deliver a determined 'public good' (e.g. even distribution and access to services). Local responses, on the other hand, are bottom-up (though often supported by the local state) and attempt to tackle challenges in situ. It has been argued that such responses are better able to address the complexity of those challenges, which processes such as rural proofing help identify.

Part 1b: Local(ising) responses

Local responses, often community-based and state-supported, have become an increasingly important means of addressing the rural service challenge. The Coalition Government has signalled strong support for this approach, built on civic 'duty and responsibility' (Cameron, 2009) and on the premise that a 'big society' should do more to address its own needs. It was noted in Chapters 1 and 5 that growth in community-based action is both a global phenomenon and something that appears to have deep roots in many rural areas. That there is a propensity towards self-help within some rural communities is evidenced by a proliferation of community-based projects and also by the lineage of neighbourhood planning in England, which clearly builds on a tradition of rural appraisal and parish planning (Gallent, 2013; Parker, 2014).

Rural Community Councils

A detailed discussion of community-based action, built on a platform of social capital, was provided in Chapter 5. Here we consider the lineage and nature of local response in relation to service challenges. Much has been written on rural community action over the years, linked in many cases to the work of the Rural Community Councils (RCC) in England. These should not be confused with civic authorities. Rather, they are voluntary support groups generally constituted to assist with community development. The first RCC was established in Oxfordshire in 1921, and by 1986 each county in England had its own RCC, which have been described by Leavett (1985) as independent voluntary organisations which bring '[...] together a wide range of other voluntary bodies, local authorities and statutory agencies to promote and support voluntary action to alleviate the social and economic problems of rural communities'. The RCCs now operate under the umbrella of Action with Communities in Rural England (ACRE), which is now part of a 'Rural Community Action Network' (RCAN). They aim to provide direct advice and support to community organisations; stimulate community action, particularly voluntary action, and encourage good practice; to develop and manage demonstration projects; provide professional support and advice to key service providers; and to raise awareness of rural issues and influence decision makers (see ruralkent.org.uk).

During the 1970s, attention was drawn to a growing number of local, community-led, initiatives, notably in the area of rural transport, which responded to the service rationalisations of the late 1960s (see Clout, 1972). Into the 1980s and 1990s, a great deal of community effort was expended on appraisals of local need with a view to either lobbying for additional investment, levering in grant support or building consensus behind local actions (Parker, 2014). Much of this effort was supported by the RCCs under the auspices of Action with Communities in Rural England (ACRE) and equivalents in Scotland and Wales. The RCCs were often re-branded as, for instance, Action with Communities in Rural Kent (see Gallent and Robinson, 2012a). These community support bodies emphasised the clear benefits to communities of helping themselves in circumstances where public investment was unlikely (Robinson, 1990), frequently working through parish or community councils.

Parish and community councils

These councils have been hubs for social entrepreneurial action for a number of decades (Gallent and Robinson, 2012a), receiving a small amount of local authority funding (the parish 'precept', derived from council tax receipts raised within the parish) and being more or less proactive on local matters depending on the inclinations of their elected and co-opted members. These councils are a product of the Local Government Act 1894 (Jones, 2007) in England and later Local Government Acts (1972 and 1973 respectively) affecting Wales and Scotland. They tend to cover one or a small cluster of villages. They are commonplace in rural areas (being labelled 'parish' councils in England, and 'community' councils in Wales and Scotland) but rare in cities. Market towns have their own 'town councils', which are the equivalent to the parishes, and these all sit beneath district or borough councils (or sometimes unitary authorities) in the local government hierarchy. This means that they are one step down from the planning authority. They had, until the Localism Act 2011, no planning powers themselves but exercised some influence over local decision-making via their local ward member or through direct lobbying.

Generally, across the UK, successive governments over the last twenty years have forwarded the view that civic engagement at the local level is an effective mechanism of improving the social, economic and environmental conditions of communities, though 'community leadership' under Labour (1997–2010) and now through the localism agenda (2010–2015?) under the Conservative–Liberal Coalition. During both periods, the community or parish councils in rural areas were viewed as instruments of these political ambitions.

They were already, however, active on a number of fronts. Some of the earliest engagement by rural actors in general, and the parish/community councils in particular, in planning matters was in response to a perception of growing design homogeneity in rural settlements (e.g. the delivery of monotonous, placeless red-brick housing) and consequent erosion of the distinctive character of some places (Owen *et al.*, 2011; Owen, 1998). During the 1990s a range of community-led planning approaches emerged including 'Village Design Statements' (VDS) (which focused on protecting, maintaining and enhancing the physical characteristics of villages) and Village Appraisals (which were community-led SWOT analyses of socio-economic or environmental conditions, or base-line assessments of various community issues), which in turn could lead to various types of action-plan, sometimes forming part of a broader 'parish plan'. Evaluations of these initiatives often revealed a confusing array of tools that were either narrowly focused on land-use planning or broader in scope, addressing issues of community wellbeing (Owen and Moseley, 2003). The former were occasionally promoted by self-interested or concerned individuals, whereas the latter were more inclusive, engaging and mobilising more of the community, but often championing the priorities of more affluent sections of those communities (Moseley 2002: 400), prioritising design aesthetics over arguably more pressing social and service problems.

The initiators of these efforts were either the parish councillors, key individuals within a community or community support groups, including the RCCs (Owen, 2002; Gallent, 2014). Their efforts tended to receive a mixed response from local government. Hence, during the 1990s, sporadic bottom-up initiative enjoyed variable success; that success being dependent on the energy of individuals (or the broader stock of community 'social capital') and the extent of local government support which might, for example, be expressed through grant-giving or through planning decisions that were supportive of community initiative.

The new localism of the 1990s: vital villages and market towns

Central government at this time came to view such initiatives as a traditional strength of rural communities, asserting that when 'people [...] do not expect Government to solve all their problems for them [...] local decision-making is likely to be more responsive to local circumstances' and arguing that '[...] improving quality of life in the countryside starts with local people and local initiative' (DETR and MAFF, 1995: 16). Drawing on the first significant review of rural policy since the Second World War, the Conservative Government under John Major advocated the devolution of greater power and responsibility to parish and town councils. Its response to the Rural White Paper (DETR and MAFF, 1995) gave official support to an existing reality of rural self-help. The promotion of localism, announced in all but name in 1995, then continued through a long period of Labour Government. After 1997, parish plans came to be seen as barometers of local opinion and aspiration. A variety of programmes were launched, including the Vital Villages and Rural Social and Community Programmes, which gave advice on – and directly supported – the undertaking of appraisals and production of parish plans. By 2004, roughly 1,000 villages had received grants of £5,000 for the preparation of plans. But this was only a fraction of eligible parishes, and overall plan coverage remained patchy. In Tonbridge and Malling District, for example, only half of its parishes had produced a plan or design statement by 2007, making it difficult to use these pieces of 'local intelligence' in higher-level planning (Gallent et al., 2008). Lack of coverage made it difficult to identify the most urgent needs or the best locations for service investment; only complete coverage could provide such guidance.

Similar community-led appraisals were supported and conducted in market towns. It was noted earlier in this chapter that the market towns had played a historic service function in the countryside and that this had been supported from the 1960s onwards through settlement hierarchies set out in Structure Plans. Despite a recent questioning of that approach, it is generally accepted that market towns play an important role in higher-tier service delivery and hence there is a logical interest in the vitality of these traditional service centres. This was expressed in the 1990s (and into the 2000s) through a 'Market Towns Initiative' (MTI), led by the Countryside Agency, which encouraged town councils to conduct their own 'health checks' and formulate actions plans in responses to any gaps in their service offer. The initiative began in the North West of England but was rolled out to 140 towns at a cost of £37 million. There was great enthusiasm for the local health checks and the ensuing production of plans, with residents and businesses coming together in common cause (Caffyn, 2004). However, despite a further £100 million being made available for project delivery after the health checks were completed, it was initially unclear as to whether the initiative had brought any lasting benefits beyond the creation of new partnerships. Caffyn (2004) has suggested that the MTI was another 'third way' project – that brought together local actors and delivered a bit more 'joined-up governance' in some instances – which really needed to be followed up with public investment aimed at bridging the identified service gaps. Much of the support from government during the 1990s and into the 2000s focused on bringing people and different groups together in the hope, perhaps, that local innovation would result (as people and groups pooled ideas, skills and their independent resources). One of the lasting outcomes of the MTI was the establishment of a 'Support for Action for Market Towns' network (Powe *et al.*, 2007), subsequently rebranded as the 'Towns Alive' network, which aimed to support, through advice and ideas, the revitalisation of market towns.

Localism and new 'community rights'

The propensity towards self-help and community enterprise have regularly been viewed as attributes of rural places, with government only needing to nudge communities towards innovative solutions to local dilemmas. This belief that the communitarian spirit of local groups can be kindled through a devolution of responsibility has been given new impetus by the Coalition Government's promotion of a new brand of institutional 'localism' (devolving to local government in the first instance and then to community actors), although support for such localism is neither new (see Stoker, 2004; Haughton and Allmendinger, 2011) nor the natural territory of either left- or right-wing UK governments. The former have seemed, in recent years, to be more comfortable with stronger central control, and regular intervention in all areas of national, regional and local policy. The latter have pursued a neo-liberal agenda, rolling back the state to make space for market rather than community solutions. However, both approaches appear to have led their proponents to the same conclusion: that neither the state nor the market offers a definitive solution to meeting very local needs – particularly at a time of economic crisis and fiscal deficit – and therefore supporting community effort to resolve their own problems makes a great deal of sense.

The Conservative Party really pushed the idea that communities should do more for themselves in the run-up to the 2010 General Election, arguing that a new type of 'open source planning', localised and flexible, would enable those communities '[...] to formulate a positive vision of their future development' (Conservative Party, 2010). This idea was jumped on by a number of community-planning advocates, who quickly offered visions of new-style plans (replacements for the VDS and Parish Plans) addressing a 'wide variety of issues' affecting quality of life, as well as actions that could be delivered either by communities themselves or with external assistance, especially from the public sector (Bishop, 2010: 613). The proposition was that communities would benefit from a new framework in which they could set out their needs and ambitions, and local government and service providers would take greater note of it.

The Localism Act 2011 introduced 'Neighbourhood Development Plans' (NDPs) which, if approved at local referenda and if deemed to be in conformity with the adopted local plan and national policy, can sit within the framework of the local plan and become an important consideration in the determination of planning applications. The same legislation created 'Neighbourhood Development Orders' (NDOs), which provide a mechanism for modifying the local definition of permitted development, effectively allowing a Parish or Town Council to grant itself either outline or full planning permission. How might these tools advance the service agenda? Once a 'neighbourhood' (meaning village or cluster of villages in a rural location) has an NDP in place, it will be eligible to receive 25 per cent of any money levied from a development occurring in its area, through the Community Infrastructure Levy. Hypothetically, this would provide it with the means to invest in new critical infrastructure: support for a local school, doctor's surgery and so on. Practically, however, these types of private development (yielding CIL funds) may be unlikely in small village locations, so it is probably urban neighbourhoods that will eventually benefit from this mechanism. In smaller villages, it is the NDOs that are likely to have greatest impact, with communities finding additional ways to allow new housing (see Chapter 8) in support of services or permitting development for community use: a farm or community shop, an extension to a village hall or a change of use that might not otherwise have been permissible.

The Act also allows local authorities to designate 'assets of community value' and it is expected that they will generally do so after receiving nominations from communities themselves. Once designated, a six-month moratorium on the sale or potential redevelopment of the site may follow, perhaps providing a community with the chance to raise money to purchase the asset itself. In addition, communities (i.e. designated neighbourhood forums in urban areas or Parish/Town Councils) were also handed, through the legislation, a 'right to acquire' and develop sites for community use. This is likely to mean that they are able to acquire land in public ownership (with the consent of the Secretary of State) that is evidently under-utilised and thereafter exercise an associated 'community right to build'. Together, these rights may lead to more communities substituting failing or poor private services with community enterprises in the form of voluntary-run shops, or initiate entirely new services. In the past, some local groups have come together to buy pubs and post-offices, aiming to secure the future of these services. Such actions may become more commonplace in the context of these new rights. And it is not only private services that may transfer to community ownership and control. The Localism Act also created a 'community right to challenge' the way in which public services are being provided (the obvious example being local transport; see Chapter 6) and to take control of those services and run them on a voluntary basis, with financial support from the local authority.

Whilst these changes are potentially very far-reaching, their impacts will be spatially discriminatory and the benefits will be disproportionately felt by those communities that are better organised, understand the rules of the game, and can gain access to the resources needed to acquire, build and deliver. By March 2014, more than 1,000 neighbourhoods had started drawing up NDPs, although only seven had been formally 'made' (DCLG, 2014b). There is wide expectation, however, of socio-economic bias in the pattern of successes achieved through neighbourhood planning and in the capacity of communities to start running their own services. Sutcliffe and Holt (2011) has used data on existing levels of volunteering in England as a basis of predicting participation in various opportunities presented by the localism agenda. In wealthier areas, the proportion of people engaged in voluntary activity is as high as 36 per cent whilst in poorer areas it drops to as low as 14 per cent. The research was well summarised by Peter Hall, who concluded that neighbourhood planning opportunities were likely to be taken up by '[...] well-meaning, well-educated people living in nice places – mostly rural – with time on their hands' (Hall, 2011: 60). It may transpire that the most affluent rural areas benefit most from the rights described above, which will not always be those in greatest need.

That need has been heightened in many areas by cuts in public budgets, a subject recently explored by a 'Rural Coalition' of interest groups (comprising Action with Communities in Rural England, the Campaign to Protect Rural England, the Country Land and Business Association, the Local Government Group, the Town and Country Planning Association and the Royal Town Planning Institute), which has warned that:

> Rural services will be more vulnerable than most to public finance constraints since rural service delivery, even at its most effective, is more expensive per head of population than in urban areas. Pressure to deliver more with less money will inevitably lead to further loss of local services altogether – unless communities are empowered, as 'Big Society' suggests, to design appropriate services and means of delivery building on a rural culture of self-help, that is already high.
>
> (Rural Coalition, 2012: 5)

The remainder of this chapter looks at key service areas – those traditionally provided by the state and those that are generally private – and considers how delivery of those services may occur in the future if community-based action becomes more important and, finally, if more services are delivered or accessed through digital technologies.

Part 2a: Access to key services

The broader issue of accessibility was introduced in Chapter 6. But how accessible are specific key services to the people and communities who rely on them? Quantifying accessibility is not an easy task as it depends on personal mobility and hence personal circumstance. However, the Department of Transport in England has developed measures of 'reasonable' accessibility using different travel modes, assuming that it is 'reasonable' to walk, cycle or drive to a service depending on its distance from home. Eight services – public, private and mixed – are considered to be 'key services':

1   Employment centres;
2   Primary schools;
3   Secondary schools;
4   Further education providers;
5   Primary health care (e.g. doctors' surgeries);
6   Secondary health care (e.g. hospitals);
7   Town centres (with a range of retail, banking and post office services);
8   Shops selling groceries.

Unsurprisingly, the Department for Transport (reported in DEFRA, 2014a) found that accessibility to all of these things tends to be generally lower in rural areas and particularly low in areas that are more sparsely populated, where distances to individual services or to service centres tends to be greater. Only 27 per cent of residents in villages and hamlets had reasonable access to a shop selling groceries, compared to 59 per cent in rural market towns and 62 per cent in urban areas. If an average for all types of rural areas is generated and compared to the average for all urban areas, then an 'access gap' can be calculated. This provides a simple means of comparing rural with urban service accessibility. The gap for primary schools is 4.5 per cent (so not a huge rural effect, with a great many villages retaining a small school). But the gap widens for other services: 14 per cent for shops selling groceries; 18.6 per cent for employment centres; and 38.1 per cent for further education (DEFRA, 2014a). These are of course gaps against the rural average, and access will fall sharply when urban areas are compared to the most sparsely populated rural areas.

Public services, community responses

Education and health care are clearly important public services, which rely on a continual stream of funding through tax revenues and cannot easily be substituted through community action. However, a broader set of delivery arrangements are now possible, with schools managed and operated in different ways and health care provided through a range of models, some of which are community-based. The once clear lines between 'public', 'private' and 'voluntary' provision are becoming increasingly blurred.

With regards to *health and social care*, it is often suggested that rural communities are healthier than their urban counterparts, particularly when measured in terms of life expectancy and premature causes of death. The notion of a healthier countryside is part of the draw of the 'rural idyll' introduced in Chapter 1. However, retirement and in situ ageing (Satsangi *et al.*, 2010) means that there is a growing demand for rural health care that is not being reflected in public spending. Per capita allocations via Primary Care Trusts (PCTs, now replaced by Clinical Commissioning Groups) is lowest, at £1,560, in rural Dorset (which has the highest proportion of elderly people of any English authority) compared to £2,084 in Tower Hamlets, which has the youngest population of any authority. Top-up allocations to local authorities (and focused on reducing health inequalities and promoting preventative health care strategies) show a similar pattern: rural counties such as Surrey and Rutland were allocated £22 and £28 per capita respectively in 2014/15, compared with £73 and £133 for the urban authorities of Birmingham and the London Borough of Westminster. The national average was £55 (Rural Services Network, 2013). The costs of providing rural services are higher, for all the reasons set out earlier, so there is perhaps a case for higher allocations in the countryside than those agreed in recent budgets. However, the barrier of distance is likely to remain. In 2011, 100 per cent of the urban population in England lived within 4 km of a GP. The figure for villages and hamlets was 69 per cent.

In relation to ageing, patterns of social care provision are perhaps even more significant. Domestic care for the elderly, and services such as 'meals on wheels' (the daily provision of hot lunches), is less regularly available in rural areas, again owing to distance and accessibility. Those providing these services need to travel more and are hence able to reach fewer people. This same pattern is observed for mental health workers, who spend up to a third of their time travelling in the countryside compared to a maximum of a tenth of their time in cities (Pugh *et al.*, 2007).

There is a clear health care disadvantage faced by rural people, which will be accentuated by the patterns of public spending and by the clustering of services noted earlier in this chapter. Policy-proofing exercises have rounded on this issue, with the Institute of Rural Health (2012) and others pointing to necessary innovations in health care provision (see Box 7.2).

Educational services – particularly in the form of primary schools – face very similar challenges. The plight of village schools has been an issue for

Box 7.2 Innovative means of servicing health and social care in rural areas

Airedale Tele-health Hub Yorkshire: Airedale NHS Trust covers a large area of Yorkshire where travel time to hospital by public transport can be almost two hours for a fifteen minute appointment. The tele-health hub provides remote monitoring for patients with long-term needs; online coaching to help deliver self-care; and remote secure video conferencing to provide clinical consultations. It is particularly useful in the care of elderly people living in residential care homes, where it is now used by more than 100 care homes and 75 patients in their own homes. Unplanned hospital visits for care home patients have been reduced by 45 per cent; hospital bed days by 60 per cent; average length of hospital stays by 30 per cent; and A&E visits by 69 per cent. The service is now being commissioned beyond the immediate boundaries of the trust (Rural Services Network, 2013).

Devon Neighbourhood Health Watch: Similar to a neighbourhood watch scheme for crime this focuses on volunteers who look after the health and wellbeing of the elderly and vulnerable in rural communities. The volunteers act as the eyes and ears of the GPs on the ground helping to reduce pressure on health services by alleviating rural isolation and stress. Volunteers may provide lifts, collect prescriptions and also provide home-cooked meals to vulnerable neighbours (Rural Services Network, 2013).

The Project Group, based in Oswestry, Shropshire, seeks to provide help and support to those who are suffering from mental health problems and low self-esteem, and who have been referred to the project by health care professionals. By focusing on creative activities, the project keeps its beneficiaries 'less ill' and therefore less of a burden and cost to statutory health care providers. Created as a community interest company, it has four part-time salaried staff and five volunteers, but still struggles to raise sufficient funds to keep going (CCRI, Rose Regeneration and the Rural Services Network, 2013).

Suffolk Coffee Caravan – A Rural Information Hub: The idea developed from a lack of access to information and support that was experienced by rural communities in the aftermath of the 2001 foot and mouth crisis. It provides mobile information services to rural communities lacking facilities. It provides information to vulnerable groups and equally can provide feedback to (health care) organisations that find it difficult to access these groups. Whilst being supported by all the local authorities, one of the group's key strengths is its ability to be seen to be independent of any statutory organisation (CCRI, Rose Regeneration and the Rural Services Network, 2013).

Oxford Adult Education Learning Services Dementia Project. It is estimated that 7,000 people aged over 65 in Oxfordshire have dementia and this figure is set to increase by 20 per cent in the next five years. The project brings together adult education practitioners, health and social care providers with local people to develop their understanding of dementia and dementia care. With many sufferers being located in rural communities it is hoped that such an approach will enhance the potential for effective care within the community (National Institute of Adult Continuing Education, 2012).

planners and educators for a number of years (Ribchester and Edwards, 1999). There are educational arguments for and against the retention of small rural schools, but these seem evenly balanced: their curricula may be more limited, but can be enhanced through the sharing of specialist teaching across a network of similar schools; the teaching environment may be too 'sheltered' but that same environment can nurture a stronger sense of community; and they may lack many of the facilities of bigger urban schools, but these again can be pooled and shared. The weight of economic argument seems, however, to favour rationalisation. Per pupil costs in rural areas (where there is a proliferation of smaller schools) are 77 per cent higher than in urban areas (where schools are generally larger and may have more than a single-form annual intake) (Hindle *et al.*, 2004). Yet, despite this economic argument, the rate of closure of small schools (with fewer than 50 pupils) slowed in the first decade of this century. The Commission for Rural Communities reported that only 19 rural primary schools closed between 2000 and 2006 (CRC, 2006c) and this trend, that began in the 1990s, was explained by the fact that the 'unpopularity of closure' had 'pushed planners and educationalists towards devising alternative reorganisation and support strategies' (Ribchester and Edwards, 1999: 58). They had increasingly been pooling the staff resources and facilities of local schools. However, swingeing cuts in local authority budgets since 2010 has meant that the economic viability of small schools has again become a major issue in rural areas. In February 2011, Shropshire County Council announced plans to close nine primary schools, including eight in rural areas in an attempt to cope with a £1 million funding gap in the county's school budget. At the same time, Hampshire County Council was facing similar funding difficulties and 40 small schools were being considered for closure or merger, the aim being to save £1.5 million annually.

Strategies for dealing with this re-emergent challenge have largely been untested to date, but under the Coalition Government parents are able to set up their own 'free schools', outside local authority control but centrally funded. Whether this is feasible in rural areas is unclear. There need to be buildings available in which to set up the schools and the limited funding available may be insufficient to overcome the same barriers (i.e. higher per pupil costs) faced by existing rural schools. There is of course the possibility that other community buildings (for example, a village hall) may be able to double-up as small schools, management can be provided by volunteers, and such schools can operate with a minimum of public money. This model might look like something between home-tuition and traditional classroom provision, but it is an option that may not be attractive to the majority of parents and is perhaps more suited to the most isolated of rural communities, where variations on this model already exist. For the majority of communities and schools, however, the preferred solution will be co-operation and partnership between existing public providers, with staff and facilities shared and significant cost savings achieved (Hill *et al.*, 2014). Innovation on the existing small school model has popular and political support, with schools viewed as a crucial service for rural communities.

Private services, community responses

The fortunes and fate of private services are perhaps best illustrated using a trilogy of the most basic and vital: village shops, post offices and pubs.

The village shop

Woods (2005a: 97) has observed that '[...] most rural parishes in England have neither a general store nor a village shop' and the disappearance of these private services is an expression of economic forces (with many independent shops purchased by chains and then closed as part of rationalisation exercises) and a shift in the socio-economic make-up of communities, resulting in new consumption patterns, including online shopping and a tendency for households to do their weekly shop at a supermarket, perhaps on their way home from work. General stores do not have the purchasing power needed to drive down supplier costs, and therefore carry a smaller range of pricier goods, which in turn leads more customers to use supermarkets or the internet (CRC, 2006c). On the one hand, shops play a vital role in communities; they are sometimes multi-functional and may host a post-office counter; and they provide important meeting places or opportunities for chance encounter. But on the other hand, they are ultimately micro-businesses that need to remain economically viable (see Chapters 3 and 4).

Village shops are a key private service that can be supported or substituted by community action. If a private shop closes, it may re-open as a community enterprise (see Box 7.3). DEFRA (2006) has estimated that a combination of voluntary action and modest state support could see local retail return to one-third of the villages that currently lack a shop, by 2020. The Plunkett

Box 7.3   Blockley rural community shop

Blockley in Cotswold District – a village with a population of about 2,000 – once boasted a thriving retail centre with eight pubs and twenty specialist retail outlets. By 2007, all had closed and the village was becoming a retail desert. The old village store and post office were closing, and a delicatessen/wine bar was being converted into a family home. But with support from the Rural Community Council (RCC), the villagers created a not-for-profit co-operative to run a community shop and cafe, with one full-time manager and 14 part-time staff. Within a year its turnover was £500,000.

(Action with Communities in Rural England, 2009)

Foundation is a national charity best known for helping rural communities to set up a range of community-owned co-operatives, including community shops, pubs and food enterprises. In the last 20 years there has been a sharp rise in the number of community-owned and run village shops, up from 23 in 1993 to 303 in 2012, with a further 197 in the pipeline (Plunkett Foundation, 2013). In 2013, the 303 existing shops employed some 945 staff, engaged 7,796 volunteers, and were hosting 174 post-office counters (see below). In the previous financial year, they collectively recorded a profit of £1.6 million, which was 28 per cent up on 2011 (Plunkett Foundation, 2013). Furthermore, the survival rate of these shops has been remarkably good. The Foundation reported that, to the best of its knowledge, just 13 community shops had closed since 1993. Success was attributed to five factors:

- Local people are keen to support a venture in which they have a social and economic interest, and from which no single individual profits.
- Staff costs can be reduced through the use of volunteers. Typically, a community shop will have about 30 volunteers each contributing 2 to 3 hours each week.
- The barrier of securing affordable premises can often be overcome by co-locating with existing community facilities; the village hall, church or pub. Ownership of these assets is often secured through long-term leases secured by grants or low rents can help maintain the viability of the community infrastructure.
- Community shops can take advantage of rate relief schemes for small rural businesses; 93 per cent of community shops currently do so, and also benefit from various tax advantages.
- As member-owned organisations, community shops are more likely to listen to their customers, providing the goods and services that the community needs.

(Plunkett Foundation, 2013)

Despite this success, and the benefits that community shops bring, they are neither ubiquitous nor a solution for all communities. The Plunkett Foundation notes that as many as 400 private village shops are closing each year in England and community action can only fill a small part of the retail gap left by these closures. Whether a community-led solution will follow-on from closure depends on whether the community has the capacity needed to mobilise and whether, also, it can secure the resources needed to start up its own enterprise. Much will depend on the characteristics of that community, its store of social capital, and the impetus provided by key individuals.

Post Offices

Even where services remain 'for-profit', their survival often depends on co-location. This has often been the case with postal services, co-located with

Figure 7.2

Convenience store with post office attached, Pluckley, Kent

village shops. The same strategy has been used by community enterprises. But despite co-location, the fate of post-offices has become emblematic of the wider challenges facing all private rural services. In 2000, there were just over 9,000 post offices in rural areas. By 2013, two-thirds of the network had been lost; the number of offices or postal services (e.g. a counter within a shop) had fallen to 2,865 (Rural Services Network, 2013). Much of this contraction can be explained by the switch away from traditional services: pensions and other state benefits being automatically credited to bank accounts rather than paid over the counter, and car tax being renewed online. Post Office revenue from government contracts fell sharply as a consequence of these changes, from 43 per cent in 2003/4 to 26 per cent in 2007/8 (Public Accounts Committee, 2009), requiring a restructuring of counter provision. Royal Mail initiated its Network Change Programme with the aim of reducing fixed services (post offices and sub-post offices) and switching to mobile alternatives and shared counters, operating just once or twice a week. A balance needed to be struck between cost and accessibility, and this also meant changing the way sub-postmasters/mistresses were paid. Fixed salary payments ceased and their income was linked to the value of transactions. Co-location quickly became a double-edged sword. Guaranteed income from government contracts had supported co-located shop functions, with postmasters/mistresses making enough to run the counters and the shops. The loss of fixed income – salary and service contracts – therefore threatened the viability of both. The

Network Change Programme led to the closure of 840 rural post offices, and although the rate of closure has slowed there is a continuing tendency for offices to shut once the postmaster/mistress retires or chooses to move on. There are simply too few people willing to take on these marginal businesses.

When that happens, there is always a chance that community enterprise will step in; working with what is now 'Post Office Local' to maintain some minimum service level. For example, the small Kentish village of Wittersham lost its post office, which was being operated out of a private house, when the householder decided to give up the service. A survey of community needs conducted by the Parish Council identified postal services as vital 'community infrastructure' (Rural Services Network, 2013). The solution was to run a sub-post office one morning each week from an existing community centre, employing the services of a postmaster from a neighbouring village.

Pubs

The third and final private 'service' that is held up as crucial to the wellbeing of rural communities is the pub. The viability of pubs – 'public houses' – across the UK is being undermined by both changing patterns of sociability and by the rising value of residential property in many locations. If fewer people are spending money in pubs and those pubs would be more valuable as private houses then the case for closure and conversion seems clear-cut. Walker (2014) claims that 1,300 pubs close in the UK each year and whilst this is not only a rural phenomenon, the loss of village pubs is thought to be especially significant given that they may be the only – or one of very few – focal points for community life. Hence there has been a recent drive to save rural pubs.

The National Planning Policy Framework (NPPF) for England points to the need for local and neighbourhood plans to 'promote the retention and development of local services and community facilities in villages such as [...] public houses' (DCLG, 2012: Para. 28). More broadly, it asserts that in order

> ... to deliver the social, recreational and cultural facilities and services the community needs, planning policies and decisions should: a) plan positively for the provision and use of shared space, community facilities (such as local shops, meeting places, sports venues, cultural buildings, public houses and places of worship) and other local services to enhance the sustainability of communities and residential environments; b) guard against the unnecessary loss of valued facilities and services, particularly where this would reduce the community's ability to meet its day-to-day needs; c) ensure that established shops, facilities and services are able to develop and modernise in a way that is sustainable, and retained for the benefit of the community; and d) ensure an integrated approach to considering the location of housing, economic uses and community facilities and services.
>
> (DCLG, 2012: Para. 70)

Figure 7.3

The 'Black Horse', Pluckley, Kent

This statement clearly has broad purpose and points to the statutory planning system's role in protecting and enhancing all types of service, rural or urban. But amongst the trilogy of services discussed here, it is pubs that have perhaps received most attention from planning. Broadland District Council in Norfolk has recently updated its Local Plan, retaining a policy which requires that:

> [...] in parishes outside of the Norwich urban fringe, proposals for the change of use or redevelopment of a public house will not be permitted unless: a) there is an alternative public house within reasonable walking distance; and b) it can be demonstrated that the public house has been marketed for a reasonable period of time and that no reasonable offers have been received to purchase or let the property, and that it can be demonstrated that the business is not economically viable.
>
> (Broadland District Council, 2011: 20)

Similar policies are now contained in a great many rural local plans that cover areas in which pubs are at threat from residential conversion. Sometimes, this is enough to ensure that a pub is retained. But in other instances, communities need to be more proactive, identifying pubs as 'assets of community value' (ACV) under the Localism Act 2011, and perhaps even acquiring that asset and running it for community benefit (Box 7.4).

Box 7.4 Community action, supportive planning and pubs

- **The Dolphin in Bishampton and Throckmorton** Parish Council in Worcestershire had changed hands several times and was under threat of closure. Having been designated an ACV, a community group sought the support of the Parish Council who had just acquired the village shop. A £300,000 loan was secured from the Public Works Loan Board and the pub was acquired by the community on the basis of a robust business plan to pay back the loan.
- **The Highbury Barn in Great Cornard** was in 2012 threatened with demolition by its owners the brewery chain Punch Taverns, who wanted to redevelop the site. Great Cornard Parish Council, supported by a 310-signature petition protested against the proposal. In the light of local concerns, Babergh District Council used an Article 4 Directive to effectively remove Punch Taverns' permitted development rights for the site (these effectively allow the site to be redeveloped without needing to apply for planning permission). Punch Taverns appealed this decision but lost with the Planning Inspector finding in favour of the local council as the existing building was '[...] a significant local heritage asset which makes a positive contribution to the street scene and local distinctiveness' and provides '[...] a traditional landmark for community and social interaction'.
- In 2008 the landlords of the **George and Dragon pub in Hudswell**, a small village of about 240 people situated in the Yorkshire Dales National Park, were declared bankrupt and the pub closed. The community established a not-for-profit company called Hudswell Community Pub Ltd, and because planning policy prevented the asset from being de-licensed within two years of closure, there was time to raise the money needed to purchase the pub. Most people in the village contributed a minimum of £500 each in a community share issue. The pub was bought by the community (through the company). Other services – library, meeting rooms, wi-fi hotspot, and children's centre – have been co-located at the pub, providing a multi-functional community facility. Furthermore, ten allotments were created on land acquired with the pub. These are rented out and enhance the viability of the enterprise.

Uttlersford District Council in Essex was an early supporter of ACVs and, on the advice of its Parish Councils, has designated 170 ACVs, 17 of which are public houses (Walker, 2014).

Part 2b: Information and communication technology

Finally, the 'judicious introduction of ICT' is often viewed as a vital means of solving the rural service conundrum (Moseley, 2000: 415), having the potential to deliver against a number of key public (health and education) and private (shopping) needs. But the role and value of digital services is contested. On the one hand, they create opportunities for greater connectivity, but may threaten the face-to-face contact that sustains community life. That said, Wellman (2001) argues that digital technology tends to augment physical social interaction rather than supplant it, playing down fears that the internet and smartphones will ultimately bring technological isolation and social meltdown. On balance, ICT is generally good for rural areas, helping to sustain social networks (Abas *et al.*, 2009) as well as offering new ways to access some services. However, as in the case of Post Offices and the replacement of counter services with online alternatives, there is a danger (perhaps already realised) that:

> [...] telematics could also render many village and small town service outlets redundant [...] For 30 years attempts have been made to stem the motor-car-driven loss of service outlets 'up the urban hierarchy' from villages and small towns to larger towns and out-of town complexes. It would be ironic if the next 30 years witnessed a struggle to stem the Internet-driven loss of services 'down the urban hierarchy' to 20 million outlets at the individual household level.
>
> (Moseley, 2000: 430–431)

Research into ICT provision has tended to view it as generally a good thing, whilst drawing attention firstly to a 'digital divide' (between areas that have good wired and wireless connection and those which do not) and secondly to 'digital differentiation' (between groups who tend to make use of the technology and those who are less comfortable with it) (Riddlesden and Singleton, 2014; Townsend *et al.*, 2013). Longley *et al.* (2008) have shown that knowledge and use of ICT is spatially differentiated: there is a greater concentration of the 'e-unengaged' in rural areas whilst the 'e-experts' seem to congregate in cities. One interpretation of this might be that access is more impeded in rural areas, owing to lower broadband speeds. However, in the US, LaRose *et al.* (2007) demonstrate that 'digital differentiation' remains even when the 'digital divide' (in infrastructure) is closed, so the explanation must lie in demographic differences, in relative poverty and in educational attainment.

In the UK, the rural population is highly mixed meaning that differentiation in patterns of potential use, between urban and rural areas, is likely to be limited. People are able to make use of the technology so long as they have good access. However, the communications infrastructure is exclusively provided by private companies who have a vested interest in servicing those markets where there is both high demand and where the costs of provision are less. With wired services, costs correlate to the distance between a customer's home and the digital exchange. The speed of broadband services is affected by that same distance. Generally, this means that slower rural broadband services are more expensive to deliver than faster urban services. Because companies enter the most profitable markets first, Broadband UK has estimated that the cost of providing access to the final 10 per cent of potential UK customers (living in the least accessible locations) will be three times greater than it was to provide services to the first two thirds of customers (Townsend *et al.*, 2013). Market reality is interpreted in some quarters as market failure, and in 2009 the UK government launched a 'Universal Service Commitment' guaranteeing a nationwide minimum connection speed of 2Mbps by 2012, thereby distributing more evenly '[...] the benefits of broadband, including the increasing delivery of public services online' (DCMS, 2009: 27).

Because of the costs involved, the target proved too ambitious. Government now has a new 'superfast' broadband delivery target to be achieved by 2017 and funding is being channelled to local authorities via Broadband Delivery UK (BDUK) to support infrastructure upgrade. In England alone, some £294.8 million has been allocated to local authorities to accelerate the delivery process (DCMS, 2013). However, big differences in broadband speed remain between the most urban of urban areas and the most isolated of rural areas. Figures from DEFRA (2014a: 69) show that the current urban average is 14.8Mbps compared to 4.4Mbps in sparse hamlets or isolated dwellings. All town/urban speeds are above 11Mbps; all village/hamlet speeds are below 6Mbps. And these speed data do not, of course, account for the fact that nearly half of all rural households and businesses have very slow or no fixed connections (Townsend, 2013).

Poor ICT infrastructure is today compounding the accessibility challenges faced by communities at the widest point of the digital divide. Rural households are often being deprived of information, missing out on e-retail opportunities, struggling to communicate with utility companies (or paying more for services), unable to access online leisure, and – crucially – disconnected from new social networks (see Box 7.5).

Conclusions

Rural situations produce very specific service challenges, which are a function of isolation, physical distance, low(er) population densities and hence more fragile service markets. Some parts of the countryside are unattractive to private enterprise: insufficient profit can be extracted from retail activities or even from investment in infrastructure such as ICT. Post offices, village shops and pubs have been closing in many rural areas as their markets continue to weaken, perhaps as a result of restrictive planning, population change or loss, and a popular shift in consumption habits. Therefore, there seems to be a corrective role for the public sector, which needs to step in and deal with this 'market failure'. However, the cost of doing so is often prohibitive. The cost of providing equivalent access to standard public services for rural populations – health care and schools – is often astronomical compared to the cost of supplying such services in relatively dense urban centres. Stepping in

Box 7.5 The value of ICT connectivity

- Public information is increasingly being delivered online, the aim being to achieve significant savings for key providers such as the NHS (Booz and Company, 2012). However, this strategy will leave behind late adopters, disproportionately located in rural areas;
- E-commerce continues to expand, putting pressure on traditional retail activities, both in market towns and villages. Internet-based retail will benefit some households, but accentuate problems for others. Many other services, besides retail are now online – e.g. banking, buying insurance, or booking holidays – and physical access to these things is reducing. Such goods are frequently cheaper when bought over the internet. This means that non-connected households, many of which are in rural areas, face both access and increased cost difficulties.
- Gas, water, electricity and telecom providers now deal with their customers online; offering 24-hour access to internet-based services as well as discounts against standard rates. Those who are not connected tend to be poorer and suffer the double burden of poorer access and higher costs;
- Leisure and entertainment activities are increasingly being offered online, and such services often need very-high broadband speeds. These are more difficult to achieve in rural areas, further reducing the appeal of such areas to the young (Townsend *et al.*, 2013).
- The internet offers opportunities for online social interaction and a means of overcoming isolation. As it becomes more ubiquitous, other traditional forms of staying in touch – the post and land-line telephones – may become increasingly irregular or costly and may ultimately cease. Generally, the internet is a way into new networks, often leading to face-to-face contact or involvement in real-world activities. Poor infrastructure may lock some people out of these activities.

to support private services is not feasible, although government has done so in the case of broadband, viewing it as a critical infrastructure for economic growth. But the survival of private and public services is viewed as vitally important for rural communities and both, to some extent, can be delivered as social goods. This is true not only of post-office counter services, village shops and pubs, all of which can be co-located and operated as community enterprises, but also health care and education, aspects of which (including soft social support, intelligence gathering, the acquisition of buildings, and the management of assets) can be led by communities themselves. In the face of market disinterest and a lack of available public investment, community action has a big part to play in the supplying of key services in many rural communities. But the potency of such action depends on the social context in which it is rooted: whether communities have the social capital (the right mix of skills, knowledge and resources) needed to progress sometimes complex and time-consuming projects, which must be led by key individuals with the drive and tenacity needed to bring these projects to fruition.

Summary

- Rural service decline, which has been a long-standing characteristic of rural areas, continues unabated, often caused by a combination of efficiency-of-scale arguments of larger more concentrated provision and the additional costs associated with rural service provision. This means access to services for the rural population is often more difficult and expensive (both in monetary and time terms) than for the urban population.
- There are new and alternative ways of delivering rural services that might offer partial solutions, but sustaining provision has become very much dependant on locally derived social capital.
- ICT offers some partial solutions, but cannot deliver universal access as there remain differences both in the physical access to the technology and user take-up.
- New and innovative ways of delivering services to vulnerable rural residents will continue to be sought as the state continues to withdraw from direct delivery.
- But changing and differentiated access to rural services will remain a challenge for both policy makers and some rural residents alike.

Key readings

- ACRE (2012) *Making a Difference: 25 Years of Community Action*, available from www.acre.org.uk/cms/resources/25-anniversary-doc-final.pdf.
  (Reviews how community action in rural England has developed over the last 25 years.)
- Powe, N., Hart, T. and Shaw, T. (eds) (2007) *Market Towns: Roles, Challenges and Prospects*, Routledge: London.
  (Provides a detailed overview of the challenges and opportunities facing market towns including a discussion of their role in rural service provision.)

Key websites

- Action with Communities in Rural England (ACRE): www.acrc.org.uk/.
  (A lobbying organisation with broad interest in the wellbeing of rural communities.)
- DEFRA Rural Proofing Guidance: www.gov.uk/rural-proofing-guidance.
  (Sets out the rationale and process of considering the impact of public policy on rural areas.)
- Rural Services Network: www.rsnonline.org.uk/.
  (Dedicated to the understanding and promotion of appropriate rural service levels.)

8 The rural housing question

One or many questions?

In Britain, the definitive 'rural housing question' is frequently presented as one of *supply*: too few homes being provided for the rural population in a context of strong external demand in the 'consumption countryside' (see Chapter 2). Many parts of that countryside are attractive to migrant households who import their income and wealth and, by bidding against one another and out-bidding poorer buyers, push property prices beyond the reach of 'local' households. A combination of supply constraint due to restrictive planning and demand pressure caused by commuters, retiring households and second-home buyers is often thought to be at the root of the rural housing question.

The reality however, in Britain and elsewhere, is that there are in fact many rural housing questions. This point is made by Satsangi and colleagues (2010: 3): at the root of everything are the processes of counter-urbanisation and social change described in Chapter 5, and these processes have spawned a number of related questions over the quantity, quality and location of housing. Rural housing markets have adjusted as a result of population movements, causing prices to rise and producing inequalities in terms of general housing access and the quality of housing available to particular groups of buyers and renters. Underpinning these market adjustments are the broader economic shifts outlined in Chapters 3 and 4, which have left remnants of the occupational communities in lower paid and sometimes seasonal jobs, and unable to compete with salaried migrants. Low pay renders home-purchase and private renting unaffordable in many instances, pushing the lowest income groups into social housing where available, or forces them to move to larger market towns – completing the social-reconfiguration of the countryside which began in the 1960s. These are perhaps the general facets of the rural housing question; they are the outcomes of gentrification and could occur in a similar fashion in an urban setting, wherever a rent gap is recognised and exploited. But more particular are the new patterns of rural consumption that began 50 years ago: retiring to secluded rural locations and buying second homes in pretty villages, and leaving the latter unoccupied for extended periods. The new seasonal or retired consumers of rural housing, alongside daily commuters, have introduced new behaviours, tastes and opinions to much of the countryside. They have often brought with them a conservative attitude to rural change and development, opposing anything that is deemed 'detrimental to the character of the village' (Newby, 1979: 167) or to the preservation of property values. It was noted in Chapter 5 that social change has been accompanied by new *expectations*; expectations of what the countryside should be, which are fundamental to planning and housing discourse and outcomes.

In Britain at least, the demand pressure expressed by households moving to rural areas may compound supply pressures, particularly as some of those households become politically active and resist the development of villages, partly out of self-interest and partly because the re-profiling of need that results from counter-urbanisation suggests a reduced requirement for social housing, or cheaper private housing to buy or rent. New residents, or their twenty-first century successors, do not need these things and their presence may mask underlying poverty amongst a minority of lower paid, less affluent groups.

Such pressures are not unique to Britain and in the first part of this chapter we examine the general drivers of the housing challenge alongside a number of socio-political outcomes and remedial frameworks and options. This is followed, in Part 2, by a return to the British situation and to recent responses.

Part 1: Questions of demand, supply and response

On the demand side

Across much of the globe, urbanisation is thought to be at the heart of housing supply difficulties. The pace of that urbanisation means that governments, and their various partners, struggle to deliver the infrastructure, the services and the new homes needed to support concentrated population growth. Therefore, the housing question is largely an urban one – as Friedrich Engels observed in Germany almost 150 years ago (Engels, 1872). But urbanisation is as much a trigger for rural housing crises as it is for crises in major cities. This was true during the Western industrial revolution and remains true today, as many cities in the global South and in south-east Asia experience growth rates far in excess of those recorded in Germany and other European countries from the mid to late nineteenth century. Urbanisation sets in train a range of immediate and longer-term processes that impact on near urban and remoter rural areas. In the first instance, cities expand beyond their current boundaries, consuming vast swathes of previously undeveloped land and overwhelming near-urban agriculture. At the same time, it sucks people from the countryside, depriving many rural areas of their youngest and most able residents. In its early phases, urbanisation will drive rural depopulation. This happened in Europe nearly two centuries ago and has been happening in much of the global South for the last 50 years. But when sections of the new urban population begin to prosper and governments attempt to 'contain' urban growth within fixed boundaries (see the 'green belt' discussion, Chapter 10) – causing pressures on urban resources, especially on things like housing and schools – those sections (now a prospering middle class) start to abandon the city. Improved transport for commuting facilitates this process. Initial rural depopulation becomes eventual repopulation, as witnessed in the United States. The economic crisis of 1929 to 1932 hit the remaining population of rural America particularly badly, triggering sudden depopulation in many agricultural areas and the disintegration of communities, leaving abandonment and dereliction in its wake. This situation persisted for the next three decades, but the arrival of post-war prosperity and the building of America's freeways precipitated an urban flight and gentrification of the newly accessible countryside.

Greater accessibility combined with post-war prosperity to produce a sharp rise in demand for rural living. In response, planning was tightened in various countries in order to protect the openness of rural areas from urban encroachment (Hall, 1974). However, not all planning systems sought to defend the countryside. After decades or even centuries of abandonment, the response in many places was to welcome new investment. The historic practice of keeping a second home was revived in many parts of Europe, especially in the south and in Scandinavia, and seemed to offer an economic lifeline to some areas and a respite from their isolation (Gallent, 2014). In Canada and the US, rural resort development was actively encouraged with second home and 'vacation cottage' ownership becoming a feature of local economic strategies. However, the combined effect of key demographic, social and economic changes was to alter radically the nature of rural housing markets. Demand, triggered by absolute population increase, outstripped housing supply; social change brought new people – new 'housing classes' – to the countryside; and economic change deprived local people of any capacity to compete against 'adventitious' homebuyers from nearby cities (Robertson, 1977). Property that was previously worth very little suddenly became highly sought after as constrained rural housing markets began to overheat.

The disequilibrium of overheating markets

A market overheats when there are key impediments to supply, despite rampant demand. During the early years of counter-urbanisation, rural housing markets were weak or stagnant. Demand for private purchase was low, with most existing residents in privately rented or tied accommodation. There was also a surplus of housing, often in a derelict or dilapidated state, in many places across Europe and North America. This situation persisted in France, for example, into the 1990s (Buller and Hoggart, 1994). When demand rises, 'the market' absorbs 'prime' property: whatever is most sought after by those adventitious buyers seeking to exploit the urban–rural rent gap. The first target might be the scattered former farmsteads, which have some land attached and can be upgraded to prestigious residences. The next (or simultaneous) target may be prime village property; larger homes with 'period features'. And the next target may be the bulk of village property, either smaller workers' cottages or more substantial homes. When these become targets of adventitious purchases, tied workers or private renters may be displaced to public housing alternatives, if available, or are forced to move away. Thereafter, once there has been full absorption of the current stock, pressure grows behind the weight of demand to open up new sites for development. Some executive housing is permitted and it may look likely that an imminent supply response (to rising prices) will restore market equilibrium. However, the restoration of that equilibrium is never an easy or simple task.

In urban areas, the supply of land will quickly be depleted and there may be few opportunities to respond to strong demand. Consequently, prices in the most sought-after urban locations – parts of New York City's lower Manhattan or key streets in London's Kensington and Chelsea – become truly astronomical, driven upwards by investment pressure. But in the countryside, it is often political rather than physical constraint that keeps prices high. On seeing the first batch of executive homes built, newcomers to the countryside – residing in their prime village houses or farmsteads – may experience anxiety over the likely effect of new building on both the value of their own homes and the character of the village they live in or near. After all, they have invested not only in a home but in a way of life and in villages that should be '[...] picturesque, ancient and unchanging' (Newby, 1979: 167). At that point, the debates that take place in community councils seldom lead to the opening up of farmland for development, in response to demand (and the needs of displaced local households), but rather to the rejection of further development for the sake of 'protecting the countryside': a laudable aim, with sometimes dire social consequences.

This imagined narrative, however, is not played out in the same way everywhere. Two countries that stand in stark contrast to one another in terms of attitudes to rural development (and therefore in the nature of their respective housing challenges) are the Republic of Ireland and England. Since independence from the UK, and following centuries of subordination under sometimes malevolent landlords, Irish planning has come to reflect a very different attitude to land and property rights and to the function of rural areas. The right of people considered 'local' to benefit from the land has become paramount, and has resulted in a proliferation of single-dwelling developments across the Irish countryside. Critics have drawn attention to the environmental cost of this development free-for-all, but it needs to be understood in the context of past depopulation and as an expression of hard-won land rights. Those rights are now defended in Ireland by IRDA (The

Irish Rural Dwellers' Association; see Scott, 2012). In England, although there have been episodes of displacement from the land, particularly as a result of the enclosures, there has never been the same level of animosity towards landed interests – certainly no lasting revolution or lurch towards republicanism. Attitudes towards development, therefore, are rooted in nostalgia rather than in militant struggle. Ireland gets its sporadic development; England its blanket curfew on all sorts of rural change.

Land-use planning can be a key impediment to supply, but only in instances where social and political discourse leads to a restrictive brand of planning. This can happen where powerful interests monopolise that discourse, where the need for additional development is unacknowledged or ignored, or where other considerations regularly outweigh the socio-economic case for new housing.

A spectrum of demand pressures

It was suggested in Chapter 5 that the countrysides of many advanced industrial nations have become more socially complex; in a word, more 'plural'. This plurality extends to the 'property classes' that come together within housing markets (Weber, 1968). Property classes are distinguished by their life-stage (and hence earning potential or accumulation of wealth), by current and past tenure (and hence by stored equity in property), by wealth accumulated through other means (for example, inheritance), by employment prospects and borrowing potential, and by family support. In composite, these characteristics create relative advantage or disadvantage. The young person, with no past history of home ownership, no accumulated wealth, few skills and no family support will, quite clearly, suffer access disadvantage relative to the forty-something buyer about to withdraw accumulated equity from a house in Milan, who has substantial savings from a professional city job, and whose family already have a portfolio of urban and rural property. Peter Saunders (1984) has argued that past history of ownership has become a key determinant of housing advantage. This is perhaps more true today than it was 30 years ago, as prices (especially in Britain) have soared, and the deposit required to purchase a home is now far too large for any first-time-buyer to find without substantial family support, or a very high salary. Current owner-occupiers form a privileged 'domestic property class' and can easily outgun those in rented homes who are trying to make the jump to home ownership. This is essentially what happened in the 1960s when external demand for rural property began to take off, and rural renters (who were in a majority at the time) with 'low income and low wealth' lost out to 'more prosperous groups' (Shucksmith, 1990) as their landlords sold their homes to incoming buyers. The same divisions persist today; in a constrained private market where prices are high and rising, wealth and income will determine the general pattern of housing access.

Historically, the rural poor were tenant farmers or labourers (see Chapter

5). This has changed today and it is often unskilled workers serving the tourist economy (see Chapter 4), renting their homes and earning only a minimum wage, who struggle to access decent housing in locations close to their work. They may also find themselves in competition with migrant workers for jobs and accommodation. In the southern US, Mexican migrants – legal and illegal – compete with US citizens for unskilled farming and service jobs. In Western Europe, migrants from the eastern accession states have arrived in large numbers. The majority have headed to the major cities, but many have taken up rural jobs in farming and in construction (see CRC, 2007).

In relation to the operation of housing markets, it is useful to distinguish between those groups who express 'effective' demand (because of their relative economic power) and those that have clear *need*, but are unable to compete in a more aggressive market. The former now comprise second-home buyers, investors (usually in holiday lets), lifestyle downshifters/changers (fleeing from cities and urban lifestyles), salaried migrants looking for more space, and retired households. The latter are a much smaller group, with particular wealth, income and life-stage characteristics. Although certainly a generalisation, it is young first-time buyers or renters who often experience the greatest housing access difficulties in many rural areas. This group includes young singles and couples, occasionally with children. They are either Pahl's (1975) *rural poor* or his successors to the agricultural workers who were once housed in tied accommodation. But if these groups are considered in the context of urbanisation and counter-urbanisation – and the displacement and replacement of population examined in Chapter 5 – the obvious question is this: why have these groups remained in the countryside? Ford and colleagues (1997) have categorised young people in rural areas as 'committed stayers', 'committed leavers', 'reluctant leavers' and 'reluctant stayers'. Genuine housing stress occurs when young people who are committed to remaining in a rural area cannot meet their needs in either the private rented sector or through home-purchase, and where state alternatives are unavailable. The difficulties faced by this group are expressed through a range of inadequate *in situ* housing solutions: living with friends or relatives for prolonged periods, making use of short-term seasonal lets (and then facing periodic episodes of homelessness), or finding themselves in temporary forms of accommodation (including caravans). Similarly genuine stress is also evident when young people leave rural areas, not because they wish to but because they cannot match their housing and employment needs (Rugg and Jones, 1999).

Both the 'committed stayers' and the 'reluctant leavers' presumably had some reason to remain in their village; including family connections and support networks, employment opportunities and a desire to stay put for a range of social reasons. Arguably, a *housing need* exists where it is housing access *alone* that causes individuals to endure the hardship of staying or of leaving. But without socio-economic anchors in a place, the existence of that need may be doubted. Without job opportunities, in particular, the case for lower income groups having a genuine housing need may appear weak; and consequently, so too is the case for intervening in market processes to deliver different forms of non-market state housing. Such intervention could leave the 'rural poor' stranded in gentrified villages without job prospects or the support of social networks. Some commentators have argued that intervention is necessary in order to maintain the right of poorer households to live in the villages of their birth: to protect an 'ancestral right' (Gallent and Robinson, 2012b). But the 'rights' argument seems unnecessary. The need to create opportunities for lower-paid service sector workers to remain in rural areas is not merely an individual need, but a broader socio-economic need. Without a cross-section of people living in the countryside, rural economies cannot function and key services cannot be provided. This has been observed on the Norfolk coast in England. This area has seen widespread gentrification since the 1960s, followed by retirement, second-home buying and *in situ* ageing. Many villages are entirely devoid of young people; but yet there is thriving demand for services, particularly care and social services for the elderly residents. Those providing these services need to travel considerable distances from more affordable market towns, adding to their living costs and making it difficult to provide those care and social services when and where they are needed (Gallent *et al.*, 2002). Without young people, the service economy will falter, and farms and tourism businesses will be starved of their labour supply – or accessing their labour will be more difficult, costlier, and lead to unsustainable patterns of car dependency and rural commuting.

The need for accessible housing in the gentrified countryside is not a need encountered solely by individuals; it is a wider socio-economic need. This is an absolutely crucial point. The countryside benefits from having a mixed profile of housing types and opportunities; but it should also be acknowledged that delivering *universal access to market housing* – in particular, in specific hotspots – is impossible. Bramley and Watkins (2009) draw a distinction between 'within area', 'outward' and 'inward' affordability in rural areas. Affordability is often expressed as a ratio between lower quartile earnings and lower quartile rents or house prices, inferring the likelihood of those on the lowest incomes being able to meet rent costs or service a mortgage: that is, 'enter' the housing market via one of these two routes. Bramley and Watkins define 'within area' affordability as the percentage of households in a village/neighbourhood able to enter the housing market in that village/neighbourhood; 'outward' affordability as the percentage of the same households able to enter the market in the wider area; and 'inward' affordability as the percentage of households in that wider area able to enter the market in that village/neighbourhood.

These distinctions are important as they express the essential problem of housing access and affordability in a gentrified countryside, in which key villages have been the targets of adventitious purchasers. A typical situation, confronting planning and public policy, is for a hot-spot village to have a misleadingly high level of within area affordability (as the village is populated by wealthy households) and therefore high outward affordability (the same wealthy households could enter the market anywhere), but *low* inward affordability: a high proportion of people not already in that village would not be able to buy or rent there.

Two issues arise from this 'village/neighbourhood' perspective on affordability. First, the concentration of wealth in gentrified villages masks *in situ* housing poverty (experienced by a small number of committed stayers). And second, it is far easier to address that poverty in the wider area than in the hot-spot villages where rents, property prices and land values are likely to be high and all forms of development will be contested.

On the supply side

The social transformation of many countrysides has altered dramatically the context for a supply response to rural housing problems. Concentrated wealth leads to a questioning of the case for additional development; it is often suspected that gentrifiers are seeking to protect property values through their opposition to additional housing. But in many instances, it is obvious that such housing would simply attract further inward investment from adventitious buyers and do little or nothing to calm prices or help lower income groups. Housing is therefore rejected on amenity grounds, because prized views (across open fields) or the character of a village might be threatened. The rejection of new housing is only problematic in instances where that housing would have otherwise met an identified need: perhaps because starter homes were being proposed, some with restrictive occupancy covenants, or the intention was to build some form of public housing.

Although additional market housing in heavily gentrified and extremely desirable locations is unlikely to assist with housing access, more specific interventions – building particular types of housing to help the 'rural poor' – has become a key battleground in rural planning. Thirty years ago, David Clark (1984) set out five possible responses to rural housing access challenges. The first was a *general market response*: open up more land for development and try to restore equilibrium between supply and demand. The second was a *density market response*: whenever permissions for new housing are given, use the planning system to encourage building more housing on less land at higher densities. The third, fourth and fifth were all *non-market, targeted responses*: give priority, through planning, to meeting the needs of the 'rural poor'; build housing for that group on agricultural land; and give priority to third-sector and community groups within the planning process.

The two market responses were acknowledged to be blunt instruments. Under circumstances of strong external demand, which was at the root of access difficulties, building more open market housing would merely attract additional external investment. The density response might well have the same effect, and would also alter the pattern of rural settlement. The idea that demand should simply be absorbed through further development clearly ignores the highly specific nature of rural housing need. In some instances, a supply response may be appropriate – for example, in addressing *in situ* growth in larger settlements or market towns, or in-migration triggered by economic growth – but in small village locations, which have been bearing the full force of gentrification, a much more targeted, non-market response would seem appropriate. Therefore, in countries affected by the meta-narrative of counter-urbanisation and restrictive planning, the answer to the rural housing question is often a highly targeted one, focused on addressing the specific needs of lower-income groups. This is achieved in a variety of ways, as the overview of the British situation in Part 2 will demonstrate, and is rationalised as a proportionate response set against the need to protect the wider countryside from development pressure.

However, in rural areas affected by continuing depopulation, there is often a very different supply response. Attempts to diversify and strengthen economic activity are regularly led by national governments or supra-national organisations (see Chapters 2 to 4). There is often a focus on tourism and the (infra) structural development of peripheral rural areas. If these attempts have any effect on local economies, they may be followed by private speculation on (potential) rising land values and the development of housing for adventitious purchasers. This all adds up to 'property-led rural regeneration' (Gkartzios and Norris, 2011; Norris *et al.*, 2013) aimed at promoting development and re-population.

There are broadly two situations in which land-use planning is called upon, by the state or by communities, to tackle rural housing challenges. The first is where the forces of counter-urbanisation make a general supply response undesirable, as it would risk environmental degradation without solving underlying market tensions or access difficulties. This is the general situation in many near-urban areas in lowland England, especially in the south. The second is where economic stagnation and depopulation suggest an approach to rural regeneration (with a view to assisting the current population) that may include an element of general house building. There are certainly poorer rural areas in England where such a response might be appropriate; though across England, Scotland and Wales *both* situations are regularly encountered. The broader issue of rural regeneration is addressed in Chapter 4. The remainder of this chapter focuses mainly on the first situation, in which targeted responses to market pressures are developed under the watchful eye of sometimes highly conservative rural communities.

Part 2: Britain – from targeted assistance to community action

Part 2a: The broad picture

Almost 100 years ago, Savage (1919: 1) called upon government to urgently address the dual issues of housing under-supply and deteriorating stock quality across rural Britain, arguing that without enough good-quality homes for farm labourers, the agricultural economy would be seriously weakened. A few years later, in 1926, the Council for the Preservation of Rural England (CPRE) was formed (followed by the Association for the Protection of Rural Scotland in 1927 and the Campaign for the Protection of Rural Wales in 1928) and began campaigning for tighter planning controls over rural development as a means of maintaining the open character of the countryside. It published its own rural housing review (Morris, 1932) which drew attention to the particular issue of 'ribbon development' extending out from villages and small towns. Later in the same decade, Government's Central Housing Advisory Committee produced another report, which, like the earlier Savage Report, focused on the economic need for additional housing and for improvements to the current stock.

Throughout the twentieth century, concerns over housing supply and its links to economic wellbeing have intertwined with a desire to control development and protect rural spaces from encroachment. During the inter-war period, a developmental agenda dominated, marked by investment in new rural housing in support of the farming sector (Satsangi et al., 2010: 34). But this agenda slowly began to give way, in the decades after the Second World War, to rural protection as the countryside seemed threatened by the dual enemies of urban expansion and counter-urbanisation. The first threat was countered by urban containment and the creation of statutory green belt in the 1950s (Elson, 1986; see also Chapter 10) and the second by a more restrictive interpretation and implementation of land-use planning by rural authorities. Local planning responses are examined in the next section; but first we look briefly at components of counter-urbanisation in Britain.

In his influential book *No Homes for Locals?* Mark Shucksmith suggests that '[...] the essence of the housing problem in rural areas is that those who work there tend to receive low incomes, and are thus unable to compete with more affluent 'adventitious' purchasers from elsewhere in a market where supply is restricted' (Shucksmith, 1981: 11). Whilst Shucksmith's adventitious purchasers remain a feature of British rural housing markets, many wealthier buyers today are in fact the successors to the incoming gentrifiers of the 1960s and 1970s, and consider themselves 'local'. Much past academic and policy effort has been expended on defining 'localness' for the purpose of rural housing analysis and allocation (see Gallent and Robinson, 2012b) but this is perhaps no longer a useful distinction, as resident populations in all sorts of places are mixed and transient. Rural residents whose families first came to a particular village thirty years ago are now as 'local' as any other group based anywhere in Britain. The only useful distinction, for the purpose of unravelling rural housing issues, is the one that can be made between *housing classes* – between those expressing effective demand and those disadvantaged in the housing market but with a clear need. Ensuring that the latter group can access housing will have wider socio-economic benefit.

That said, it is also the case that there is an original 'rural poor' in the countryside and a successor 'rural poor', which includes new economic migrants from outside of the UK. There are also original and successor prosperous groups: those whose families arrived half a century years ago, and those whose flight from urban living has been more recent. These distinctions may underpin competing claims on resources and become a source of conflict, for example with households whose families moved to a village 20 years ago critical of the arrival of 'newcomers' today. Arguably, this amounts to mere parochial tension. The most pressing issue centres on those market pressures that restrict housing access. But attempts to link particular pressures (i.e. different types of adventitious purchaser, listed below) to specific impacts is a difficult task. In many rural areas, market competition between different prosperous groups, in a context of inevitable planning constraint, means that the market no longer functions for the 'rural poor'.

However, there is a great deal of concern in Britain over *specific* types of market pressure in rural areas. A study of second-home buying in England for the Countryside Agency in 2002 – drawing on a survey of 119 planning, housing and National Park Authorities (NPAs) – found that nearly 90 per cent of housing and planning officers considered retirement, commuting and second-home purchasing to be having a 'major impact' on local housing access particularly in small village locations and market towns (Gallent *et al.*, 2002: 25–26). For some authorities included in the study, commuting and demand from 'permanently resident' economically-active 'migrants' dominated the housing market: elsewhere, retirement was the critical issue; and in other markets, second-home buying and seasonal pressure (e.g. homes converted to holiday lets) was thought to be generating the biggest impacts, accentuating market disadvantage. Views on the perceived significance of different demand components are shown in Table 8.1, and are drawn from this same study of English rural authorities. The study revealed that retirement and commuting were judged to be either 'very significant' pressures or 'significant' pressures (66 per cent and 76 per cent respectively).

Table 8.1 The components of external housing pressure (n = 119; Gallent *et al.*, 2002)

Component	% very significant	% significant	% minor	% not at all significant
Retirement	31.0	35.3	25.0	8.6
Commuting	25.6	51.3	20.5	2.6
Purchasing for private holiday use	12.5	32.1	30.4	25.0
Purchasing for holiday letting	8.8	26.3	33.3	31.6
Purchasing for rental investment	6.1	28.1	48.2	15.0
In-migration to take up employment	4.5	33.0	43.8	18.8
Teleworking	1.0	3.9	51.5	43.7

Figure 8.1

Village housing, Woodchurch, Kent

Fewer officers felt that second-home purchasing – either for private use or letting – was a 'very significant' pressure.

These different reasons for acquiring rural housing (or sources of demand) 'make' the market, and may also make that market inhospitable to low-income, low-wealth groups – especially in hot-spot areas. Three conclusions can be drawn. First, that *market allocation* is problematic because it fails to deliver the wider socio-economic benefits of broader, mixed access. If this conclusion is accepted then fundamental market controls might need to be implemented that would interfere with private property rights. This view is usually rejected because of its broader economic implications. Second, *particular types of demand should be discouraged* and resisted in order to dampen market pressure as a means of broadening access opportunities. This conclusion necessitates an examination of the rights and wrongs of different forms of housing consumption; a task undertaken below. And the third conclusion is that *market processes need to be bypassed* in order to broaden housing access, accepting that the market is open and essentially untameable.

The majority view on housing market pressures is that nothing much can be done to alter that market: it behaves in much the same way as all markets, and any attempt to restrict access would have profound economic consequences. Parallel debates are now happening in London: should international investment buyers be stopped from purchasing flats along the Thames? What would this mean for the housing market or for London as a centre of international commerce? And what would it mean for the UK in the context of an open European market?

In rural areas, similar debates centre on **second homes**. For some commentators, these represent a fundamental social inequity and bring havoc to rural communities by replacing permanent residents with seasonal ones, causing schools to close and rural shops to fail (Monbiot, 2006; see also Gallent *et al.*, 2005). But for others, they merely express new ('hetero-local') ways of dwelling (Halfacree, 2012) and may bring social benefits to more isolated communities (Rye, 2011) through the introduction of new skills, ideas and connections, which enhance community capacity and social capital (Gallent, 2014). For some, second homes are inherently bad; for others, they need to be seen in their local context with their social impact carefully weighed up. Another important point about second homes in Britain is that their ownership and use is not endemic. In some countries, having some sort of small cottage or hobby farm at the coast or in the rural interior is extremely common practice. Ownership rates in Scandinavia exceed 20 per cent, and they top 10 per cent in southern Europe. Across Britain the figure is between 1 and 2 per cent, meaning that overall they are not a large component of housing demand. But the local pattern is highly variable. There are significant concentrations in London, many owned by Russian or Chinese investment buyers (Rees, 2014). In rural areas, second homes tend to cluster in accessible beauty spots: in Cornwall, on the Norfolk coast, in parts of upland Wales, in the Lake and Peak District National Parks and in the Cotswolds. Similar concentrations are found in Scotland, with owners often drawn to areas of scenic beauty; their second homes providing bases for trekking or rambling holidays.

Many of the same factors that entice second-home buyers to rural areas also draw in **retired households**. Retirement to the countryside has been a major component of counter-urbanisation since the 1960s, with the 'early-retired' in particular choosing more distant and sometimes isolated destinations before heading to more accessible ones in later retirement, as they become less mobile (Satsangi *et al.*, 2010). Although it has often been observed that wherever there are second homes there are retired households, the latter seldom face the same criticism. Retiring is something that people need to do; buying second homes is not. But because retirement is a larger component of counter-urbanisation, its price impact on rural housing is greater. And whilst second homes may undermine local services due to under-use (displacing, it is claimed, permanent residents), the elderly may place great additional use pressure on those same services.

The **teleworking** in Table 8.1 might be regarded as part of broader **lifestyle migration**. When it first became a notable feature of rural housing pressure it was treated separately (Clark, 2001) but remote working, at home via the internet, expresses a composite of economic and social choices – getting out of the city to find more space at a lower cost, spending more time with family, commuting less, and securing a different lifestyle in a potentially very different environment. For the purpose of understanding the mixed nature of housing markets, it is worth differentiating between **teleworking**, **lifestyle shifting** and urban flight supported by **commuting**, but the reality is that all these components are part of general counter-urbanisation by economically active households. Perhaps a key difference between the groups is that teleworkers and lifestyle shifters tend to be more footloose and will head to a greater range of locations. Commuters, on the other hand, are frequently more reliant on good transport infrastructure. They choose rural locations close to bigger market towns that have 'parkway' stations from which they can start their daily trip to urban jobs. Gallent and Robinson's (2012a) study of rural communities engaged in parish planning in Kent drew a picture of affluent commuter villages circling the town of Ashford, with its High Speed train link to London (36 minutes). More than 50,000 residents exit the borough in the mornings, heading to jobs either in London or other major centres (Gallent and Robinson, 2012a: 59). These sorts of commuter areas can be thought of as an extension 'of the wealthy suburbs of the south east into the surrounding rural areas' (Lowe and Ward, 2007; see Chapter 3).

Broadly, counter-urbanisation has brought a mix of old and younger people to the British countryside, with and without families, economically

Figure 8.2

House in Kentish vernacular style, Appledore

active or retired, seasonal and permanent. As Newby observed in 1979, these groups have been drawn to both near urban and remoter rural areas, to new developments and to older village housing. They are motivated by lifestyle factors, by the perceived qualities of rural places, by investment opportunity, and by relative housing costs. Their arrival has spearheaded a fundamental social-reconfiguration of rural spaces, transforming what were once stagnant housing markets, and presenting land-use planning with a challenge that has become *the* litmus test of the effectiveness of Britain's planning systems. We look now at land-use planning's general response to these demand pressures before examining targeted interventions, intersecting this discussion with a brief examination of the changing politics of development in the countryside.

Part 2b: Planning for housing in rural areas

Key settlements and land allocations

The 1947 Town and Country Planning Act, introduced in Chapters 1 and 2, was part of a broader post-war framework for the comprehensive planning of future development. It was grounded (i) partly in the 1940 Barlow Report, which set out the need for national control over strategic and local development; (ii) partly in the 1941 Uthwatt Report, concerned with nationalising development rights and capturing the uplift in land values on the grant of planning permission; (iii) partly in the 1942 Scott Report on land utilisation in rural areas and the special privileges given to farming; and (iv) partly in the later 1945 Dower and 1947 Hobhouse Reports, which paved the way for the creation of National Parks (see Chapter 10). The broader framework, which included the 1947 Agriculture Act, sought to protect land for farming by restricting development in the open countryside and by handing local authorities *discretionary* powers over decisions affecting the extent and siting of that same development. Elected County and County Borough Councils became gatekeepers, able to decide where development should happen in accordance with their own plans. This system operated until 1971 when an incoming Conservative Government created a two-tier system of development and structure plans. The subsequent 1972 Local Government Act tasked County Councils with preparing the latter (dealing with strategic priorities rather than local detail) and dropped development plan responsibilities down to the level of district and borough authorities.

Right from the very beginning 'land utilisation in rural areas' was to be restricted, with agricultural land protected from urban encroachment. Some areas, including National Parks, were given specific 'amenity' protection, irrespective of the quality of agricultural land. The restriction of rural development is orchestrated at the levels of strategic and development planning. At the *strategic* level, County Councils were, by the 1970s, distinguishing between 'key' and 'non-key settlements' (Cloke, 1979), using strategic powers to steer the provision of rural services, including *public* housing, to service centres. There was a public-sector rationale behind this: cluster the services and achieve economies of scale (Sillince, 1986: 176) and the advantages of agglomeration (Chapter 3). In order to strengthen the market for services at key settlements (and counter the thinning and stretching of service markets), a ring-fencing of non-key settlements occurred, affecting *private* development. Many County Councils adopted settlement hierarchies for planning purposes: most service investment would go to the higher tier; a little, but not much, to tier two; and absolutely nothing elsewhere. In some areas, the result was the containment and protection of gentrified villages within their existing development envelopes, and an accentuation of social change. This was the case in the 1984 Warwickshire Structure Plan (Sillince, 1986: 181). But in declining and depopulating rural areas, lower tier settlements in need (and wanting) further development were deprived of it, and key settlement policy in those cases amounted to settlement rationalisation (Cloke, 1979).

The directing of development away from small village locations was also a feature of local *development* planning from the 1970s. County Structure Plans remained in place until 2004, when they were replaced – as the main strategic tier – by Regional Spatial Strategies (see Chapter 2). At that point, elements of key settlement policy were 'saved' in some local plans, but these same plans had always been concerned with the *allocation of land* for housing in a way that respected political sensitivities and contributed to some kind of spatial rationale. They also, from the 1980s, adhered to a presumption against development in the open countryside set out in planning policy guidance.

Land allocation for housing is a primary responsibility of local planning. It has technical and political dimensions. On the *technical* side, Strategic Housing Land Availability Assessments (SHLAA) involve officers of a local authority sieving through a long list of possible sites in order to identify those with clear housing potential. This is done in a systematic, evidence-based way, taking account of primary (physical) and ownership constraint. On the *political* side, there is a parallel process of consultation, lobbying and decision-making in council chambers, which may see sites included or excluded from allocation in a local plan. The technical process provides officers with a basis on which to offer advice to politicians, but it is the politicians who ultimately take decisions.

When development 'issues and options' are brought to Council Members, as part of the plan-making process, the technical SHLAA becomes political. It is also at this point that the public have been traditionally consulted on the different options for future development, and politicians come to understand more clearly the sensitivities that their decisions need to reflect if they are to remain in office (this has changed recently, as we shall see below in the case of Thame). At this stage in the local planning process, options for a broad 'spatial development strategy' are presented, typically ranging from concentration in one or two larger centres to a more even spread of development across many locations. There are often variant options too, linking to specific local opportunities: for example, a strategy to locate development along key transport routes or to direct it to a well-known brownfield site. Stratford-upon-Avon District Council reached this stage in 2007 and consulted the public on six different options. Of a total population of 28,000 people, 164 responded to the consultation. These favoured a dispersal of development to smaller towns and service villages (i.e. key settlements), though there was also support for additional new homes on the edge of Stratford itself (Stratford-upon-Avon District Council, 2010). Rural discourse often favours the status quo: concentration and the avoidance of smaller villages. There is seldom an appetite for new settlements in rural areas, unless this option provides long-term protection for smaller villages. In Stratford, the consultation (and local protest) illustrated a weight of feeling against an 'eco-town' six miles from the main town on a former airfield site.

Planning and general housing supply

Supply pressures, affecting small village locations, 'reinforce the inequitable consequences of a free-market allocation and lead to socially regressive distributional consequences' (Shucksmith, 1981: 13). However, it is wrong to view the 'regressive distributional consequences' in villages as a simple outcome of land scarcity. Whilst planning debate in the 2000s focused on the need to release land on the basis of market signals (that is, respond to evidence of declining affordability), with criticism levelled at the way in which local politics – rather than planning *per se* – may inhibit development for reasons of NIMBY-inspired redlining (Barker, 2004: 33), this debate never addressed the specific circumstances of gentrified villages. Kate Barker's *Review of Housing Supply* (in England) led to a revision of planning policy for housing (then PPS3; DCLG, 2006b), which required authorities to maintain an 'on-demand' supply of land above their core allocation – in case prices spiked and more land was needed. This same mechanism exists within the NPPF (DCLG, 2012): it is broadly accepted that, at a strategic level, not having enough land available to maintain housing supply will affect local prices and threaten housing affordability and access. But in neither 2006 nor 2012 was the issue of overall land allocations and supply billed as a solution to the particular problems in small village locations.

A year after the Barker Review, the government's own assessment concluded that in a given market, 'large increases' in construction will have a significant impact on affordability (that is, on the percentage of households able to afford rents or secure a mortgage), but increases in 'construction have to be *large*' (ODPM, 2005: 48) – and such large increases cannot be matched against defined price spikes (in, for example, a specific village). Chasing price signals and affordability targets is something that can only be done strategically, and big supply increases are not going to happen in and around villages. No one wants it to happen and the planning system is stacked against it.

Figure 8.3

1990s housing, edge of Shadoxhurst, Kent

Meen (2005, p. 968) further suggested that supply has an indirect effect on prices: as the stock of dwellings builds up, prices slowly adjust (displaying elasticity in relation to supply). This presumes that levels of demand remain unchanged during the intervention (construction), yet over-supply will affect the desirability of a location. It is likely that NIMBY opposition to new housing, in small villages, is fuelled not by a fear of general price adjustment but by the more severe 'amenity' effects of new development, rendering their village less attractive – to themselves and prospective buyers.

The reality in many rural areas is that development is directed away from villages, to either key settlements or market towns (Cameron and Shucksmith, 2007). Even where more land is opened up for housing, this has no effect on the most sought-after villages. Gallent (2009: 276) has previously argued that 'unless smaller village locations are earmarked for and accept a share of new allocations [post 2006, and now post the 2012 framework], this strategic approach may be seen as a missed opportunity for rural areas'. However, whilst it is true that some land needs to be made available for development in villages, this will command a premium price and quickly be swallowed up by the market. No affordable housing will be built on the site unless there is a *targeted* intervention. Market mechanisms do not provide a solution for the most intractable rural housing challenges faced by gentrified villages. Achieving broader access will require the delivery of non-market housing.

Non-market housing: the 'right to buy'

However, in these same locations the only source of non-market housing – council homes built in the interwar and immediate post war periods – was sold to sitting tenants after 1980, through the Right to Buy, and has since been sold on to incomers often at spectacular profit. In the 35 years since the Housing Act 1980, the stock of council housing in rural areas has dwindled. Jones and Murie (2006) provide an analysis of the impact of the Right to Buy, suggesting that its effect in rural areas has been magnified by a general insufficiency in housing supply. They argue that '[...] incomers buying former council housing and other housing for second homes are an important force especially in tourist areas or areas within commuting reach of large employment centres' (Jones and Murie, 2006: 93) and that, more generally, the loss of council housing in areas of planning constraint (limiting the supply of new homes) is problematic. Using a study of South Ayrshire in Scotland, Jones and Murie show that as the amount of council housing in villages reduces, applicants for this housing spend longer on waiting lists and are less likely to secure a home from a local authority. It is also suggested that the Right to Buy has had a far greater effect in rural areas than in many urban areas where the quality of council provision (e.g. in pre-fabricated high-density and sometimes unattractive developments now blighted by social problems) was more variable. Indeed, '[...] in rural settlements where the supply of council housing is more constrained and the quality less variable the impact of the Right to Buy has been to reduce access to social rented housing for low income households' (Jones and Murie, 2006: 94). Nationally, nearly 1.66 million council homes were sold between 1979 and 2004 (Jones and Murie, 2006: 60), and the percentage of households renting from local authorities in the UK fell from 29.2 per cent in 1981 to 13.0 per cent in 2003 (Jones and Murie, 2006: 53). Data on the extent of rural sales were reported by the Affordable Rural Housing Commission (ARHC, 2006): whilst the cumulative percentage loss of urban council housing stood at just over 30 per cent (of the 1979/1980 stock) by 2004/05, the figure for rural areas exceeded 36 per cent (ARHC, 2006: 68). Fewer council homes were built in the countryside and more have been sold. And because the supply of new housing association homes has been outstripped by the rate of Right to Buy sales, the percentage of all social housing in rural districts (of England), fell by 24 per cent between 1980 and 2005 compared to a loss of 18 per cent in urban areas (ARHC, 2006: 68).

Whilst key settlements and market towns will get their share of land allocations, and may see a mix of new development, small village locations in many areas have been starved of new housing. General supply responses can help with 'outward affordability', but will not address 'inward affordability' stresses: they may calm prices across a market, but not in hot-spots. In many places, prices have risen and the villages have become socially exclusive. The socio-economic divisions between key and non-key settlements can be startling, perhaps nowhere more so than in areas washed over by green belt, where intolerance to development is greatest (see Chapter 10). There would seem to be two possible courses of action: first, do nothing and be content with the mix of housing available in service centres; or second, see socio-economic value in including a mix of households in all sorts of rural settlement, and therefore intervening in a targeted way.

The politics of development

Several references have already been made to the politics of development in rural areas. The populations of gentrified villages tend to be conservative, and this is reflected in the attitudes to development found in Parish Councils and in some rural local authorities. However, many communities are acutely aware that general supply responses do little to widen housing access. They are simply a magnet for external investment. For this reason, there is seldom an appetite for speculative building and, in some instances, there is an expressed preference for highly targeted development over which communities have a degree of control (Gallent and Robinson, 2012a, 2012b). The popular answer to the rural housing question posed in this chapter is thought to be one of targeted, community-led, intervention.

Targeted interventions: planning and affordable housing

In parallel with their policy of selling council homes to sitting tenants, the New Right governments of the 1980s believed that local authorities should facilitate the provision of affordable housing by other agencies and groups (a mix of private, third-sector and community actors), and that the way in which provision was taken forward should encourage innovation and cost-efficiency. A Housing White Paper in 1987 carried forward this idea:

> The future role of local authorities will essentially be a strategic one, identifying housing needs and demands, encouraging innovative methods of provision by other bodies to meet such needs, maximizing the use of private finance, and encouraging the new interest in the revival of the independent rented sector.
>
> (HM Government, 1987: 14)

The Housing Act of the following year handed local authorities an enabling role and curtailed grant funding to housing associations, encouraging them to find new ways of funding affordable housing schemes. The most important mechanism to emerge from this context was the procurement of affordable housing through the planning system – putting policies in local plans that required 'developer contributions' in the form of a number or percentage of affordable housing as the price payable for planning permission. This approach has meant that affordable housing has been concentrated in areas of general supply and where land values are greatest, and hence where the value of development makes it possible for a 'contribution' to be delivered without undermining development viability (Newhaven Research, 2008: 3).

'Planning and affordable housing' mechanisms represent an evolution of the idea that the benefits from the uplift in land values, produced by the granting of planning permission, should be captured for wider community benefit (Satsangi *et al.*, 2010: 127). Land on which there is permission to build housing is far more valuable than agricultural land (Evans, 1991: 854) and the idea that value uplift should be taxed or otherwise captured originates in the 1941 Uthwatt Report. Several consequent mechanisms were brought forward for achieving this goal, but none proved workable and all had a dampening effect on economic growth. The last attempt to capture uplift in a general way, through a national approach, was scrapped in 1979. Yet the goal of developing a *local* approach, that would be sensitive to market conditions, remained; and an opportunity for achieving that goal was created through local planning. The 1971 Town and Country Planning Act in England and Wales allowed authorities to enter into 'agreements' with applicants, imposing 'obligations' on developers and requiring them to deliver planning gains in return for planning permission: the gain was treated as a local levy on land value uplift and negotiated in the context of local market conditions, using discretionary powers. Similar powers were created in Scotland through the 1972 Town and Country Planning (Scotland) Act. A few authorities, during the 1970s and 1980s, used this approach to procure affordable housing, but developers were resistant to such sizeable gains fearing that they could become a standard requirement. It was only in the context of the 1988 Housing Act, and the stipulation that the need for affordable housing should indeed be a 'material planning consideration' (set out in government circulars on planning and affordable housing in 1991 and then in new planning policy guidance on housing in 1992), that this approach entered general use.

It works in the following way. 'Local needs policies' are set out in local plans, backed up by evidence (usually provided by consultants) that an area has an affordability problem and that a portion of future housing should be for social rent or discounted purchase (shared equity). Development control decisions need to be in conformity with that adopted plan, and hence planning officers will negotiate for affordable housing contributions within proposed market housing schemes. The policies contained in the plans become one of a number of public tests that have to be passed before development can proceed. Developers will factor the cost of meeting that test (that is, providing the affordable homes) into the price paid for land. In practice, the cost of affordable housing is shared between landowners (who accept a lower sale price) and developers (who accept a narrower profit margin). In situations where the market for housing is strong, sites will have a greater capacity to absorb planning gains without adversely affecting viability; and expected profits, for landowners and developers, will provide enough of an incentive for development to proceed. But in situations where the market is weaker, affordable housing requirements may de-incentivise development or threaten viability. For that reason, agreements drawn up in a strong market may be renegotiated if that market dips before development goes on-site.

Agreements to provide affordable housing as a planning contribution are now based on Section 106 of the 1990 Town and Country Planning Act in England and Wales or Section 75 of the 1997 Town and Country Planning (Scotland) Act. Depending on market conditions – and evidence of need – contributions typically ranged from 20 to 50 per cent of all units in England (Crook *et al.*, 2006: 354) and between 5 and 25 per cent in Scotland (Newhaven Research, 2008: 27–30) before the global financial crisis that began in 2008. The mechanism provided a key funding stream for affordable housing, usually then managed by housing associations or other 'registered providers' of social housing. Roughly half of all affordable housing in Britain has been funded in this way over the last decade. However, because this approach relies on land being opened up for development by private enterprise, its contribution in rural areas is confined to locations receiving new market development: market towns or other service centres. If that is where new housing is directed, it will also be where new affordable housing through Sections 106 or 75 is procured. It is not the type of targeted approach that would potentially widen housing access in small village locations.

Targeted intervention: rural planning exceptions

Planning obligations demonstrate how land-use planning can be used to intervene in market processes and extract community benefit. A proportion of *potential profit* is diverted and the planning system becomes a tool for widening housing access. This also happens in the more targeted case of 'rural planning exceptions'. For all of the reasons listed earlier in the chapter, development in rural areas is not generally permitted on agricultural land outside of village envelopes. This means that land adjoining villages has an agricultural use value, which is far lower than housing use value. In market hot-spots, the land cost element of building houses is likely to be a large part of total development costs. If that land element could be removed (or significantly reduced) and if the development were undertaken by a not-for-profit provider (removing the profit margin), then the total cost of building new homes, to be protected in perpetuity through occupancy controls or managed and let by an housing association or a community trust, would be reduced. The homes would be affordable and accessible to lower-income groups.

This is how the rural planning exceptions approach works (see Gallent and Bell, 2000). The framework for its operation was first set out in England and Wales in 1991; an equivalent in Scotland was put in place through Planning Advice Note 74 (see Shiel *et al.*, 2007), although there appears in Scotland to be limited enthusiasm for policy approaches, such as exceptions, that are '[...] unlikely to provide an answer to the scale of the site availability problem which has become evident, and are not particularly compatible with the preference for a plan-led system and a clear policy framework on houses in the countryside' (Shiel *et al.*, 2007: 46). These same criticisms have been levelled at exceptions in England and Wales for 25 years. They circumvent development planning and make only limited contributions to housing need. However, they can make important contributions where they are most needed, in those difficult-to-reach small village locations. Exceptions schemes seldom comprise more than a dozen or so new homes, meaning that if all rural authorities in Britain progressed a handful of such schemes each year, there would probably be fewer than a couple of thousand homes procured in this way. But they could well be in market hot-spots: villages in the lower part of the settlement hierarchy, which have been redlined for several decades and where counter-urbanisation pressures are strong. These sorts of targeted intervention create opportunities for housing access where none previously existed, distinguishing them from general market and supply responses, which are confined to bigger service centres.

Community-based housing solutions

Rural exceptions have, in the past, been taken forward by a partnership comprising community councils, a landowner (willing to release land at only slightly above agricultural value), the local authority, a housing association and a 'rural enabler', whose task it was to broker the exceptions deal between the different partners. Communities have played a significant role in exceptions schemes, with development only progressing where there was broad community support. It is now generally recognised that communities have a key role to play in addressing their own needs. Land-use planning is able to support community initiative, as in the case of exceptions, but it is in mixed communities that agreement has to be reached on the need to progress schemes that broaden access to housing and deal with specific local needs. General market approaches are facilitated by the release of land for housing in bigger population centres. It is vital that enough land is released in these locations as they play an important role in rural economies and are home to much of the rural population. However, community-based action and planning, supported by land-use planning, is the key to negotiating the social complexity and anxieties of small village locations. Recent government reviews and initiatives – including the Affordable Rural Housing Commission in 2006 and the Market Towns Initiative – have tended to focus on strategic rural sites. The slightly more recent Taylor Review (2008) – undertaken by the then Member of Parliament for Truro and St Austell in Cornwall – placed more of an accent on villages, arguing for a new accommodation between the needs of rural communities, the environment and economic wellbeing. Sustainable rural communities enjoy the wider socio-economic benefits of more equitable housing access.

At the same time as the Taylor Review, there seemed to be heightened interest in 'Community Land Trusts' (CLTs) as a model for addressing the needs of smaller rural communities. Rodgers (2007) describes these as not-for-profit organisations, initiated and controlled by community groups, whose primary purpose is to hold 'land or other assets so as to promote the social, economic and environmental sustainability of a specified geographical community'. They have an extensive history in urban areas, where 'development trusts' (see Bailey, 1990) have sometimes been established in areas of economic decline, taking ownership and control of land and buildings with the aim of bringing various benefits to their community. Where they exist in rural areas, they represent a desire for self-determination and self-help, signalling also the inadequacy of public-sector responses to particular needs, including the need for accessible housing or soft community infrastructure (Moseley, 2000). Whilst communities have been active in exceptions schemes, they have often been critical of the way 'social housing' has been allocated (Gallent and Robinson, 2012b) and have argued for more direct control over defining need and providing homes. They are able to do so when land or assets (including buildings) are gifted to, or purchased by, the community and when land-use planning supports a community's re-use of that land or asset.

This famously happened on the Isle of Gigha in Scotland (Satsangi, 2014). In that case, land reform in 2000 gave community groups the opportunity to purchase land from private owners. The island and its built assets were put up for sale in 2001. A Community Land Steering Group was set up, supported by the Scottish Land Fund, and purchased the island a year later for £4 million using public sector loans and grants (Satsangi, 2014: 124). The Steering Group became a Trust and within three years of acquisition sold six plots of land to local families on which to build new homes and entered into an agreement with a housing association to build eighteen affordable homes for local need on the island (Satsangi, 2014: 124). Back in England, discussion focused, prior to 2010, on how resources might be directed to villages so as to enable community acquisition. The obvious mechanism seemed to be Section 106: cash contributions from developers *in lieu* of on-site contributions might be channelled to CLTs. The Community Infrastructure Levy (CIL) might be used in the same way. There appeared to be no legal impediment to such transfers, and the Quirk Review (DCLG, 2007) into the 'community management and ownership of assets', underscored the wider community benefits accruing from asset-building and ownership (DCLG, 2007: 10), where those assets are retained in perpetuity for community use. However, a Land Trust would encounter the same price barriers as any private individual trying to access housing in a gentrified village. It would need to acquire or borrow substantial sums of money to purchase assets and then recover its costs through market rents. Without some sort of philanthropic or grant support, the Trust would not be able to deliver community benefits. Alternatively, these community efforts need to be supported by land-use planning, making 'community exceptions' to the re-use of buildings or land for housing, thereby reducing costs in the same way that the regular exceptions approach has since 1991.

In the late 2000s, there was evidence however that some local authorities were reticent about working with Trusts because of their lack of a proven track-record and unclear governance arrangements (Gallent, 2009). In Devon, the *County Council* entered into an experimental partnership with the Community Land Trust at High Bickington, offering to transfer land at substantially reduced cost to the community thereby enabling it to provide a mix of low-cost housing and work units. However, the *planning authority* rejected the scheme, ostensibly because it was too large, but also because the proposal – including workshops – was not what they had expected from the site. The High Bickington example perhaps expressed a broader lack of trust in CLTs, and more generally in communities, to take control and guide the development of assets (Moran, 2007). However, much has changed in the last five years and much of what the CLT model was trying to achieve – social entrepreneurialism, community control and responsibility – has now been mainstreamed within the Neighbourhood Planning framework in England.

Neighbourhood planning and the community focus

The arrival of 'neighbourhood planning' in England was noted in Chapter 1 and outlined in greater detail in Chapter 2. It was described in Chapter 5 as a practical expression of the broader governance shift that has, in recent decades, meant a rescaling of power and responsibility in order to respond to social complexity, in rural areas and elsewhere. Politically and practically, the Neighbourhood Planning agenda in England addresses three critical challenges. The *first* is a perceived 'democratic deficit', or gap between the traditional structures of authority and the level at which actual problems are experienced and addressed. The *second*, closely related to the first, are the general challenges of development planning and particular issues around the selection and allocation of sites for housing. And the *third*, also closely related to the first, is the extent to which communities can own local solutions, tailoring them to problems defined by a community. Put simply, Neighbourhood Planning deals with issues of democratic renewal, tries to make strategic planning more acceptable, and provides a supportive context for the generation of social goods through communitarianism.

In relation to the first, the Conservative–Liberal Democrat Coalition has presented local planning conflict (centred on housing development) not as a reaction to development *per se*, but rather a natural response to the manner in which decisions have been traditionally reached, with government side-stepping local people, and presenting sometimes enormous and fundamental changes as *fait accompli*. NIMBYism is a measure of the 'democratic deficit' (Sturzaker, 2011) in local decision-making rather than, as others would have it, evidence of vested private interest, resisting development for personal and selfish reasons. Indeed, government has pointed to the opportunity of trans-forming a generation of NIMBYs into much more welcoming IMBYs (In My

Back Yard [please]) through democratic renewal at a local and neighbourhood level, replacing the 'big state' – which flourished under previous governments – with a 'big society'. Such claims have been greeted with some scepticism, as it is doubtful that all NIMBY sentiment is rooted in democratic deficit. Opposition to development has multiple causes, and the local environment for planning is often 'messy' and complex, without necessarily being parochial.

In relation to the second, the goal of democratic renewal feeds into attempts to alter general planning orthodoxies. Earlier in this chapter, it was noted that in the past, land allocations have been guided by a technical process that sieves for possible housing sites before a political process kicks in that requires *consultation* on fixed options. Communities tended to be consulted on professional advice, and had little sway over local detail. Neighbourhood Development Plans (NDP) give communities much more direct control over the siting of housing, although the amount of housing that needs to be planned for is set within a district or borough plan. One recent example of community control over general allocations can be found at Thame in the District of South Oxfordshire. Thame Town Council was faced, in 2010, with a Core Strategy from the District Council that proposed that 600 new homes should be built on a single site on the edge of the town (of a total allocation of 775 homes over 15 years). In the same year, the Town Council approached the District to express its desire to become one of England's first Neighbourhood Planning frontrunners. This was accepted and the town became an earlier recipient of a grant to produce a neighbourhood development plan. In the summer of 2011, the process of putting together the plan began and the town engaged the assistance of private consultants. The consultants worked towards the production of the residents' vision. Initially, the plan for a single site with 600 homes was opposed. Eventually, the residents' identified enough land for 2,000 homes in and around the town; seven sites were eventually designated in the Neighbourhood Development Plan, for 775 homes. The plan went to referendum, was approved by 75 per cent of voters, and went before the District Council in the summer of 2013 (Cook, 2013). The Town Council's website notes that originally an allocation was 'decreed', that 'protests fell on deaf ears' but the Localism Act gifted an 'opportunity for self-determination' (www.thametowncouncil.gov.uk). Other neighbourhood groups have been reviewing the Thame experience and it may come to pass that site allocations become a key feature of future Neighbourhood Development Plans.

Lastly, and in terms of delivering community-based solutions, the Upper Eden Neighbourhood Development Plan provides a good example of how communities are seeking to tailor the planning system to their own needs. The plan has a strong focus on housing, containing policies that '[...] are aimed at making sure that the opportunities that exist for local people to build to solve their own housing problems are positively supported through the planning process' (Upper Eden Community Interest Company, 2012: 6).

Four key policies (that amend policies contained in Eden District's local Plan) are set out. The first alters the approach to rural exceptions. The District's Plan stipulated that exceptional permission for affordable homes could only be given on sites close to existing dwellings. This had limited the number of suitable sites in small villages. Upper Eden's NDP relaxes this rule, saying that site suitability will be judged only on visual impact. Moreover, the possibility of converting outbuildings to residential use is opened up by the same policy – but again, only for local needs. The second policy allows new housing on farms, for use by 'family members, holiday letting or renting to local people' (Upper Eden Community Interest Company, 2012: 16). In supporting the policy, it is argued that Eden Council's own plan is 'silent' on the 'flexible use' of farm dwellings, and it is the intention of the NDP to create a framework in which the farming economy, and those reliant on it, will be supported. The third policy is concerned with older residents, requiring that affordable homes in certain named villages are reserved for local older people and designed appropriately. The fourth policy addresses a key concern in rural settlement planning, acknowledging that non-key settlements can be starved of vital development. The NDP notes that Eden District's Core Strategy contains an LSC (Local Service Centre) de-designation policy: every two years, the local authority will review its settlement hierarchy and can remove LSCs on the basis of their judged sustainability. The NDP contends that top-down assessments of sustainability are flawed and, in opposition to the Core Strategy, allows for the development of single dwellings for local need (on exception sites) in de-designated service centres. In terms of defining 'local need', all the standard socio-economic criteria apply (i.e. length of occupancy and employment need) but discretion is also handed to Parish Councils, who can provide 'written support' for an individual's claim to be 'local' – raising the spectre, perhaps, of a much more personalised approach to housing allocations (Gallent and Robinson, 2012b).

Neighbourhood Planning in England may appear to address some of the tensions surrounding rural housing discussed in this chapter: broader land allocations may become more acceptable, reducing the conflicts and delays that sometimes slow housing supply (see Barker, 2004); and local frameworks for land-use planning may be altered in ways that make it easier to deliver targeted, non-market, housing in small village locations. Whether or not these measures are sufficient to fully answer the rural housing question – and can deal with the rampant gentrification that afflicts many parts of the countryside – is not yet known, but they at least provide some glimmer of hope at the end of what has been a very long tunnel.

Summary

- Housing challenges in rural areas are associated either with rural decline and depopulation or with counter-urbanisation and gentrification. These processes present a number of problems for rural communities, requiring either a general development and investment response or targeted assistance with the aim of widening housing access.

- This chapter has focused on counter-urbanisation and gentrification, arguing that addressing 'housing need' in small villages does not benefit only individuals, but has a much broader socio-economic purpose. It ensures the sustainability of those communities and the vitality of rural economies.
- The key problem facing many rural areas is the 'inward affordability' of some gentrified villages. These have become centres of concentrated wealth from which the 'rural poor', or others on modest incomes, are excluded.
- Land-use planning has been primarily concerned with 'outward affordability'; with the general supply of housing at a district or borough level. At the same time, strategic planning has tended to redline smaller villages, concentrating services in key settlements and directing housing growth to those same locations.
- Land allocations for housing are often contentious, and there have arguably been insufficient efforts to widen housing access in villages.
- A targeted approach to rural housing provision seems vital, with community-based planning perhaps the appropriate scale and vehicle for addressing very local needs. The Neighbourhood Planning approach in England seems to be providing useful opportunities for community groups to take control of local planning. It parallels efforts elsewhere, notably in Scotland, to give communities a greater power over the use of land for local benefit.

Key readings

- Satsangi, M., Gallent, N. and Bevan, M. (2010) *The Rural Housing Question: Communities and Planning in Britain's Countrysides*, Policy Press: Bristol.
  (A comprehensive account of rural housing challenges across Britain, looking at all aspects of demographic change, migration, planning and housing policy.)
- Taylor, M. (2008) *Living Working Countryside: The Taylor Review of Rural Economy and Affordable Housing*, DCLG: London.
  (A government commissioned review of the housing–economy–society linkages in rural England.)
- Shucksmith, M. (1981) *No Homes for Locals?* Gower Publishing: Aldershot.
  (Old but seminal, setting out the essence of the rural housing access problems in Britain, drawing on research in the English Lake District.)
- Marcouiller, D., Lapping, M. and Faruseth, O. (eds) (2011) *Rural Housing, Exurbanisation, and Amenity-Driven Development: Contrasting the 'Haves' and the 'Have Nots'*, Ashgate: Farnham.
  (Broad coverage of key topics including counter-urbanisation and the transformation of housing markets, but focused on the US.)

Key websites

- English Rural Housing Association: www.englishrural.org.uk.
  (A leading 'researcher and developer' of rural housing: its site also has links to a broader array of housing organisations working in rural areas.)
- Rural Housing Trust: www.ruralhousing.org.uk.
  (A housing company specializing in rural shared ownership schemes, providing many examples of these on its website.)

Part 4

Environment and landscape

··

9 Environment and landscape

A changing relationship with the 'natural' world

Earlier chapters in this book have charted how views on the countryside have changed over the last one hundred years or so and have been reflected in changing patterns of land use and development in rural areas and approaches to rural planning. For example, many rural areas, particularly in contemporary Britain, are no longer seen as being focused on agricultural production but are increasingly viewed as areas of consumption in which the quality of the environment provides an attractive setting for leisure and tourism and for living (see Chapters 2 and 4). These shifts reflect our changing relationship with the 'natural' world, which is often considered as the key feature of rural areas. They illustrate the differing and often conflicting environmental perspectives that lie at the centre of some of the most hotly contested rural planning debates. As in other spheres that have been considered so far, it is apparent that these perspectives are constantly evolving in response to both local and global forces and that the early twenty-first century is a time of particularly intense change.

This chapter starts by illustrating how the economic, social, cultural and political forces discussed in the earlier chapters interact with the natural environment to shape the rural 'landscape' and wider 'environment' we know today. Key contemporary challenges for environmental policy and planning arising from landscape and environmental change will then be distilled. The second part of the chapter will examine rural landscape and environmental change in Britain and illustrate how local and global challenges are manifesting themselves here. The chapter sets the scene for a more detailed examination of environmental policy and planning responses in Chapter 10.

Part 1: Landscape and environment: key concepts and challenges

Humans and the natural world

It is important at the outset to explore what is meant by the terms environment and landscape and human relationships with them. There is perhaps a popular conception that the countryside represents the natural environment, the antithesis of the worldly or 'human' environment of towns and cities. However, as with other widely held views of rurality, this understanding does not live up to scrutiny, for there are few places in the world today that can truly be described as natural, or completely devoid of human influence. Over the past century there has been a growing understanding that man is an integral part of the natural world and is a major force in shaping environmental change at both a local and a global scale.

At a local scale, it is now appreciated that although underlying natural structure and natural processes do significantly determine the diverse character of the rural environment, the countryside we see today is perhaps more accurately described as landscape. Zonnefeld defines landscape as a 'part of the Earth's surface, consisting of a complex of systems formed by the activities of rock, water, plants, animals and man and that by its physiognomy forms a recognisable entity' (Zonnefeld, 1990: 55). Put more simply, 'landscape = habitat + man'. The determinants of landscape character are summarised in Box 9.1 and include the physical backdrop of geology, topography, climate and natural processes that determine the underlying form of the landscape, and the soils and natural flora and fauna of places. Overlying this is the accretion of millennia of human activity, which Meinig describes as 'the symbolic expressions of cultural values, social behaviours and individual actions worked upon particular localities over time' (Meinig, 1979: 6). Rural areas have traditionally exhibited very evident 'time depth' in their landscapes and this historic 'patina' reflecting past cultures has frequently been a key element in their human appeal. For example this is reflected in UNESCO's inclusion of cultural landscapes as a category meriting World Heritage Site designation. Beyond physical characteristics, increasing recognition is being given to the variability of personal perceptions or experience of landscape: the individual or culturally determined appreciation of place that influences different peoples' responses to and use of different landscapes (Gray, 2003; Moore-Colyer and Scott, 2005). These perceptions can change over time and vary quite markedly between people. For example prior to the Romantic Movement and its idealised notions of 'natural history' and 'natural beauty', much of society tended to view the wild and untamed countryside with fear. Today it is likely that a farmer will view the landscape in a very different way from a natural scientist or urban residents who often retain Romantic sensibilities and hold dear an image of

Box 9.1 What is landscape? (Derived from Countryside Agency, 2006: 5 and
Moore-Colyer and Scott, 2005: 508)

Experience (visual and sensory)	How people perceive the landscape influences how it is used or valued.
Land use	The present day pattern of land use, such as settlement, farming, energy production and forestry.
History/culture/archaeology	Landscapes have also been shaped by past patterns of human activity.
Wildlife (biodiversity)	The types and abundance of plants and animals, determined by the physical backdrop of the natural environment and by economic and social factors.
Natural form	Geology, landform, river and drainage systems and soils shape the land and its 'usefulness' for agriculture and other human functions.

rural areas that is far from historic or present reality and who can find the current pace and direction of landscape change in the countryside extremely challenging.

Human activity has not only influenced the diverse visual and cultural character and appeal of rural landscapes but also the underlying ecological dynamics that shape their development over time. Since Alfred George Tansley first coined the term ecosystems in 1935 (Tansley, 1935: 299), the close interaction between man and ecological change has been the focus of much scientific enquiry. Wider public appreciation of the connections grew during the 1960s following, for example, the publication of *Silent Spring*, Rachel Carson's emotive account of the impact of modern farming methods on the ecology of the American countryside (Carson, 1962). The book's publication is often presented as a milestone in the development of the modern environmental movement. More recently, developing ecological understanding, including increasing acceptance of many aspects of James Lovelock's *Gaia Theory* (Lovelock, 1979), has highlighted the wider significance of anthropogenic forces in shaping global environmental change. Earth System Science – which envisages the Earth as an integrated ecological system in which local and global forces interact (Steffen *et al.*, 2004) – has emerged as an important field of trans-disciplinary research and has led to calls for the designation of a new geological epoch – the 'anthropocene' – which started around 200 years ago with the development of industrialisation. Some of the key concepts and concerns of Earth System Science were captured in the 2001 Amsterdam Declaration, which summarised the conclusions of four major global scientific research programmes – the International Geosphere-Biosphere Programme (IGBP), the International Human Dimensions Programme on Global Environmental Change (IHDP), the World Climate Research Programme (WCRP) and the international biodiversity programme DIVERSITAS (See Box 9.2). Some of the data upon which the Declaration was based is shown in Figure 9.1. Although these conclusions are subject to much ongoing debate and critical scrutiny, they are now accepted by many international agencies and national governments and are important influences on the direction of rural planning activity. Some of the key environmental challenges that are the focus of current attention are introduced below.

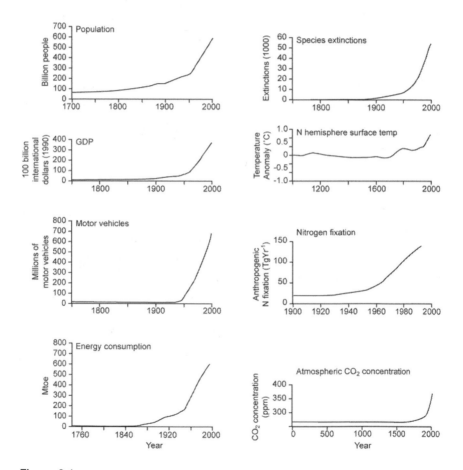

Figure 9.1

Examples of data that informed the Amsterdam Declaration on Earth System Science
Source: Steffen *et al.*, 2004

Box 9.2 Extracts from the Amsterdam Declaration on Earth System Science (IGBP, 2001)

Research carried out over the past decade has shown that:

- The Earth System behaves as a single, self-regulating system comprising physical, chemical, biological and human components. The interactions and feedbacks between the component parts are complex and exhibit multi-scale temporal and spatial variability. The understanding of the natural dynamics of the Earth System has advanced greatly in recent years and provides a sound basis for evaluating the effects and consequences of human-driven change.
- Human activities are significantly influencing Earth's environment in many ways in addition to greenhouse gas emissions and climate change. Anthropogenic changes to Earth's land surface, oceans, coasts and atmosphere and to biological diversity, the water cycle and biogeochemical cycles are clearly identifiable beyond natural variability. They are equal to some of the great forces of nature in their extent and impact. Many are accelerating. Global change is real and is happening *now*.
- Global change cannot be understood in terms of a simple cause-effect paradigm. Human-driven changes cause multiple effects that cascade through the Earth System in complex ways. These effects interact with each other and with local- and regional-scale changes in multidimensional patterns that are difficult to understand and even more difficult to predict. Surprises abound.
- Earth System dynamics are characterised by critical thresholds and abrupt changes. Human activities could inadvertently trigger such changes with severe consequences for Earth's environment and inhabitants. The Earth System has operated in different states over the last half million years, with abrupt transitions (a decade or less) sometimes occurring between them. Human activities have the potential to switch the Earth System to alternative modes of operation that may prove irreversible and less hospitable to humans and other life. The probability of a human-driven abrupt change in Earth's environment has yet to be quantified but is not negligible.
- In terms of some key environmental parameters, the Earth System has moved well outside the range of the natural variability exhibited over the last half million years at least. The nature of changes now occurring simultaneously in the Earth System, their magnitudes and rates of change are unprecedented. The Earth is currently operating in a non-analogue state.

The pace and direction of landscape change

The European Landscape Convention (ELC) notes

[...] that developments in agriculture, forestry, industrial and mineral production techniques and in regional planning, town planning, transport, infrastructure, tourism and recreation and, at a more general level, changes in the world economy are in many cases accelerating the transformation of landscapes.

(Council of Europe, 2000, preamble)

The Convention acknowledges the complex and interrelated drivers of landscape change (see Figure 9.2). It reflects a general, underlying concern over the current pace and direction of landscape modification. This concern is not simply focused on perceived threats to the character of highly valued cultural/heritage landscapes but also on what is considered an erosion of the quality of landscapes more generally, associated with contemporary social, economic and environmental processes. The ongoing and global nature of such concerns is reflected in the Florence Declaration on Landscape, which was signed in 2012 by a wide range of international agencies and experts who expressed 'deep concern about the degradation of landscapes worldwide due to industrialisation, urbanisation, intensification of agricultural processes and other threats and risks caused by global change' (UNESCO, 2012: 1). The Declaration was seen as a staging post towards the introduction of an international landscape convention that would draw attention, in particular, to

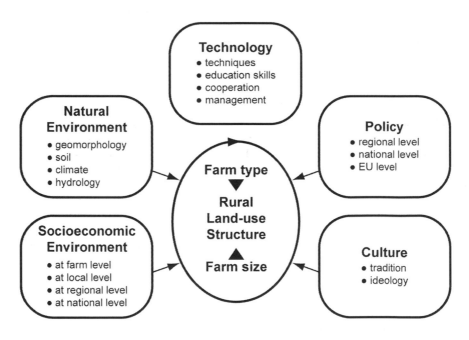

Figure 9.2

Driving forces of rural landscape change
Source: Brandt et al., 1999: 81

the social significance of landscapes in terms of cultural heritage and local identity, and also their role in supporting economic activity and job creation. It was envisaged that a new international landscape convention would follow in the footsteps of the ELC in encouraging action at all levels to promote landscape protection, improvement and enhancement and encourage a forward-looking and creative approach to future landscape change. The ELC advocates a holistic approach to landscape planning that includes natural science perspectives, but it has its roots in the arts and social sciences and therefore puts considerable weight on the aesthetics and cultural value of landscapes, aiming to bring together expert and public perspectives in the practice of landscape planning. The implications of these developments for rural planning are discussed in Chapter 10.

Habitat loss, degradation and fragmentation, and changes in biodiversity

As noted above, changes in the physical character of the countryside are mirrored by changes in its ecology, and research indicates that many of the driving forces that are shaping rural areas today (illustrated in Figure 9.2) are also having negative impacts from an ecological perspective. Prominent among ecological concerns is the loss, degradation and fragmentation of habitats and the consequences of this for biodiversity. For example, one of the most obvious features of rural landscape change is the increasing urbanisation of the countryside. This includes direct shifts from rural to urban land-uses such as housing and industry and more diffuse urban land-take associated with roads, railways and other forms of infrastructure such as water supply and waste disposal. Coupled with the development of more industrialised farming methods many countries have experienced a significant loss in the area of natural or high biodiversity value habitat, which has traditionally been a feature of rural areas. In addition, those areas that remain often face degradation associated with human activities, for example in the form of light, noise and chemical pollution as well as direct physical disturbance associated, for example, with leisure pursuits. To compound matters fragmentation of habitats is an increasing issue. As natural habitats shrink and become more distant from one another, the scope for movement and genetic exchange is reduced, thereby threatening the long-term viability of species and raising the risk of crossing 'tipping points' where a collapse in species numbers or a major readjustment in ecosystem dynamics could occur. Taken together, these factors are thought to account for extinction of many of the species highlighted in Box 9.3. It has attracted much global attention and led to the signing of the United Nations Convention on Biological Diversity (CBD) in 1992, which, as we shall see in Chapter 10, is now a key driver of environmental policy and practice in rural areas.

Box 9.3 Some key features of late-twentieth-century/early-twenty-first-century landscape change

Element	Nature of change	Nature/landscape impacts
Woodland Location/magnitude of cover. Extent of active management.	Woodland increased from a low of 5 per cent of the UK land area in the 1920s to over 12 per cent in 2014. Majority non-native coniferous planting, but growing proportion of native broad-leaved woodland in recent times. Increasing proportion of woodlands being certified as being sustainably managed, amounting to 44 per cent of all UK woodlands in 2014.	Coniferous plantations of less value to native wildlife and inconsistent with natural/historic landscape character.
Boundaries Extent and condition of hedgerows, walls, and ditches.	Rapid reduction in boundary features between 1950 and 1990 associated with farm mechanisation. Position now stabilised and possibly increasing.	Loss of cover and habitat for flora and fauna, contributing to species decline. Loss of historic landscape features.
Agriculture Extent of cultivated land area, changing patterns of arable/grassland farming, stocking density and number, size and type of farm.	In 2013, 70 per cent of UK land area still in active agricultural use. Long-term trend of increasing use of intensive production methods, reducing number of farms and increasing size of farm units. Since 1980s declining area under crops and growth of area of uncropped land and land covered by agri-environmental schemes.	Loss of cover and habitat from larger field sizes. Fertilizers/pesticides/time of harvesting linked to species loss, diffuse pollution, soil erosion, flooding. Together with modern farm buildings, alter historic landscapes.
Settlement and development Changing settlement size/structure (morphology). Changing pattern of land-use.	Population/household increase, expanding cities, towns and villages, New Towns, out-of-town shopping and business development, expanding road/motorway network and global communication infrastructure e.g. satellite masts and airports. Growing energy supply network – power stations, wind farms, and power lines. Mineral extraction supporting urban growth, increasing demand for waste disposal/management sites.	Physical, noise, light intrusion, fragmentation of countryside, reduced 'tranquil' areas results in loss of biodiversity, changes historic character, homogenises landscapes.

Element	Nature of change	Nature/landscape impacts
Semi-natural habitats Extent and condition of semi-natural habitats.	Increasing development/recreational pressure on many semi-natural landscapes. Pattern of damage and decline of many designated sites but some recovery of status evident for protected sites 2008–2012/13.	Reduced size of undisturbed 'core' areas reduces resilience of species and habitats. Reduced sense of 'wilderness'.
Historic features Changes to historic landscape features and buildings.	Decline in the 'status' of many traditional farm buildings either due to dereliction or 'poor' conversion. Loss of ponds associated with intensification of farming methods. Damage to some historic land features due to deep ploughing.	Loss of wetland and other habitats/species. Erosion/loss of local landscape character.
River/coastal management Location and extent of river and coastal engineering. Condition of river corridors.	Increasing use of hard defences in coastal and river management in post-war years, but recent shift to 'soft' solutions. Significant new development on flood plains. Gradual improvement in river water quality, but soil erosion and diffuse pollution (particularly associated with modern farming methods) is a new focus of concern. Invasive, non-native flora and fauna significant in many river corridors.	Loss of habitat/native species associated with hard engineering methods and invasive non-native species. Reduced value as wildlife corridors. Introduction of urban elements to rural landscapes.

Climate change

Of all contemporary environmental concerns climate change is arguably the most significant for both rural and urban areas. It has certainly received extensive attention in recent years leading to criticisms from some quarters that it is deflecting attention from other pressing environmental issues (Verissimo *et al.*, 2014). The core findings supporting concern about human-induced climate change – from Earth System Science and key datasets – are outlined in Boxes 9.2 and 9.3 above. The issue has spawned wide-ranging research, most notably by the Intergovernmental Panel on Climate Change (IPCC). This has included work developing understanding about the drivers and potential consequences of climate change, exploring mitigation activities that could help reduce human-induced global warming and adaption strategies that might help people to cope with the effects of climate change.

The conclusions to emerge from the IPCC's Fifth Assessment of Climate Change are more cautious than previous assessments and indicate that, by the end of the century, global warming of between 1.5°C and 2.0°C and a sea level rise of between 0.26 and 0.55 meters are likely. Rises considerably in excess of this are however possible if the Greenland Ice Sheet disappears (which would result in warming of more than 4.0°C and a 1-meter sea-level rise). The assessment also predicts that the contrast in precipitation between wet and dry regions and between wet and dry seasons will increase. Interestingly, the findings are particularly more cautious in predicting increased storminess and incidence of extreme events and emphasise that experience is likely to vary greatly between regions in this, and in other, respects (IPCC, 2013). The Fifth Assessment is helpful for those interested in rural planning as it involved a new examination of the implications of climate change for rural areas. This concludes that major impacts will be felt through changes in water supply and availability, food security and agricultural incomes. Increased incidence of drought is envisaged for some regions making agricultural production increasingly marginal. In addition, increased inter-annual climate variability is likely to make crop yields and agricultural incomes more uncertain and vulnerable to climate-related shocks. Coupled with rapidly rising crop demand associated with global population growth, all aspects of food security are potentially affected by climate change including food access and price stability. In addition it is envisaged that climate change will bring shifts in agricultural production of food and non-food crops. For example increases in energy supply from renewable resources and cultivation of biofuels are anticipated. These may bring new employment to rural areas but may also have negative impacts in terms of landscape change and competition for scarce land resources (IPCC, 2014).

Resource scarcity

Access to key resources such as food, timber, water and minerals that are obtained mainly from the rural environment has been a matter of central concern in countryside planning for generations. Previous chapters have charted the rise and fall of interest in this issue and have highlighted how this has mirrored for example shifts in technology, which have resulted in increased agricultural production, and variations in resource availability. Global dynamics have also been central to changing attitudes with shifting patterns of international trade and political relationships and the impact of war reflected in the increasing or decreasing weight of national resource security arguments in rural debates and approaches to environmental planning in rural areas. In the early twenty-first century there is no doubt that concerns about resource scarcity and national security are growing again as a result of the rapid rise in resource demand referred to above. Together with climate change, there is perhaps no other issue that illustrates more clearly how the fortunes of rural areas are so closely connected to global developments and the forces of globalisation and how perceptions of the natural world are evolving once again.

Current concerns about resource security are tied to global population growth. The United Nations 2012 projections estimate that the global population will grow from 7.3 billion in 2013 to 9.6 billion in 2050 and 10.9 billion in 2100. These figures compare to a global population of 2.5 billion in 1950 (United Nations, 2013). Running alongside population growth there is also anticipated to be a significant expansion in economic activity and income levels for many, and with growing affluence, further increases in resource demands are envisaged. The resource and wider environmental implications of these developments are self-evident and are captured in a very visible way in Figure 9.3. It is anticipated that this growth in demand will be felt across the spectrum of resources and there is particular concern about

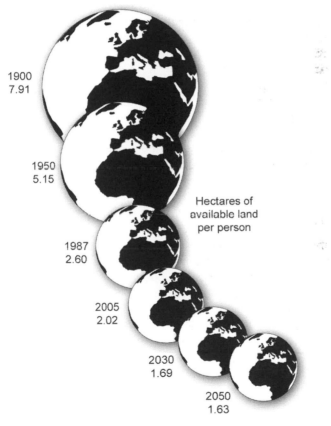

1900
7.91

1950
5.15

Hectares of available land per person

1987
2.60

2005
2.02

2030
1.69

2050
1.63

Figure 9.3

Our shrinking Earth
Source: UNEP, 2007: 367

water resources in many parts of the world. Of more general concern however is food and this has prompted fears about rising food costs and the ability to feed such a large population. Although the bulk of population growth and increased resource demands will be in developing countries, major impacts on global markets and commodity prices are envisaged affecting all parts of the world. Recent research by the Food and Agriculture Organisation of the United Nations (FAO, 2009) suggests that the required increase in food production can be achieved if the necessary investment is put into agricultural research and technological development and competition for alternative uses of rural land and production systems – for example, the use of agricultural feedstock for bio-fuels – is curtailed. What is evident from the above however, is that a new era of productivism can be envisaged for the countryside (perhaps recast with ecological modernisation, eco efficiency or smart production associations) and economic, social and environmental arguments in favour of local self-sufficiency and sustainable use of environmental resources may, once again, become increasingly significant. In this context it can be envisaged that conflicts between cultural, ecological and sustainable resource-use perspectives on the rural landscape and environment may increase, and finding a balance between different views will be a growing area of rural planning concern.

Part 2: The British experience

Britain (or rather the UK here, as it is the nation as a whole that is subject to a variety of international environmental agreements) is by no means detached from the challenges associated with the global landscape and environmental changes outlined above, but as in other parts of the world their impact on rural areas reflects the particular economic, social, environmental and political context. In this respect, it is perhaps correct to say that many of the issues discussed are particularly acutely felt in the UK and they engender particularly intense debates as they challenge deep-seated UK cultural tendencies (dating back to the Romantic movement) to idealise the 'natural beauty' and 'natural history' of the countryside (see Chapter 1). Growing understanding about the inevitability and pace of twenty-first-century landscape change, rapid loss of biodiversity, global warming and resource scarcity fly in the face of the popular image of rural areas (held by much of its very urbanised population) as being unchanging and seemingly detached from the pressures of the modern world. The discussion below provides an insight into the UK's experience of some of the key contemporary landscape and environmental challenges facing rural areas. How this is reflected in contemporary rural planning is explored further in Chapter 10.

Landscape change

As noted in Chapter 2 (Figure 2.5) the UK is one of the most urbanised countries in Europe and although urban development still accounts for less that 10 per cent of the land area, human influence is clearly evident almost everywhere. The interaction between human activities and underlying environmental conditions over past millennia is reflected in the very diverse rural landscape we see today. Part of the current UK picture is presented in the landscape categorisation that was published by Natural England in 2013, which divides England into 159 distinct National Character Areas (NCAs). The map (which is difficult to reproduce here because of the complex and subtle shading) shows how geo-diversity, biodiversity, cultural and economic activity combine to produce areas of distinctive landscape identity. Each area is accompanied by a descriptive area profile. This is the latest iteration of landscape character assessment work that has been undertaken by government agencies covering all parts of Britain. The work is intended to provide a resource for understanding the current landscape context of different localities. It aims to inform a range of environmental/rural planning activities including the preparation of local plans and their sustainability appraisals and landscape and visual impact assessments associated with development proposals. The NCA profiles not only describe the key environmental characteristics of each area including the ecosystem services they provide, but also chart historic and recent landscape change, drivers of future change and highlight opportunities for environmental enhancement. This work forms part of the UK's response to the European Landscape Convention and its ambitions to promote greater sensitivity to landscape distinctiveness. It also reflects the ELC's underlying concern for the impact of recent landscape changes on the quality of the British landscape and promotion of environmental protection, improvement and enhancement activities.

As discussed in previous chapters, important drivers of change in rural Britain over the past fifty years include:

- population and household growth (Chapters 1 and 8);
- a rapidly expanding economy (Chapter 3);
- a transformation in agricultural practices associated with the 'productivist' policies of government (Chapter 4);
- increasing affluence (Chapters 5 and 8);
- increasing car ownership (Chapter 6);
- increased leisure time; and
- new patterns of leisure activity (Chapter 4).

These factors have not only had implications for the socio-economic context of the British countryside but have also resulted in significant physical change in the rural environment. Box 9.3 provides an overview of some of the key elements of UK landscape change in recent times and the impacts that have been a focus of concern. The Tranquillity Map in Figure 9.4 provides an illustration of their cumulative effects. The map was produced for the Campaign to Protect Rural England using a series of GIS layers related to environmental variables that a social survey identified as adding to or detracting from people's perceptions of tranquillity. This included data on patterns of urban development, roads and railways and light pollution. More than two-thirds (72 per cent) of respondents in the CPRE survey identified 'tranquillity' as the quality they most valued in the English countryside. The map illustrates how widespread 'urban' influences are now thought to be

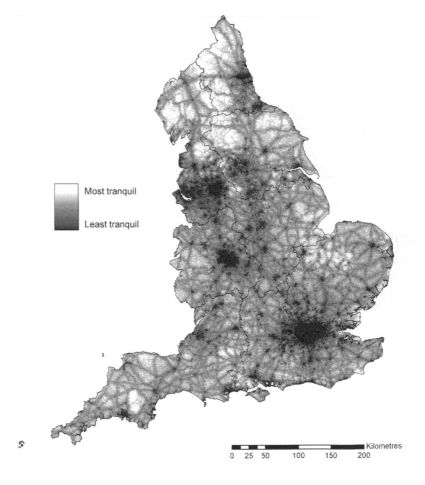

Figure 9.4

Tranquillity map of England, 2007
Source: CPRE, 2007

across England, but also that there are distinct spatial differences in experience. For example, long-standing concern has been expressed about the deterioration in the environmental quality of urban fringe areas (e.g. Hoskins, 1955) and the Tranquillity Map illustrates the scale of intrusion around urban centres. This rural/urban transition zone is the area of the countryside where human 'pressures' and landscape changes are most intense. Here the concentration of uses such as waste disposal facilities, sewage works, transport infrastructure, office parks, recreation facilities, education and health establishments, farm shops, and 'bad neighbour' land uses such as scrapyards, combine with often marginal agricultural activity to produce what some see as a disorderly, unplanned and unappealing rural scene. It contrasts with more formalised urban development patterns within towns and cities and the managed farmland landscapes of the wider countryside, which, it has been argued, align better with English preferences for neatness (Lowenthal and Prince, 1965).

The urban fringe case is interesting as it illustrates how landscape debates are closely bound up with aesthetic appreciation and subjective responses to particular environmental qualities that tend to be culturally determined. In the UK, public debate has often been based around a Romantic view of an 'ideal' landscape, which condemns the new and is backward-looking in its emphasis. As a consequence it has been argued that there is a tendency towards the 'Disneyfication' of British landscapes, either through attempts to fossilise them at a particular point in their development (in valued landscapes such as the Lake District) or by the application of normative treatments that have little local relevance in areas perceived to be of lesser quality, leading to banal landscapes and deterioration of local distinctiveness and meaning (Rackham, 1986). Shoard (2002), for example, believes these views have informed recent environmental 'improvements' (such as the Community Forest Programme) in urban fringe areas aimed in part at hiding urban intrusions and she highlights the lack of appreciation of the distinctive cultural history and value of these 'Edgelands'. Gallent et al. (2006) develop this line of thinking and present a range of alternative aesthetic perspectives on the urban fringe, which place value on its functional and disorderly qualities. A key point here, which applies beyond the urban fringe as well, is that these are living landscapes that are authentic to their time and place. Another related line of argument was very eloquently expressed by Nan Fairbrother (1970) in her book *New Lives New Landscapes* in which she encourages the British public to embrace landscape change as a reflection of contemporary lifestyles and as an essential feature of landscapes themselves, which through their very nature are constantly evolving. Trying to achieve a balance between sensitivity towards the past, the present and the future is of course challenging. These issues, however, are central to thinking about the future of rural landscapes particularly when set against some of the environmental challenges discussed below.

Habitat and species loss

The diversity in the character of the British landscape is also reflected in the rich diversity of British habitats and the species of flora and fauna that they support. Britain's coastal, estuarine and heathland habitats are of global significance and the country is also particularly noted for its veteran trees, ancient woodlands and chalk rivers. However, as in other parts of the world, Britain has experienced a decline in biodiversity in recent times. The Tranquillity Map shown in Figure 9.4 reveals the fragmentation of the English countryside that is thought to be contributing to the loss of habitats and species. Britain is a signatory to the United Nations Convention on Biological Diversity and as part of this is required to develop indicators to monitor progress in relation to internationally agreed biodiversity goals and targets. The 2013 UK Biodiversity Indicators (JNCC, 2013) provide an overview of trends and reveal a mixed picture (see Table 9.1). It can be seen that a number of indicators show long-term improvement (indicated by ticks in Table 9.1). These include areas of increased conservation activity such as conservation volunteering, sustainable fisheries management, air and marine pollution and the extent of protected sites. These are paralleled by a long-term improvement in the ecological status of seabirds and of wintering water birds and in the plant species richness of arable and horticultural land. However, the decline in populations of farmland birds (indicated by a cross in Table 9.1) is continuing and, as this is often seen as a key indicator of wider ecological health in the countryside, further decline here is significant. The continuing rise in the prevalence of invasive species across terrestrial, freshwater and marine habitats is also worth noting as it highlights some of the difficulties inherent in biodiversity planning and management. Many of the species we see as common today are not native to Britain. The sycamore is an example that might well be classed as an invasive species if it were introduced today. Grey squirrels do receive this classification but taking effective action to halt their spread and reduce their numbers raises lively economic and ethical arguments as well as questions of practicality.

There is no doubt though that the general arguments supporting conservation and enhancement of biodiversity attract widespread public support. A report published in 2013 (Cracknell *et al.*, 2013) indicated that 139 not-for-profit environmental organisations in the UK had a combined membership of 4.9 million and that 1 in 10 UK adults was a member of an environmental group. Biodiversity and nature conservation activity was the major focus of attention accounting for 45 per cent of expenditure. In contrast (and of relevance to the discussion below) climate change accounted for just 7.3 per cent. UK government agencies cite ecological, economic, emotional and ethical arguments as underpinning their work related to biodiversity and it is notable that, as with National Character Area activities, increasing reference is being made to the ecosystem services that biodiversity delivers. The reasons for this are explored further in Chapter 10. However, as with many aspects of public policy, while broad arguments may receive

Table 9.1 Overview of UK biodiversity indicators 2013 (JNCC, 2013)

Indicator number (strategic goal/number), title and measure where applicable			Long-term change	Short-term change
A1. Awareness, understanding and support for conservation			Under development, no interim measure(s) available	
A2. Taking action for nature: volunteer time spent in conservation			☑ 2000–2012	☒ 2007–2012
A3. Value of biodiversity integrated into decision making			Under development, no interim measure(s) available	
A4. Global biodiversity impacts on UK economic activity/sustainable consumption			Under development, no interim measure(s) available	
A5. Integration of biodiversity considerations into business activity			Under development, no interim measure(s) available	
B1. Agriculture and forest area under environmental management schemes	B1a. Area of land in agri-environmental schemes	B1a(i). Higher-level/targeted schemes	☑ 1992–2012	☑ 2007–2012
		B1b(ii). Entry-level type schemes		☑ 2007–2012
	B1b. Area of forestry land certified as sustainably managed		☑ 2001–2013	⇔ 2008–2013
B2. Sustainable fisheries			☑ 1990–2011	☑ 2006–2011
B3. Climate change adaptation			Under development, no interim measure(s) available	
B4. Pressure from climate change			Not assessed	Not assessed
B5. Pressure from pollution	B5a. Air pollution	B5a(i). Area affected by acidity	☑ 1996–2010	☑ 2005–2010
		B5a(ii). Area affected by nitrogen	☑ 1996–2010	☑ 2005–2010
	B5b. Marine pollution		☑ 1990–2011	☑ 2006–2011
B6. Pressure from invasive species	B6a. Freshwater invasive species		☒ 1960–2008	⇔ 2000–2008
	B6b. Marine invasive species		☒ 1960–2008	☒ 2000–2008
	B6c. Terrestrial invasive species		☒ 1960–2008	☒ 2000–2008

Table 9.1 continued

Indicator number (strategic goal/number), title and measure where applicable		Long-term change	Short-term change
B7. Water quality		Not assessed	Not assessed
C1. Protected areas	C1a. Total area of protected areas: on land	☑ 1950–2013	☑ 2008–2013
	C1b. Total area of protected areas: at sea	☑ 1950–2013	☑ 2008–2013
	C1c. Condition of A/SSSIs	=	☑ 2008–2012/13
C2. Habitat connectivity	C2a. Broad-leaved, mixed and yew woodland	=	=
	C2b. Neutral grassland	=	=
C3. Status of habitats of European Importance		=	☒ 2007–2013
C4. Status of threatened species	C4a. Status of priority species	☒ 1970–2010	☒ 2005–2010
	C4b. Status of species of European Importance	=	☑ 2007–2013
C5. Birds of the wider countryside and at sea	C5a. Farmland birds	☒ 1970–2011	☒ 2006–2011
	C5b. Woodland birds	☒ 1970–2011	☑ 2006–2011
	C5c. Wetland birds	☒ 1975–2011	☒ 2006–2011
	C5d. Seabirds	☑ 1970–2012	☒ 2007–2012
	C5e. Wintering water birds	☑ 1975/6–2010/11	⇔ 2005/6–2010/11
C6. Insects of the wider countryside (butterflies)	C6a. Semi-natural habitat specialists	☒ 1976–212	⇔ 2007–2012
	C6b. Species of the wider countryside	⇔ 1976–2012	⇔ 2007–2012
C7. Plants of the wider countryside	C7a. Change in plant species richness (arable and horticultural land)	☑ 1990–2007	☑ 1998–2007
	C7b. Change in plant species richness (Woodland and grassland)	☒ 1990–2007	☒ 1998–2007
	C7c. Change in plant species richness (boundary habitats)	☒ 1990–2007	☒ 1998–2007

Table 9.1 continued

Indicator number (strategic goal/number), title and measure where applicable			Long-term change	Short-term change
C8. Mammals in the wider countryside	C8a. Bat populations		☑ 1999–2012	⇔ 2007–2012
	C8b. Historical pipistrelle bat roost counts		☒ 1977–1999	Not assessed
C9. Genetic resources for food and agriculture	C9a. Animal genetic resources	C9a(i). Native sheep breeds	=	⇔ 2001–2007
		C9a(ii). Native cattle breeds	=	☑ 2001–2007
	C9b. Plant genetic resources – enrichment index		☑ 1960–2012	☑ 2007–2012
D1. Biodiversity and eco-system services (marine – fish size classes in the North Sea)			☒ 1953–2011	☑ 2006–2011
D2. Biodiversity and eco-system services (terrestrial)			Under development, no interim	

☒ = decline/detioriation ☑ = increase/improvement ⇔ = no change

general favour, it is the detailed delivery of biodiversity ambitions where tensions arise and where the legitimacy of what might seem 'well-meaning actions' is called into question.

As with the landscape concerns discussed above, attitudes and approaches to nature conservation are inevitably culturally determined and do change over time as alternative perspectives open up lines of critique and debate about the most fundamental assumptions on which traditional views are based. For example, a core feature of nature conservation activity over many generations in Britain as elsewhere has been the designation of protected sites. The 2013 UK Biodiversity Indicators show that the area covered by special protection such as Site of Special Scientific Interest (SSSI) status has grown steadily since 1950 and there has been a marked jump in recent times associated with new marine area designations. Terrestrial designations now cover more that 6 million hectares, approximately 25 per cent of the UK land area. Most protected areas are however, individually quite small, the exception being the UK's estuaries where sites are generally extensive. The increase in protected area coverage is presented as a positive development in the 2013 UK Biodiversity Indicator report and while there is no doubt that it is likely to bring biodiversity benefits, the approach does raise some difficult questions. For example, designations have built up over generations and the current stock of protected areas does not necessarily constitute the ecologically coherent network of sites that contemporary ecological understanding aspires to. In many cases areas have been protected on the basis of rarity rather than their intrinsic value in the wider scheme of things. Similarly, protection is often related to preserving what we have inherited as a result of historic agricultural or industrial practices that may be uneconomic in the current context and far from the ecological richness of the more natural conditions they replaced. There has also been growing scientific understanding of the 'island dilemma', which indicates that species reduce in number and abundance with the size of habitat and with increased distance between habitats. Overall therefore despite continuing international and national support and indeed legal requirements, it is now widely recognised that nature conservation designations are insufficient on their own and may ultimately represent a self-defeating zoo mentality, as wildlife in zoo-like islands amidst a hostile sea cannot survive very long (Gilg, 1996). Thus the scientific justification for site designation as the cornerstone of nature conservation activity is increasingly open to challenge from a range of different angles. Running alongside this, a new paradigm has emerged, which believes that to tackle biodiversity decline and go some way to *reversing* the rate of species extinction shown in Figure 9.1, biodiversity should be a consideration in all areas and attention should shift from a defensive 'preservation and protection' mentality to a creative 'improvement and enhancement' approach. This is a key message of the Convention on Biological Diversity and it is opening up new lines of argument very much akin to Nan Fairbrother's *New Lives New Landscape* thinking. The cross-connections in planning for the future of Britain's rural landscapes and environment are illustrated in the NCA profiles, which seek to integrate science- and social-science-based perspectives. Taking the Northern Thames Basin NCA profile (which covers a large part of London's urban fringe) as an illustration, it is evident that quite a dramatic change of the landscape and environmental character is being proposed. The following extract is an example of one of the opportunities it is promoting, with biodiversity concerns very much to the fore:

> [...] manage, restore and significantly expand the wetlands of the river valleys of the NCA including wet grassland, valley woodlands, flood plain woodlands, non-coastal grazing marsh, fens, rush pasture, swamp and valley mires which, with their high water tables are important sites for over-wintering wildfowl, adding significantly to biodiversity and landscape character. Restore river valley minerals sites to wetlands and washlands and seek opportunities to restore the natural geomorphology of rivers where this will significantly enhance their biodiversity and landscape contribution, including the conservation and planting of a new generation of riverside willows.
>
> (Natural England, 2013b: 50–51)

A similar line of thinking is reflected in The Wildlife Trusts' proposals that have been put forward as a response to the new high-speed rail link between

London and Northern Britain. In *A greener vision for HS2* (The Wildlife Trusts, 2014) they set out ambitious ideas for large-scale nature restoration along the proposed route.

Climate change

One of the key developments that is also challenging traditional approaches to landscape and nature conservation planning in Britain is climate change. It is likely that over time, the environmental change that is predicted will have significant implications for the visual appearance of the British countryside and its ecology, as habitats and species adjust to new conditions. In addition to an underlying momentum of natural adjustment, efforts to mitigate human-induced climate change and adapt to its consequences are also likely to shape the future of the countryside, and they are bringing a new dynamic to landscape and environmental debates including arguments for a much more forward-looking approach.

The latest UK Climate Projections (DEFRA, 2009) investigated three greenhouse gas emissions scenarios and what they might mean for changes in climate by 2080. Taking the central estimate, it is anticipated that the UK generally will experience warmer and wetter winters, hotter and drier summers, sea-level rise, and more severe weather. Across the UK summer (June, July, August) temperatures could rise by between 3°C and 4°C, while summer precipitation is projected to reduce by between –17 per cent and –23 per cent and winter precipitation to increase by between +14 per cent and +23 per cent. Regional variations are evident as indicated in Figure 9.5. This is particularly so in relation to sea-level rise, which in London is predicted to increase by 36 cm by 2080 while in northern areas this rise will be offset to some extent by isostatic uplift that is still occurring following the last ice age.

Following publication of the 2009 UK Climate Change Projections, government agencies have considered what they may mean for the UK's landscape and environment. Discussion of the implications of climate change features for example in all the NCA profiles produced by Natural England. A general overview is set out in Box 9.4, which summarises the key findings from work undertaken for Scottish Natural Heritage looking at likely changes in Scotland. The research suggests that direct impacts of climate change for Scotland will include increased flooding, a changing pattern of habitats and loss of land to the sea as a result of sea-level rise. Mitigation measures designed to slow the impact of climate change will promote increased production of renewable energy in the form, for example, of wind farms and biomass production, and efforts to increase the amount of carbon locked up in the landscape by increased woodland cover and peatland restoration. In terms of adaptation, flood management measures are envisaged as a significant aspect of future landscape change both in terms of increasing hard river and coastal defences in some areas and restoring natural flood plains and more natural approaches to flood control in others. Woodland and forest

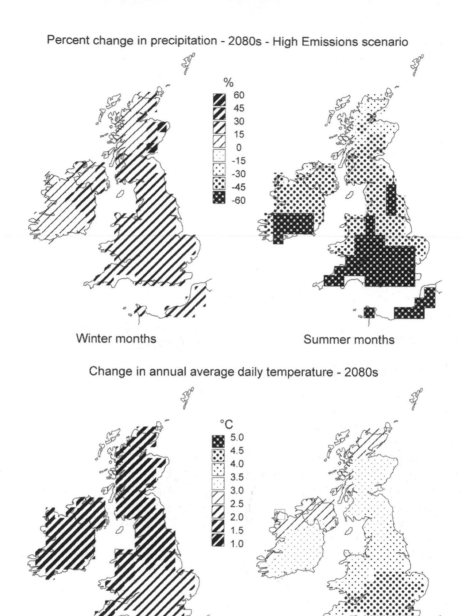

Figure 9.5

UK climate change projections, 2009
Source: DEFRA, 2009

management practices are also expected to alter, for example, in terms of species selection and there is likely to be an intensification of agricultural production and development in lowland areas. Overall, lowland areas are expected to see the most evident landscape changes while in upland areas the effects are likely to be more subtle with the exception of increased wind-farm developments. Interestingly, the report suggests that the most noticeable landscape effects will be associated with mitigation and adaption measures rather than through the direct impacts of climate change.

Box 9.4 Predicted effects of climate change on Scottish landscapes (Scottish Natural Heritage, 2011: 3)

Direct changes

- an increase in coastal flooding and erosion, including loss of low-lying areas of land to the sea as a result of sea level rise, larger waves and storm surges;
- an increase in river flooding, erosion and slope instability;
- effects of changing patterns of rainfall, increases in storm damage and a potential increase in pests and disease on trees and forests.

Impact of measures to mitigate climate change

- wind-farm developments, introducing large modern structures into many upland and some lowland and coastal landscapes;
- cultivation and processing of biomass in the form of short-rotation coppice and energy crops;
- tidal and marine-based renewables;
- micro-renewables;
- carbon storage including woodland expansion and peatland restoration.

Impact of measures to adapt to climate change

- engineered responses to the threat of coastal and riparian flooding;
- sustainable flood-management responses, including restoring natural flood plains, increasing woodland planting in river catchments, and allowing coastlines to change as sea levels rise;
- changing patterns of woodland and forest management, and the expansion of habitat networks;
- intensification of agricultural production in lowland areas, with changes in crops, new buildings and greater use of irrigation;
- policies designed to steer development away from locations where there is an increased risk of flooding or storm damage;
- expansion of outdoor recreation and tourism, particularly during the summer months.

This latter point illustrates once again the significance of human activity, and in particular government policy, in shaping the evolving character of the landscape and wider environment. It is evident from recent UK experience that policy approaches do vary and that the pace and extent to which climate change adaptation and mitigation measures are implemented will be influenced by shifting political and to some extent public opinion. At a UK level, this is evident in the shift in government's approach to climate change in recent times. From a position where the Labour government established a new Department of Energy and Climate Change in 2008 (DECC) and supported the production of the *Stern Review on the Economic of Climate Change* (Stern, 2006), which made the case for early mitigation and adaptation action, it is evident that the Coalition government is much more reserved in its approach with some of its DECC Ministers being open about their climate change scepticism. This change in stance is perhaps reflected in the scale of public-sector spending reductions for climate-change-related action and in government support for shale gas exploration in the face of increasing concerns about UK energy security and interest in the economic growth potential that the 'fracking' industry could offer. A DECC report into the development of shale gas and oil in the UK published in 2013 illustrates this position.

> Shale gas and oil could provide the UK with greater energy security, growth, jobs and tax revenue. The Government is encouraging safe and environmentally sound exploration to determine our shale potential.
>
> (DECC, 2013: 3)

Such a stance may seem surprising given widespread international acknowledgement of climate change as one of the most pressing environmental issues facing twenty-first-century society. However, it illustrates some of the challenges in translating this understanding into practical action. Scepticism is one end of a spectrum of views on climate change that make planning for climate change problematic. This is revealed by a survey of UK public attitudes that identified a range of barriers to public engagement with climate change (Lorenzoni *et al.*, 2007). These included uncertainty and lack of trust in information sources, perceptions that climate change is a distant threat and other things are more important, and fatalism and 'drop in the ocean' feelings linked to a sense of helplessness due to the global scale of the problem. It is evident therefore that the long-term, diffuse and intangible nature of climate change contrasts with the much more locally visible and tangible concerns associated with landscape change and biodiversity discussed above. Landscape and biodiversity issues also benefit from a long history of public interest and government and charitable action in Britain. Unlike climate change concerns they are well embedded in public consciousness and have a long-established architecture of legislation and organisational infrastructure and clear (but not unproblematic) lines of action associated with them. In contrast climate change is very much the 'new boy on the block' and its place in the overall picture of landscape and environmental planning in Britain is still at an early stage of development.

Resource scarcity and security

The same is not entirely true about resource scarcity. For although contemporary concerns about this are closely bound up with global population growth and associated increased resource demand rather than war, they do continue a long-standing concern about resource security. Indeed as has been discussed in earlier chapters, up to the 1970s this was a fundamental driver in shaping approaches to rural planning in Britain and the form of its countryside (see Chapters 2 and 4). The stance on shale gas and oil mentioned above has connections with these concerns and highlights a particular interest in energy security at the present time. The UK government's 2012 Energy Security Strategy (DECC, 2012a) explains the reasons for this, including the country's declining reserves of fossil fuels, which is making it increasingly dependent on imports at a time of rising global demand and increased resource competition. Interestingly, the report reveals that war continues to be an issue. For example, it makes reference to risks associated with international conflicts including those in the Middle East where it notes that the closure of a major energy choke point such as the Strait of Hormuz (between Iran and other Gulf states) would affect 20 per cent of global oil and one-third of global liquefied natural gas (LNG) supply. The potential development of shale gas and oil in the UK reflects these considerations. This may all seem far removed from rural Britain but if one looks at a map (Figure 9.6) of where these reserves are, the significance becomes apparent. Large parts of the British countryside, including much of southern and central England and a band running across central Scotland are now thought to offer potential for gas and oil production associated with underlying shales through the process of 'fracking'. The landscape and environmental concerns related to fracking have been well publicised. The British Geological Survey indicates that these can include methane gas emissions, contamination of groundwater and increased seismic activity as well as built development associated with extraction. The government believes that with proper environmental practices these risks can be minimised but not unsurprisingly there has been significant local opposition in areas where trial drilling has been proposed. Fracking it seems may be a very good example of the types of conflict that could increasingly emerge in future as growing resource demands bring new uses to the countryside that challenge established notions of the form and function of rural landscapes and raise a complexity of environmental issues.

Food is another area where national security concerns are growing once again. In 2014 the House of Commons Environment, Food and Rural Affairs Select Committee published a report on *Food Security* which indicated that the UK is now only 68 per cent self-sufficient in the foodstuffs that can be

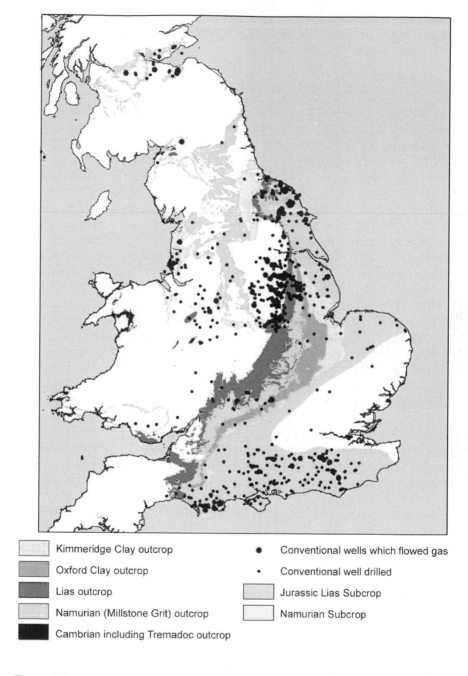

Key:

Kimmeridge Clay outcrop

Oxford Clay outcrop

Lias outcrop

Namurian (Millstone Grit) outcrop

Cambrian including Tremadoc outcrop

● Conventional wells which flowed gas

• Conventional well drilled

Jurassic Lias Subcrop

Namurian Subcrop

Figure 9.6

UK shale gas and oil potential
Source: DECC, 2012b

produced here and that there has been a steady decline in self-sufficiency over the last 20 years. The report presents the case for immediate action to plan for changing weather patterns and increases in global demand for food. Among its many recommendations it promotes sustainable intensification: producing more food, on a finite amount of land, in a sustainable way and the use of new technologies (including a review of approaches to genetically modified [GM] crops) to improve productivity. It also makes the case for changing the type of crops that are produced and notes in particular how the UK's livestock and dairy sectors are highly dependent on imported feeds and that there is a need to improve this position with more production at home. These pressures and arguments challenge the reality that has set in over the last 50 years, of the countryside as space for consumption. The next 50 years may see a (neo)-productivist revival.

Concluding remarks

This chapter has illustrated how the economic, social, cultural and political forces discussed in earlier chapters interact with the natural environment to shape the rural 'landscape' we know today. Far from being static and amenable to 'preservation', rural landscapes are in constant flux and are shaped by both local and global drivers of change. This has always been the case, but as we progress further into the twenty-first century, particularly significant environmental challenges are evident associated with the trajectory of human development. These have attracted international attention and calls for the designation of a new geological epoch – the 'anthropocene'. Key areas of concern include:

- *The current scale, pace and direction of landscape change,* which is trans-forming the face of rural areas and eroding their cultural heritage and natural beauty.
- *Habitat loss, degradation and fragmentation and changes in biodiversity,* which reflect the ecological consequences of current patterns of human development in rural areas and the global reduction in areas of high biodiversity value.
- *Climate change* associated with global warming, and predicted impacts on sea-level rise, climate variability and water and food availability.
- *Resource scarcity* associated with global population growth and increasing demands for food, timber, minerals and other resources that are mainly sourced from rural areas resulting in increasing competition for rural space.

These issues are acutely felt in Britain as they challenge deep-seated cultural tendencies (dating back to the Romantic movement) to idealise the 'natural beauty' and 'natural history' of the countryside and to see rural areas as serene and unchanging.

Summary

- Rural areas are on the front line of a changing relationship between human activity and the 'natural world'. The pace of change, linked to population growth, has accelerated markedly during the twentieth century.
- 'Earth System Science' has emerged as a frame for understanding new relationships and the interplay between key challenges.
- These key challenges are rooted in habitat change and loss, climate change, and in resource scarcity and security. Britain feels these challenges acutely. How they are addressed is a source of considerable debate and conflict.
- Responses can generate a great deal of conflict. Fears over resource scarcity and security have been used to justify support for shale gas and oil extraction in Britain, which itself has the potential to bring habitat loss and contribute to climate change.
- Finding ways to address key environmental and landscape dilemmas that work across all components of the 'Earth System' has become a major challenge for public policy and rural planning.

Key readings

- Council of Europe (2000) *The European Landscape Convention*, Council of Europe: Brussels.
  (Text of the Landscape Convention, which summarises current international approaches to landscape planning and enhancement.)
- Steffen, W., Sanderson, A., Tyson, P.D., Jäger, J., Matson, P.A., Moore III, B., Oldfield, F., Richardson, K., Schellnhuber, H.J., Turner II, B.L. and Wasson, R.J. (2004) *Global Change and the Earth System: A Planet Under Pressure*, Springer: New York.
  (A useful introduction to Earth System Science and the research underpinning proposals for the Anthropocene.)

Key websites

- Intergovernmental Panel on Climate Change: www.ipcc.ch/.
  (Provides access to all IPCC reports and associated resources.)
- Millennium Ecosystem Assessment: www.millenniumassessment.org/en/index.html
  (Provides access to all Millennium Ecosystem Assessment reports associated statements and Green facts summary.)

- National Character Area Profiles for England: www.gov.uk/government/publications/national-character-area-profiles-data-for-local-decision-making/national-character-area-profiles
  (Provides access to all NCA profiles plus useful case studies showing use of NCA profiles in different contexts.)
- UK Biodiversity Indicators: http://jncc.defra.gov.uk/page-4229
  (Provides access to 'UK Biodiversity Indicators in your Pocket 2013'.)

10 Environmental policy and planning

••

Integrating the environment into rural development

The previous chapter explored how the economic, social, cultural and political forces discussed in this book have interacted with the 'natural' environment to shape the rural 'landscape' and wider 'environment' we know today. It also examined how human perspectives on the 'natural' world have changed over time, encompassing a range of distinct and often conflicting viewpoints that have made planning for the rural environment a subject of intense debate. Chapter 9 illustrated this through a discussion of four key contemporary landscape and environmental challenges that are prominent at both a global and a UK level:

1   the pace and direction of landscape change;
2   habitat loss, degradation and fragmentation and changes in biodiversity;
3   climate change; and
4   resource scarcity.

This chapter picks up these themes and examines how they are reflected in the development of environmental governance, policy and planning for rural areas.

This final thematic chapter starts, however, by examining how evolving understanding of the natural world and humans' relationship with it have resulted in changing approaches to environmental governance and are increasingly informed and directed by international agreements and laws, particularly those flowing from the United Nations and, for European countries, from the European Union. These agreements and laws reflect and promote a number of key policy principles including those related to Sustainable Development, but also increasingly to the *Ecosystem Approach* and the related rise of *Ecosystem Services* thinking. Core ideas associated with international environmental policy principles and concepts are explained. The second part of the chapter then considers how the international legal framework and associated environmental planning and management paradigms are reflected in current environmental governmental bodies and environmental designations related to each of the four landscape and environmental challenges introduced in last chapter.

Part 1: Approaches and instruments for governing the rural environment

The importance of international environmental law

Chapter 9 introduced the developing understanding of human/environment interactions and humans' place within the natural world. It also drew attention to the increasing pace and scale of environmental change affecting rural areas, associated with current patterns of human development, and the growing appreciation of the *global connectedness* of environmental issues. The global, or international, nature of many environmental challenges is now broadly recognised, as is this need for an international response:

> It is a trite observation that environmental problems, though they closely affect municipal laws, are essentially international and that the main structure of control can therefore be no other than that of international law.
>
> (Sands and Peel, 2012: xxii)

This recognition goes back a surprisingly long way and was first reflected in a series of bilateral treaties (for example between the UK and France, and the United States of America and the UK) related to the control of fisheries, which were agreed in the second half of the nineteenth century (Sands and Peel, 2012). These treaties were pioneering in recognising the adverse economic and social as well as environmental consequences of uncontrolled human 'exploitation' of the natural world and the need in certain cases for transnational arrangements to deal with these. With their focus often on the sea and wild fish stocks, they were early examples of an appreciation of the value of what later became known as the Global Commons; those internationally shared ecosystems and their shared natural resources that fall outside national jurisdictions. Today these are defined in international law – the High Seas; the Atmosphere; Antarctica; and, Outer Space – but it is recognised that concepts raised by the Global Commons apply more widely due to the global inter-linkages between all ecosystems. The debate surrounding the Global Commons and the introduction of international agreements guiding their use gathered pace in the mid twentieth century. This was spurred by the publication in 1958 of ecologist Garrett Hardin's theory of the *Tragedy of the Commons* (Ostrom, 2008), which concluded that without appropriate action, the common pool resources they offered would be exhausted by utility-maximising individuals rather than conserved for the benefit of all.

International effort to respond to this concern and collectively manage the global commons has played a significant part in informing the development of the wider pattern of international environmental law and in establishing its legitimacy in guiding national and local level policy and planning. The extent of its influence is significant and it has a prominence that is not evident in most other rural planning issues discussed in this book. This development has also taken the debate about environmental policy and practice beyond local and national arenas to a global scale. The following sections provide an overview of some related areas of development and the main international agreements and associated principles and concepts that are shaping governance of the environment and rural development today.

The environment as a 'governance' issue

The tragedy of the commons analysis was based on the assumption that human behaviour is always ultimately guided by efforts to maximise utility for the individual in question. Like many theories, this is hypothetical or 'idealised', rather than a true description of complex and contingent reality. However, the tragedy of the commons thinking illustrates the difficulty of achieving sustainable management of the environment and highlights not only the need for trans-national action but also the need for non-market actors such as states to intervene to shield the environment from the operation of market forces. It must be noted, however, that state actors are not necessarily configured at a scale that is appropriate for dealing with environmental issues and are also by no means neutral in their approach to environmental matters. Foucault, for example, concludes that the main motivations for governmental interventions in market forces in today's increasingly neo-liberal world is the growth of national economies and the development of competitive capacity (Foucault, 1991). The biological wellbeing of the population, access to sufficient nutrition and the availability and health of a nation's natural resources may also be viewed as important as they are obviously foundational components for economic growth and competitiveness (Sairinen, 2002).

While early environmental policies were based on a 'command and control' approach – where a central authority such as a national government legislated in relation to resource use – this has increasingly been seen as an inadequate approach to achieving sustainable environmental management as a result of the issues discussed above. As Ostrom (2010) observes, groups of rational individuals have come together, organising themselves to address what they see as deficiencies in the established practices of environmental planning and management. Such groups have been prominent in challenging the adequacy of state activities. Many operate at a national or sub-national scale, but some are global in their reach such as the WWF and Greenpeace. Global networking among local-level groups is also increasingly evident. Overall, such groups are now very influential in environmental planning and management decision-making at all scales. These developments are part of an emerging pattern of environmental governance. Evans (2012) describes environmental governance as the steering of relations between society and the environment in a situation where market forces are blind to most of the values and services offered to societies by their immediate and more distant environments. Governance studies have for a long time charted the rise of various forms of political co-ordination related to resource management. These not only span the conventional public–private divide but also involve 'tangled hierarchies' and parallel 'power networks' involving state and non-state actors such as private corporations and non-governmental organisations (NGOs) (Jessop, 1995: 311). These observations reflect a wider shift from government to governance. In an increasingly post-modern and globalised world, the role of the nation state is diminishing in the face of market forces and the need for international collaboration (Rhodes, 2007). In short, the same shift of power away from national governments towards corporations and NGOs influencing the global agri-food system and the trajectory of agriculture, discussed in Chapter 4, also applies to governance of the environment at both international and more local scales.

Environmental issues tend to constitute particularly challenging governance problems due to their typically slow emergence over extended time frames; their tendency to concern and connect geographically distant areas and to involve causes and consequences that are unequally distributed among populations; and finally, because they tend to be mostly formulated by scientists but frequently engage and involve wide ranges of stakeholders with different beliefs, value systems and interests (Mickwitz, 2003; Evans, 2012). Climate change, for example, represents a trans-boundary problem, requiring collective action by national governments with potentially conflicting interests. Similarly, issues such as diffuse pollution of water resources involve conflicting interest groups at a more confined scale, such as farmers, fertilizer companies and local inhabitants who require clean drinking water.

Uncertainty in environmental understanding and its governance implications

Due to their complexity, both trans-boundary and locally confined environmental issues almost always involve a range of scientists from different fields such as geology, chemistry and hydrology, for example, and due to their 'ontologically distant' and accumulative nature evade direct and certain detection, forcing scientists to rely on proxy indicators and models that integrate and accumulate uncertainty. This complexity and compounded uncertainty renders most environmental issues highly politicised, further compromising the ability of national governments to derive legislative standards based on straightforward scientific evidence. In many cases, it is now recognised that it makes more sense to enable the involved stakeholders such as scientists, interest groups, local inhabitants and concerned businesses to collaborate, or at least contribute to negotiations, for the best possible management outcomes (e.g. Radaelli, 1999; Fisher, 2005).

Given the challenging and uncertain nature of most environmental issues, arriving at optimal environmental standards in any given context is not an easy feat. Ansell and Gash (2007: 547) suggest that collaborative governance is also a solution in this context as it has the potential to transform the often adversarial relationships between policy makers and recipients, or different stakeholder groups, into more co-operative ones in policy negotiations. The levels of uncertainty in understanding related to environmental issues is therefore also one of the factors that has encouraged the shift away from state 'managerialism' of the environment, in which decision-making power is concentrated in the 'centre' – often a ministry of a national government and public agencies – and where decisions are made unilaterally or through closed decision processes, typically relying on evidence from agency experts (see Chapter 2). Instead, there is increasing acceptance that high levels of uncertainty necessitate more collaborative and deliberative approaches to environmental planning and management. Drawing from research on collaborative planning, Innes and Booher (2003, 2004) suggest a model of collaborative governance that is multi-dimensional, integrating a dialogue between citizens and planning/policy officials that enables learning by all participants that closely informs action. Significantly, in this type of collaborative planning the polity interests and the citizenry co-evolve. The central contention is that effective participatory methods are reflexive, integrate a diverse range of stakeholders, and involve collaborative learning that produces innovative outcomes (Fritsch and Newig, 2012). In the face of high levels of uncertainty, the aim of such reflexive governance is to build shared knowledge and heuristics for collaborative action (Innes and Booher, 2003). Reflexive governance systems are complex and often difficult to establish due to the need to integrate a very diverse range of stakeholders who need to be able to collaborate on an equal basis but the reward is a system that can be adaptive instead of stalemated. It can build societal capacity and produce innovative responses to seemingly intractable problems.

Key international agreements related to the rural environment

The need for international co-operation to address environmental problems, together with an appreciation of the case for more collaborative styles of governance and for mechanisms to deal with significant levels of uncertainty, are reflected in the development of international environmental law. Table 10.1 shows that today there are a number of important international agreements guiding environmental policy and planning in rural areas. Those led by the United Nations and its associated institutions such as UNESCO, are extensive in their influence and apply to many countries all over the world. A number of other agreements are also wide ranging in their international coverage such as the Ramsar Convention (which now has 168 contracting

Table 10.1 Key international agreements related to the rural environment

Environmental challenges facing rural areas	International agreements	Lead agency (where relevant)
The pace and direction of landscape change	• 1992 World Heritage Convention • 2000 European Landscape Convention	• UNESCO • Council of Europe
Habitat loss, degradation and fragmentation and changes in biodiversity	• 1971 Ramsar Convention on Wetlands of International Importance Especially as Waterfowl Habitat • 1979 Bonn Convention on Conservation of Migratory Species of Wild Animals • 1979 Bern Convention on the Conservation of European Wildlife and Natural Habitats • 1992 Convention on Biological Diversity (CBD) • 1992 Habitats Directive • 2009 Birds Directive	• Council of Europe • United Nations • European Union • European Union
Climate change	• 1992 United Nations Framework Convention on Climate Change • 1997 Kyoto Protocol	• United Nations • United Nations
Resource scarcity	• 2000 Water Framework Directive • 2004 Environmental Liability Directive	• European Union • European Union
Other	• 1985 Environmental Impact Assessment Directive • 2001 Strategic Environmental Assessment Directive	• European Union • European Union

parties) and the Bonn Convention (which has more than 120 countries involved). European nations have also been active in their international collaboration on environmental matters through the European Union (EU) and in addition through the Council of Europe, which also includes European countries that are not formally part of the EU. For those that are members of the EU, the various environmental directives that are listed are particularly important as they: require formal legislation in each state to be in place to carry forward Directive ambitions; involve set timetables for action; and can lead to financial penalties for non-compliance. Other international agreements are softer in their approach but still entail a commitment to action and monitoring and reporting of related activity.

Table 10.1, although selective, reveals that the major focus of international environmental law to date has been on wildlife conservation. Of the wildlife conservation agreements mentioned, it is worth noting that the Convention on Biological Diversity (CBD) (one of the series of very influential outputs from the 1992 United Nations Rio Earth Summit) has been particularly significant in bringing contemporary natural science and, to some extent, social science understanding together and setting a new direction in this area. Important ideas incorporated in the CBD (and made a legal requirement in the EU under the 1992 Habitats Directive) include an obligation to consider biodiversity *everywhere* not just in sites designated for 'nature conservation' purposes. In addition it moves beyond a concern for species protection and conservation and calls for significant *enhancement* of biodiversity, recognising that much of what we once had has gone. Similar sentiments are also evident in the 2000 European Landscape Convention (ELC), which equally calls for more effective landscape planning everywhere and a new focus on the enhancement of landscapes in light of concerns over the global degradation of landscape quality which was discussed in Chapter 9.

The ELC clearly extends ideas set out in the 1992 UNESCO World Heritage Convention which is concerned with the protection, conservation and celebration of built and natural heritage sites and with areas that are judged to be of outstanding universal value. Table 10.1 suggests that, to date, more limited attention has been given to issues of climate change and resource scarcity but there is no doubt that recognition of the need for international action is growing in relation to both issues. In the case of the former, the 1992 United Nations Framework Convention on Climate Change, which commits signatories to internationally binding emission reduction targets, was given firmer direction through the 1997 Kyoto Protocol but it remains a hugely contentious area. The difficulties in moving forward mirror those revealed in a survey of public attitudes in the UK, which were set out in Chapter 9. These include uncertainty and a lack of trust in information sources, perceptions that climate change is a distant threat and that there are other more urgent challenges, and feelings of fatalism and helplessness when confronted with the global scale of the problem. The qualified commitment of key 'greenhouse gas' producing countries such as the USA and potential constraints implied on the activities of poorer developing countries are just some of the additional complicating factors here. In relation to resource scarcity, the European Union countries seem to have a strong track-record in international co-operation. For example, the 2000 Water Framework Directive (WFD) responds to widespread concern among EU citizens over the pollution of surface and ground water, both from urban waste water and from agriculture. The Directive requires the preparation of a comprehensive set of River Basin Management Plans, which aim to achieve cleaner European rivers and provide long-term sustainable water management based on a high level of protection for the aquatic environment. Less well-known is the 2004 Environmental Liability Directive, which establishes a framework for the prevention and remedying of environmental damage based on the Polluter Pays Principle (PPP). It is purely concerned with ecological damage covering protected species and natural habitats; damage to water and damage to soil (as opposed to damage to the aesthetics of the environment for example). PPP is an issue of growing significance in planning for all forms of development in rural as well as urban areas.

Key principles and concepts in international environmental law

The European Union is not alone in establishing the Polluter Pays Principle within its legal frameworks. It is an important concept in international environmental law more generally and it dates back to the 1970s and efforts by the Organisation for Economic Co-operation and Development (OECD) to ensure that companies paid the full cost of pollution control that was not covered by national governments. As can be seen from Table 10.2, it is one of a number of principles and concepts that have similar status and provide the underpinning philosophy upon which environmental governance at all levels is increasingly based. Many of these principles reflect the issues discussed earlier in this chapter and were confirmed or established as core dimensions of international environmental law through the Rio Declaration on Environment and Development, which was another key outcome of the 1992 United Nations Rio Earth Summit. Examination of the principles and concepts in Table 10.2 indicates that a number emphasise the connectedness of environmental issues and environmental responsibilities and the need for integration of environmental considerations across sectors, at different levels of government and at a trans-national scale. Others can be seen to have a

Table 10.2 Key principles and concepts in international environmental law (Kurukulasuriya and Robinson, 2006)

Principle/concept	Explanation
Sustainable development, integration and interdependence Principles 4 and 25 Rio Declaration	*Sustainable development*: 'Development that meets the needs of the present without compromising the ability of future generations to meet their own needs' (WCED, 1987). *Integration*: In order to achieve sustainable development, environmental protection shall constitute an integral part of the development process and cannot be considered in isolation from it. *Interdependence*: Encompasses interdependence of economic, social and environmental development; territorial interdependence across the world; and intergenerational interdependence.

Table 10.2 continued

Principle/concept	Explanation
Inter-generational and Intra-generational equity Principles 3 and 5 Rio Declaration	*Inter-generational equity*: Relates to the right of future generations to enjoy a fair level of the common patrimony. *Intra-generational equity*: Relates to the right of all people within the current generation to fair access to the current generation's entitlement to the Earth's natural resources.
Responsibility for trans-boundary harm Principle 2 Rio Declaration	*Trans-boundary harm*: States have the responsibility to ensure that activities within their jurisdiction or control do not cause damage to the environment of other States or of areas beyond the limits of national jurisdiction.
Transparency, public participation and access to information and remedies Principle 10 Rio Declaration	*Transparency*: Each individual shall have appropriate access to information concerning the environment that is held by public authorities. *Public participation*: The opportunity to participate in decision-making processes. *Access to information*: States shall facilitate and encourage public awareness and participation by making information widely available. *Access to remedies*: Effective access to judicial and administrative proceedings, including redress and remedy, shall be provided.
Co-operation, and common but differentiated responsibilities Principal 7 Rio Declaration	*Co-operation*: States shall co-operate in a spirit of global partnership to conserve, protect and restore the health and integrity of the Earth's ecosystem. *Common but differentiated responsibilities*: In view of the different contributions to global environmental degradation, there is explicit recognition that different standards, may be appropriate for different countries.
Precaution Principal 15 Rio Declaration	*Precautionary principle*: Where there are threats of serious or irreversible damage, lack of full scientific certainty shall not be used as a reason for postponing cost-effective measures to prevent environmental degradation.

Table 10.2 continued

Principle/concept	Explanation
Prevention Feature of many international agreements	*Prevention*: Seeks to avoid environmental harm. It requires prior assessment of environmental harm, licensing or authorization that set out the conditions for operation and the consequences for violation of the conditions, as well as the adoption of strategies and policies preventing harm.
Polluter Pays Principle Principal 16 Rio Declaration	*Polluter Pays*: National authorities should endeavour to promote the internalization of environmental costs and the use of economic instruments, taking into account the approach that the polluter should, in principle, bear the cost of pollution.
Access and Benefit Sharing regarding Natural Resources Principal 22 Rio Declaration	*Access and Benefit Sharing*: Indigenous people and their communities and other local communities have a vital role in environmental management and development because of their knowledge and traditional practices. States should recognise and duly support their identity, culture and interests.
Common Heritage and Common Concern of Humankind	*Common Heritage*: The natural environment, particularly the climate system, biological diversity and fauna and flora of the Earth. *Common Concern*: States and other actors should not cause harm to the Common Heritage and they share responsibility for protection, preservation and enhancement of the natural environment, particularly the proper management of the climate system, biological diversity and fauna and flora of the Earth.
Good Governance Goal 8 Millennium Development Goals	*Good Governance*: States and international organizations should: (a) adopt democratic and transparent decision-making procedures and financial accountability; (b) take effective measures to combat official or other corruption; (c) respect due process in their procedures and observe the rule of law more generally; (d) protect human rights; and (e) conduct public procurement in a transparent, non-corrupt manner.

strong moral dimension and relate to complex issues concerning responsibility for harm and equity including the concept of common but differentiated responsibility. There are also several principles that define important attributes of good environmental governance such as access to environmental information, stakeholder engagement and transparency in environmental decision-making and establishing mechanisms to anticipate and help prevent environmental harm. These mechanisms include rural planning regimes where it can be seen that the establishment of planning policy frameworks that help to prevent environmental damage and promote protection, conservation and enhancement of the rural environment are key ways in which nations can demonstrate their attention to their international environmental obligations. Other mechanisms that are very familiar to rural planners (operating the land-use planning system) such as Environmental Impact Assessment (EIA) and Strategic Environmental Appraisal (SEA) also have their origins in obligations related to international environmental law.

The list set out in Table 10.2 is derived from a Training Manual on International Environmental Law produced by the United Nations' Environment Programme (Kurukulasuriya and Robinson, 2006). The list begins with 'sustainable development' which, as highlighted in early chapters, has been an overarching concept guiding many aspects of rural development since the early 1990s. The Guide which was published in 2006 is interesting, however, in that it makes only passing reference to two other related concepts that are of growing prominence in international environmental law and which are increasing points of reference in planning for the environment. These are the Ecosystem Approach and Ecosystem Services concepts and they are introduced below.

The ecosystem approach

Many of the principles set out in the 1992 Rio Declaration reflect a paradigm shift in the planning and management of the natural environment and the resources that are derived from the functioning of component ecosystems. This shift is based on a number of premises, which recognise that:

- The sustainability of economic systems and the quality of human life is inevitably dependent on the maintenance of healthy ecosystems.
- Humans are an integral part of ecosystems rather than separate from them.
- A sectoral approach to planning and management is generally insufficient to deal with complex interrelationships and diverse stakeholder priorities.

Adoption of the Ecosystem Approach (EA) as a methodological framework for a more holistic style of planning and management that reflects these premises has been promoted in particular, but not exclusively, by the CBD and has been incorporated in a growing number of other international conventions and policy documents. Following the ratification of the CBD in 1993, the Conference of the Parties (COP) to the CBD has gradually elaborated the detail of the EA, which it has defined as:

A strategy for the integrated management of land, water and living resources which promotes conservation and sustainable use in an equitable way.

(CBD COP, 2000: V16)

Alongside this definition, the COP has developed a series of EA principles and these are set out in Box 10.1.

The EA principles carry forward many of the key concepts of international law set out above but they also encompass additional elements that reflect some of the discussions set out earlier in this chapter. There is no doubt, however, that they are complex, and various agencies have sought to provide additional guidance for practitioners on what the EA means and how to apply it. For example, the COP has developed a website with a wide range of EA related resources for practitioners (see useful web-links at the end of this chapter). This includes five points of *operational* guidance which stress:

Box 10.1 Ecosystem approach principles (CBD COP, 2000: V16)

1 The objectives of management of land, water and living resources are a matter of societal choices;
2 Management should be decentralized to the lowest appropriate level;
3 Ecosystem managers should consider the effects (actual or potential) of their activities on adjacent and other ecosystems;
4 Recognising potential gains from management, there is usually a need to understand and manage the ecosystem in an economic context. Any such ecosystem-management programme should be a priority target of the ecosystem approach;
5 Conservation of ecosystem structure and functioning, in order to maintain ecosystem services, should be a priority target of the ecosystem approach;
6 Ecosystems must be managed within the limits of their functioning;
7 The ecosystem approach should be undertaken at the appropriate spatial and temporal scales;
8 Recognising the varying temporal scales and lag-effects that characterise ecosystem processes, objectives for ecosystem management should be set for the long term;
9 Management must recognise that change is inevitable;
10 The ecosystem approach should seek the appropriate balance between, and integration of, conservation and use of biological diversity;
11 The ecosystem approach should consider all forms of relevant information, including scientific and indigenous and local knowledge, innovations and practices;
12 The ecosystem approach should involve all relevant sectors of society and scientific disciplines.

1　A focus on functional relationships and processes within ecosystems.
2　Enhanced benefit-sharing, recognising that the goods and services that ecosystems offer are essential to human environmental security and sustainability but in many instances are external to the market economy and lack proper valuation and recognition in decision making.
3　The use of adaptive management practices, recognising the constant flux of environment/human interactions and imperfect scientific understanding.
4　The importance of carrying out management actions at a scale appropriate to the issue being addressed with decentralisation to the lowest appropriate level.
5　Ensuring inter-sectoral co-operation and partnership working.

Despite these efforts, applying the EA to day-to-day environmental planning and management practice presents many challenges. However, a review of early experience concluded that there was evidence of genuine effort to engage with the EA in many countries, with the USA and UK being notable examples (SBSTTA, 2007). The most prominent areas of application were in water and marine management. This had been encouraged, for example, by international policy papers associated with the Ramsar Convention (Swiss Agency for the Environment, Forests and Landscape *et al.*, 2002) and through UNEP-funded projects promoting better management of Large Marine Ecosystems (Wang, 2004), both of which had been notable in their promotion of EA principles. Although the review revealed clear evidence of engagement with the EA, it also highlighted a range of barriers to its application and these are set out in Box 10.2.

Prominent among these barriers was limited understanding of what the concept is trying to achieve. Linked to this, some commentators have drawn parallels with 'sustainable development' and have highlighted the similarly dynamic – and often nebulous – nature of the EA concept and the widely

Box 10.2　General barriers to the use of the ecosystems approach (SBSTTA, 2007: Para. 27)

- ineffective stakeholder participation in planning and management;
- limited understanding of what the approach seeks to achieve;
- the lack of capacity for decentralised and integrated management;
- insufficient institutional co-operation and capacity;
- the lack of dedicated organisations able to support delivery of the ecosystem approach;
- the overriding influence of perverse incentives; and
- conflicting political priorities, including those that arise when a more holistic approach to planning is adopted.

different interpretations that it attracts (Kidd *et al.*, 2011). For many natural scientists, for instance, it is apparent that the key aspect of the EA is an emphasis on the application of ecological understanding to the planning and management of the environment. However, others see the EA as fundamentally an approach that draws social science understanding into this arena and believe that it is as much about distilling what is currently considered as good practice in environmental governance as it is about ecology. Indeed the barriers to use of the EA, listed in Box 10.2, indicate the significance of the governance dimensions of the EA. Many barriers relate to issues associated with institutional structures and approaches and the wider socio-economic context within which environmental planning and management is situated. The points set out reveal the difficulties inherent in breaking out of traditional ways of seeing and organising the world and delivering more integrated, participative and partnership-based planning and management of places in which environmental considerations receive a more significant (and many would argue, more appropriate) share of attention. It is this ambition that lies at the heart of the Ecosystems Approach.

Ecosystem services

One of the key ideas reflected in the EA (Principle 4 in particular) is that there is a need to take better account of the value of the ecosystem services which the environment provides. Many of these environmental services – such as beautiful landscapes, biodiversity but also ground or surface water that is of safe quality for recreational uses – fall within the realm of public goods, where consumers cannot be excluded or asked to pay directly. In rural areas, for example, the asset base and quality of these 'environmental goods' (again, see Chapter 4) is often being heavily eroded by pressures such as pollution from farming, traffic, and the need to develop land for housing, commercial and/or other built uses. The argument is that many of these (potentially) damaging activities and uses have more direct economic or more immediate and tangible social value and tend to attract greater attention and support in decision-making. Equally, protection of those services that fall within the public goods category is often difficult to justify as no direct income streams are attached to them. The development of the EA as an internationally endorsed paradigm for environmental planning and management has supported major advances in ES research and practice, which is seen as a way of counteracting these tendencies, by enabling the integrity of the environment and the wide range of services it provides to be more explicitly valued and balanced with economic priorities and other human needs.

The Millennium Ecosystem Assessment (MEA) (UNEP, 2005) was a landmark in the development of Ecosystem Services thinking and, as can be seen from Figure 10.1, it classified Ecosystem Services (ES) into *provisioning services*, *regulating services*, *cultural services* and *supporting services*. This way of conceptualising the environment in terms of a range of dynamic flows of

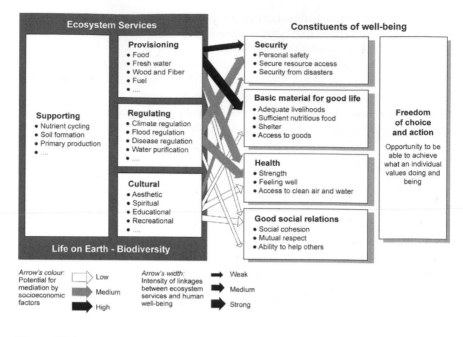

Figure 10.1

Ecosystem services and their links to human wellbeing
Source: UNEP, 2001: 5

services pays heed to the quality of habitats (ecosystems) and frames environmental values in terms of their utility to societies and, in the case of the MEA, very explicitly to human wellbeing.

The potential application of ES thinking in environmental planning and management is wide-ranging. At one end of the spectrum, and where much of the focus of attention has been so far, is the development of ES valuation techniques. This reflects a view that formal valuation facilitates more effective environmental governance particularly via means of economic governance instruments such as those related to formal Payments for Ecosystem Services (PES; see Figure 10.3). As discussed in Chapter 4, PES methods are increasingly being used, for example, to alleviate problems associated with nature conservation activities in developing countries where conservation efforts threaten local livelihoods (Branca et al., 2009). However, the problems associated with the valuation of ecosystem services are manifold. There are immense difficulties in identifying and measuring the state of a range of ES that are crucial to human and environmental wellbeing (Robertson, 2006). The challenge yet again boils down to the complexity of the environment as a governance or indeed scientific problem. The difficulty of translating the uncertainty,

Figure 10.2

Cultural ecosystem services: The Glastonbury Tor offers magnificent views of the Somerset Levels and is also described as the most 'spiritual' location in England

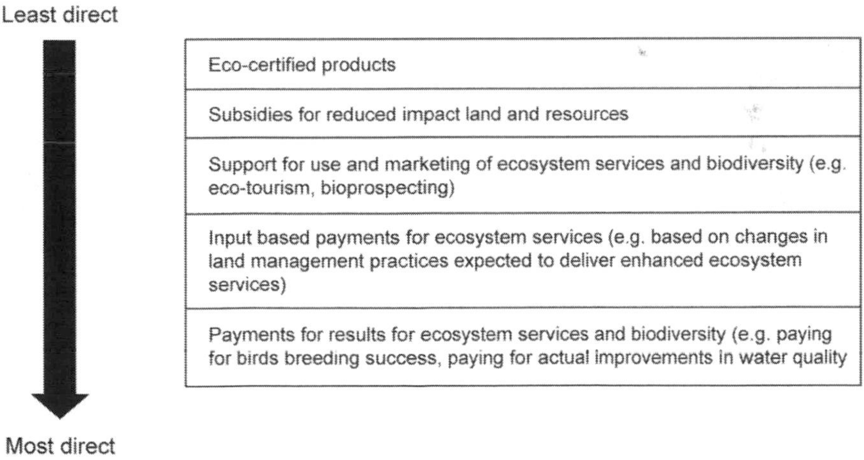

Least direct

Eco-certified products
Subsidies for reduced impact land and resources
Support for use and marketing of ecosystem services and biodiversity (e.g. eco-tourism, bioprospecting)
Input based payments for ecosystem services (e.g. based on changes in land management practices expected to deliver enhanced ecosystem services)
Payments for results for ecosystem services and biodiversity (e.g. paying for birds breeding success, paying for actual improvements in water quality

Most direct

Figure 10.3

Payments for Ecosystem Services (PES) can be seen to range from direct to indirect
Source: Dunn, 2011; adapted from Ferraro and Kiss, 2002

multiplicity and variability of natural processes and environmental values into the logic of economic valuation inevitably leads to crude simplifications in the valuation process (van Hecken and Bastiaensen, 2010). While efforts to overcome these difficulties are extensive, increasing attention is being paid to activities at the other end of the ES application spectrum involving softer, more qualitative, uses of ES concepts. These include using ES as a framework for more holistic consideration of environmental attributes and processes; extending the scope of environmental planning and management activities; and improving public understanding of the environment. For some however, the approach will always remain controversial as it is essentially anthropogenic in its focus and heavily reflective of market-orientated perspectives.

Part 2: Existing and emerging environmental governance instruments in the UK

So what does all this mean for rural planning? Much of the discussion set out above might appear detached from the coal-face of rural planning. However, the influence of international environmental law upon rural policy and practice is considerable and the following sections will illustrate the connections with reference to the UK (rather than just Britain, as the UK is often the data reporting unit and the signatory to international agreements). First, an overview of key environmental legislation and its connections to planning is provided along with an account of the administrative structures that currently guide UK environmental action. Key environmental designations, which are the focus of much attention in rural planning, are then explored and finally some new directions in environmental planning are examined. Throughout, connections to the four key contemporary landscape and environmental challenges discussed in Chapter 9 are drawn out.

Key aspects of environmental law in the UK and rural planning

Table 10.3 below provides an overview of some of the key legislation that forms the backbone for environmental planning and management in the UK today. It also illustrates how these connect to rural planning with reference to the National Planning Policy Framework (NPPF) for England. Many of the more recent Acts reflect, to a significant degree, the content of international agreements and efforts to embed these in national law. The table is inevitably selective in the legislation it mentions. A fuller picture can be obtained from Bell and McGillivray (2013) which is an excellent source on the environmental legislation that planning decisions need to abide by in England and Wales.

From a broad 'Sustainable Development' perspective the Acts and regulations listed in Table 10.3 are crucial. For example, the Planning and Compulsory Purchase Act 2004, the Planning etc. Scotland Act 2006 and the Government of Wales Act 2006, all carry forward commitments from the

1992 Rio Declaration on Environment and Development and establish Sustainable Development as the overarching purpose of planning. In all instances this is the first time that a formal purpose of planning (and in the case of Wales, a formal purpose of government more generally) has been set out in UK legislation. Similarly, a clear link to the international commitments pursuant of Sustainable Development is evident in the introductory paragraph to the NPPF. It refers, for example, to the United Nations definition of Sustainable Development and embeds the five guiding principles set out in the UK Sustainable Development Strategy – *Securing the Future* (DEFRA, 2005) – as the basis for planning activity in England. These guiding principles are:

- Living within the planet's environmental limits;
- Ensuring a strong, healthy and just society;
- Achieving a sustainable economy;
- Promoting good governance; and
- Using sound science responsibly.

Also mentioned under the 'Sustainable Development' heading in Table 10.3 are regulations that relate to Strategic Environmental Assessment (SEA) of plans and Environmental Impact Assessment (EIA) of development applications that may harm the environment. Both of these derive from international law and many see SEA and EIA as important mechanisms for increasing the attention given to environmental considerations in planning processes. The extent to which these activities do influence decision-making and deliver more environmentally sustainable patterns of development is, however, much debated (Jay *et al.*, 2007; Fischer, 2007, Glasson *et al.*, 2013).

In terms of other areas of legislation referred to in Table 10.3, it is worth noting that international obligations have not played a major part in the development of the legislation mentioned in relation to 'The pace and direction of landscape change'. Much of this derives from the post Second World War era when there was considerable popular concern in the UK over the spread of towns and cities, loss of areas of countryside to urban development, and associated enthusiasm to protect areas of 'natural beauty'. The 1949 National Parks and Access to the Countryside Act reflected these concerns and laid the foundations for activity and planning responsibilities in this area. Although this Act was also significant in establishing the legislative framework related to nature conservation, much of the legislation mentioned under the 'Habitat loss, degradation and fragmentation and changes in biodiversity' heading has been greatly influenced by more recent international agreements and is very closely connected to them. The same is true for the legislation mentioned in relation to 'Climate Change' and 'Resource Scarcity'.

What is evident from Table 10.3 is that there is a close relationship between planning policy in England, and international and national environmental law. Across all of the contemporary landscape and environmental challenges

Table 10.3 Key landscape/environmental challenges, related legislation and the National Planning Policy Framework (NPPF) for England

Environmental challenges facing rural areas	Selected UK legislation of relevance to rural planning	Related paragraphs in the NPPF for England
The pace and direction of landscape change	• Town and Country Planning Act 1947 • Environment Act 1995 • National Parks (Scotland) Act 2000 • The Countryside and Rights of Way Act 2000	Conserving and enhancing the natural environment: Paras. 109–125
Habitat loss, degradation and fragmentation and changes in biodiversity	• 1949 National Parks and Access to the Countryside Act • Wildlife and Countryside Act 1981 • Wildlife (Northern Ireland) Order 1985 and the Nature Conservation and Amenity Lands (Northern Ireland) Order 1985 • The Countryside and Rights of Way Act 2000 • Nature Conservation (Scotland) Act 2004 • The Conservation of Habitats and Species Regulations 2010	Conserving and enhancing the natural environment: Paras. 109–125
Climate change	• Climate Change Act 2008 and UK Regulations • Flood Risk Management (Scotland) Act 2009 • Flood and Water Management Act 2010	Meeting the challenge of climate change, flooding and coastal change: Paras. 93–108
Resource scarcity	• Environmental Protection Act 1990 • Water Industry Act 1991 and 1999 • Environment Act 1995 • Pollution Prevention and Control Act 1999 • Water Framework Directive Regulations 2003 • The Water Act 2003 • The Environmental Permitting Regulations (England and Wales) 2010	Conserving and enhancing the natural environment: Paras. 109–125 Minerals: Paras. 142–149
Sustainable development	• Planning and Compulsory Purchase Act 2004 • The environmental assessment of plans and programmes Regulations 2004 • Planning etc Scotland Act 2006 • Government of Wales Act 2006 • Town and Country Planning (EIA) Regulations 2011	Sustainable Development: Paras. 18–219 taken as a whole

Protecting Green Belt Land: Paras. 79–92

discussed in Chapter 9, both local plans and development management are seen as having a key role to play. For example, the NPPF makes a number of references to issues associated with 'the pace and direction of landscape change'. Planning authorities are encouraged to promote the protection and enhancement of valued landscapes and pay particular attention to protecting the scenic beauty of National Parks and Areas of Outstanding Natural Beauty (AONB) and areas of Heritage Coast. They are also asked to consider protecting areas of tranquillity that have remained relatively undisturbed from noise, as well as intrinsically dark landscapes that are remote from light pollution and prized for their recreation and amenity value. The NPPF is more extensive in its discussion of matters related to 'habitat loss, degradation and fragmentation and changes in biodiversity' and clearly carries forward CBD sentiments in not only requiring planning authorities to secure appropriate protection of the hierarchy of international, national and locally designated sites of special wildlife and geo-diversity value, but also the establishment of coherent ecological networks and broader patterns of green infrastructure. The NPPF suggests that these should be developed to secure the preservation, restoration and recreation of priority habitats and recovery of priority species and to achieve *net gains* in biodiversity, recognising the wider benefits of ecosystem services. In this context local authorities can consider designating Nature Improvement Areas in their local plans.

'Climate Change' merits a section of its own in the NPPF and includes both mitigation and adaptation considerations. In terms of mitigation, elements of the wording seem surprisingly strong given the climate change scepticism evident within sections of the Coalition government. For example, the NPPF suggests that planning authorities have a key role in securing *radical* reductions in greenhouse gas emissions by helping to increase the supply of renewable and low-carbon energy and moving to a low-carbon future by promoting energy efficiency in the location and design of new development. Clearly these sentiments have important implications for rural planning. On the face of it, the NPPF suggests active support for wind-farm development in areas and concentration of development in key settlements (see Chapters 2, 6, 7 and 8) with energy efficiency among other issues in mind. Much of the climate change discussion in the NPPF relates, however, to issues of adaptation and in particular to matters related to flood risk. Strategic Flood Risk Assessments are seen as an essential underpinning for planning activity, and using this understanding, local planning authorities are asked to apply a 'sequential test' to steer new development away from areas at risk from flooding. They are also asked to consider flooding issues in all development proposals to ensure that flood risk elsewhere is not increased and where possible overall flood risk is reduced. It is worth noting that matters related to 'resource security' form a substantial element of the NPPF and the coverage is wide-ranging. In the main, the interest is related to natural resource protection from various forms of pollution (air, water, light, noise and soil). However, there is also a clear recognition that land itself is a finite resource.

Local planning authorities are encouraged to make effective use of land by re-using previously developed sites (brownfield land) and considering agricultural land quality in decisions about the location of green-field development, recognising the value of retaining the best and most versatile agricultural land. The sustainable use of minerals receives particular attention as well. Here the emphasis is very much on making the best use of finite resources and securing their long-term conservation, recognising that minerals are essential to support economic growth and quality of life. A final point that is worth mentioning in relation to Table 10.3 is that it poses a question about whether the Green Belt aspects of the NPPF can be seen as relating to all of the environmental challenges identified. This answer is by no means straightforward and some of the issues involved are explored further in the discussion of key designations set out below.

In many respects, the NPPF (DCLG, 2012) is forward-thinking in its approach to environmental matters. It certainly reflects concepts and principles set out in international environmental law and carries them into the planning arena, making them material considerations in rural land-use planning. It may seem surprising to learn therefore that the NPPF has attracted considerable criticisms from environmental NGOs. Although it is felt by some of the leading environmental campaign groups like the RSPB that the final version is much better than earlier drafts, at least from an environmental perspective, many still feel that it remains a strongly pro-growth document. The concern from environmental groups is what weight will be given to environmental considerations in planning appeals and other decision-making in light of this pro-growth emphasis. Equally there is a very real and growing concern that following a period of swingeing cuts in public budgets, local authorities may no longer have the expertise and resources needed to protect and enhance the environment in the way the NPPF sets out (RSPB, 2012).

Key authorities responsible for the environment in the UK

Not only has the legislative framework summarised in Table 10.3 set the direction for planning policy in the UK, it has also played a part in establishing the institutional structures overseeing environmental planning and management. Figure 10.4 outlines the key governmental authorities responsible for developing and implementing environmental legislation in the UK at the present time. The pattern is complex and again the picture presented is simplified to aid understanding. For example, all government agencies departments (including DCLG) have environmental responsibilities and there are many other related bodies with quite extensive environmental remits that could have been included. Interestingly, the pattern of responsibilities in some ways is becoming increasingly complex, but in other ways there are signs of increasing integration. A major reason for increasing complexity is that many environmental responsibilities,

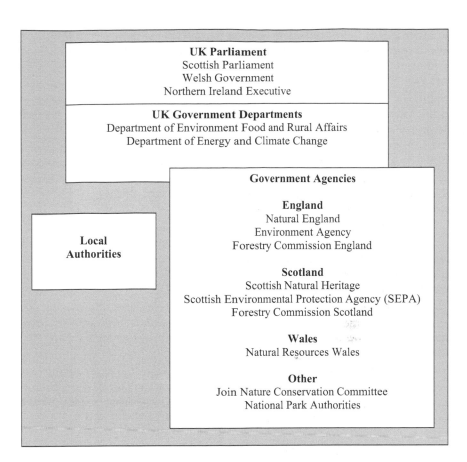

Figure 10.4

Key governmental authorities with environmental responsibilities in the UK

although co-ordinated by the UK government, are now delegated to the devolved administrations and distinct differences in emphasis and approach are beginning to emerge between England, Scotland, Wales and Northern Ireland as these responsibilities take root and develop. This situation makes the activities of the Joint Nature Conservation Committee (JNCC) for example – a long-established body integrating and supporting activity across the four 'national' jurisdictions of the UK in relation to nature conservation – even more significant.

The structure presented in Figure 10.4 highlights the inherent difficulties in developing an integrated approach to environmental planning and management in the UK, which is advocated in international agreements, as even at governmental level responsibilities are split so many ways. However, this criticism of the arrangements is well recognised, and steps towards greater integration have also occurred in recent times. For example, Natural England (which was set up in 2007) brought together responsibilities that were previously divided among at least three former agencies but it left responsibilities for water management and integrated pollution control in the hands of the Environment Agency. However, the newest agency mentioned, Natural Resources Wales – which was established in 2013 – is much broader in its environmental remit. This is reflected in its purpose, which is 'to ensure that the natural resources of Wales are sustainably maintained, enhanced and used, now and in the future'. As the newest agency in environmental administration in the UK, its set up and rationale is perhaps indicative of future developments.

What is particularly notable from Figure 10.4 is that in contrast to the complicated national-level administrative arrangements, at the local level the picture appears much simpler. It is clear that local authorities are expected to play a key role in delivering international as well as national environmental ambitions on the ground and as indicated in Table 10.3 (setting out NPPF instructions) local planning authorities are especially significant in this respect. However, as discussed earlier, since the economic crises of 2008 there are significant concerns that public spending cuts are compromising the ability of local authorities to play their part. Budgetary constraints, together with the other governance factors referred to in Part 1 of this chapter, are therefore prompting increasing partnership-working in relation to local-level environmental planning and management at the present time as local bodies seek to maintain the momentum of environmental care.

Rural planning and the designated areas system in the UK

It is also evident from Table 10.3 that the UK has a long history of environmental legislation. The picture has built up over many years and while, periodically, Acts have been introduced that consolidate and revoke previous laws (the Environment Act 1995, for example) the current legal framework remains very complex and is by no means totally coherent or reflective of current ideas about good practice in environmental planning and management. This situation adds to the contentiousness of activity in this area. The ongoing prominence of special area designations in the UK is a case in point. Traditionally, these have been perceived as a key mechanism for environmental protection and the UK has been particularly active in its designation activity. As already discussed, however, the limitations of special area designations as a means of stemming environmental degradation are now well recognised. This is why more recent international agreements such as the CBD and ELC emphasise more holistic environmental planning and management approaches that go beyond protecting areas deemed to be of special value and seek environmental enhancement more generally. However, while this may be the current environmental paradigm, much of the legal framework related to designations (at both international and UK level) remains in place and still attracts a strong body of support from different quarters. In contrast, there is as yet little evidence of a willingness to review the historic legacy of designated sites in light of current understanding and efforts to redirect action to a more wide-ranging approach are at a very early stage of development. As a consequence, environmental designations continue to be very significant considerations in rural planning in the UK. Some of the key designations are discussed below.

Designations that particularly relate to 'the pace and direction of landscape change' are shown in Figure 10.5. These include National Parks, Areas of Outstanding Natural Beauty, National Scenic Areas and Heritage Coasts (which relate to areas of attractive coastline). Perhaps the most widely known of these 'landscape' designations are National Parks, which cover areas with beautiful or relatively wild features that are seen to be of national significance (Gilg, 1996). The National Parks and Access to the Countryside Act 1949 set out the idea of National Parks as an instrument for planning and conservation in the countryside with the overall aim of safeguarding areas of exceptional natural beauty and improving means of access to the countryside. Each National Park is managed by a National Park Authority led by nationally appointed Members who include stakeholders from local authorities, representatives from external interest groups and community representatives. This Authority has responsibility for preparing both a National Park Management Plan and a development plan for the area, and for handling all planning applications within the National Park. Box 10.3 sets out the statutory purpose of National Parks and shows that the more recent National Parks in Scotland (which were established post-2000) have a broader remit than those in England and Wales. In Scotland

---

Box 10.3 Statutory purpose of National Parks in the United Kingdom

England and Wales

- To conserve and enhance the natural beauty, wildlife and cultural heritage.
- To promote opportunities for the understanding and enjoyment of the special qualities (of the parks) by the public.

Scotland

- To conserve and enhance the natural and cultural heritage of the area.
- To promote sustainable use of the resources of the area.
- To promote understanding and enjoyment of the special qualities of the area by the public.
- To promote sustainable economic and social development of the area's communities.

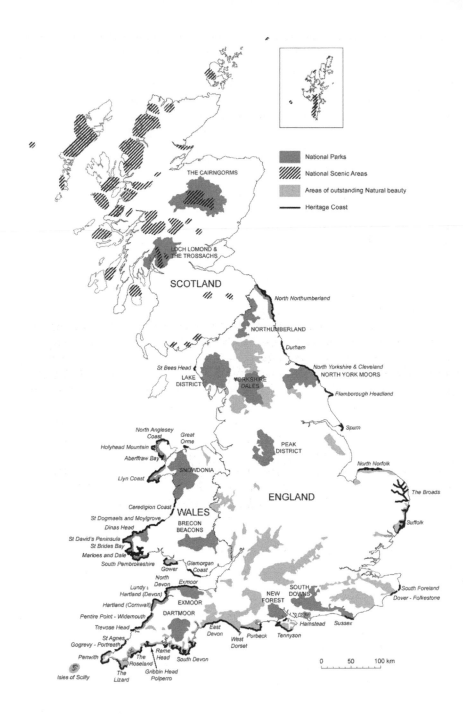

Figure 10.5

Key landscape designations

this includes sustainable use of the resources of the area and the promotion of the economic and social development of the area's communities. The Scottish variant reflects current concerns about resource scarcity and the need for thoughtful management of National Park resources. It also responds to criticism of the longer established National Parks, which have been accused of being elitist in setting environmental protection and 'quiet' recreation interests above the social and economic wellbeing of local communities (McCarthy *et al.*, 2002). Reference to recent National Park Management Plans and National Park development plans in England and Wales indicates something of a softening of this approach. Similar shifts in emphasis are evident in the planning and management of AONBs (National Scenic Areas are the equivalent in Scotland). These areas are generally smaller than National Parks and the AONB remit is more focused on the conservation and enhancement of the natural beauty of the landscape. Their management is the responsibility of local authorities and often involves formal partnership arrangements funded by a range of local stakeholders. Since the 2000 Countryside and Rights of Way (CRoW) Act, the production of AONB management plans is now a statutory requirement. Despite the strong landscape quality focus of AONB designation some of the management plans produced (such as that for the Wye Valley AONB) have been exemplary in integrating economic, social and environmental concerns and providing models of good practice in integrated countryside management (Wragg, 2000).

Perhaps less well-known but of equal significance for rural planning are nature conservation designations that seek to stem 'habitat loss, degradation and fragmentation and changes in biodiversity'. Table 10.4 lists the main nature conservation designations that can be found in the UK. These include Sites of Special Scientific Interest (SSSI), quite a number of which also attract other designations associated with international environmental laws such as those covered by the Ramsar Convention focused on wetland protection, the CBD and the associated European Union Habitats Directive (Special Areas of Conservation, SAC) and Birds Directive (Special Protection Areas, SPA), which together form the 'Natura 2000' network of protected areas. There are also National Nature Reserves managed for the protection or flora, fauna or special geological features and a range of local level designations such as Local Nature Reserves and Sites of Local Biological Importance that are designated at a local authority level, often through local plans.

Nature Improvement Areas are a recent addition to these long-standing designations and they reflect the very different focus of nature conservation activity that was heralded by the CBD. Table 10.4 indicates that there is a hierarchy in the importance of these sites and the levels of protection that they should be afforded in planning decisions reflects this. Generally, significant harm to nature conservation interests resulting from development on these sites, or close to them, should be avoided, but in some circumstances mitigation proposals and even compensation for harmful impacts may be accepted.

Table 10.4 Key nature conservation designations

Importance	Site designation and explanation
Sites of international importance	*Ramsar Sites* listed under the 1971 Convention on Wetlands of International Importance *Special Protection Areas* (SPAs) classified under the EC Directive on the Conservation of Wild Birds *Special Areas of Conservation* (SACs) designated under the EC Directive on the Conservation of Natural Habitats and Wild Fauna (the Habitats Directive)
Sites of national importance	*National Nature Reserves* (NNRs) designated under the 1949 National Parks and Access to the Countryside Act, the 1981 Wildlife and Countryside Act and the 1965 Amenity Lands Act (Northern Ireland) *Marine Nature Reserves* (MNRs) designated under the 1981 Wildlife and Countryside Act *Sites of Special Scientific Interest* (SSSIs) notified under the 1981 Wildlife and Countryside Act. Notified as *Areas of Special Scientific Interest* in Northern Ireland under the 1965 Amenity Lands Act
Sites of regional/local importance	*Local Nature Reserves* (LNRs) declared by local authorities under the 1949 National Parks and Access to the Countryside Act. Declared as *Local Authority Nature Reserves* (LANRs) in Northern Ireland. *Non Statutory Nature Reserve* established by a variety of public, private and voluntary bodies such as the Royal Society for the Protection of Birds (RSPB). *Sites of Local Biological Interest.* These sites are recognised by local authorities for planning purposes. The name and status afforded to these sites varies considerably *Nature Improvement Areas.* These sites are run by local partnerships to create joined-up and resilient ecological networks at a landscape scale.

In England, Natural England (NE) is the statutory agency mainly responsible for landscape and nature conservation designations, and in its Biodiversity 2020 strategy, it outlines the current guiding principles of designation work. Connection to the concepts and ideas discussed in Part 1 of this chapter is clearly evident:

> [...] our approach to all future landscape and nature conservation designations will:
>
> • be underpinned by the best available evidence;
> • deliver our relevant national and international commitments;

- wherever possible enhance connectivity;
- operate at a sufficient scale to facilitate coherence;
- facilitate resilience to climate change and other impacts, and accommodate change where necessary;
- wherever possible encourage stakeholder and community engagement;
- increase public understanding and enjoyment of the natural environment;
- take a holistic approach so that decisions on revised boundaries and new designations take account of existing designations;
- give priority to actions that seek to implement the Natural Environment White Paper and Biodiversity 2020;
- In fulfilling the above points, demonstrate a commitment to the ecosystem approach.

(Natural England, 2012)

An account of protective designations in the UK countryside would not be complete without reference to Green Belt. Although not conceived as an environmental designation (but rather as an 'urban containment' tool), green belts do have significant implications for the planning of rural areas surrounding major towns and cities. Perhaps the best known and most widely supported planning policy in the UK, green belts emerged as a response to vociferous campaigns against urban sprawl in the early years of the twentieth century. They were pioneered in the Greater London Plan of 1944, and the Town and Country Planning Act 1947 made formal provision for local authorities to designate green belts as areas in which very stringent controls on new development could be applied. Over the years green belts have achieved a unique place in planning in the UK. As Cullingworth and Nadin have commented they '[...] are the first article of the planning creed. They are hallowed by use, popular support, and fears of what would happen if they were weakened' (Cullingworth and Nadin, 2006: 183).

Figure 10.6 shows the current pattern of green belt designations in the UK (which it must be stressed do not cover all urban fringe areas). The NPPF sets out five purposes for the Green Belt in England:

- to check the unrestricted sprawl of large built-up areas;
- to prevent neighbouring towns merging into one another;
- to assist in safeguarding the countryside from encroachment;
- to preserve the setting and special character of historic towns; and
- to assist in urban regeneration, by encouraging the recycling of derelict and other urban land.

From these it can be seen that Green Belts address concerns about careful management of land resources and play a role in controlling landscape change. At first sight, they seem also to be a mechanism for promoting more

Figure 10.6

UK Green Belt designations

intensive, high density and energy efficient urban form. However, studies have shown that by restricting peripheral development local authorities are simply causing development to leapfrog the green belt and encouraging longer commuting journeys, traffic congestion and increased pollution (Cullingworth and Nadin, 2006; Nathan, 2007). There have also been criticisms that the policy has lacked sophistication, in that it is mainly concerned with restricting development and has traditionally exhibited little regard for the quality of the environment within green belt areas, which many view as severely degraded particularly at the urban edge (Gallent *et al.*, 2006).

To some extent, both these criticisms are addressed in the NPPF. For example, it encourages local planning authorities to retain and enhance landscapes, visual amenity and biodiversity within the Green Belt. It also supports the review of Green Belt boundaries as part of Local Plan-making processes, and in undertaking these reviews it requires local authorities to consider the consequences for sustainable development of channelling development towards urban areas inside the Green Belt boundary, towards towns and villages within the Green Belt, or towards locations beyond the outer Green Belt Boundary. Such reviews and the need for release of Green Belt land are, however, still hotly contested by groups such as the Campaign to Protect Rural England (CPRE, 2014).

The designated areas approach is without doubt the most significant governance instrument for safeguarding natural and landscape values in the UK. At present 28 per cent of England, for example, is covered by SSSI or landscape designations and a further 13 per cent lies within areas of Green Belt. This focus on designated sites has, however, come under sustained criticism from many quarters. From an environmental perspective, Selman (2009) identified a number of lines of critique. These include: the elitist nature of designation processes, which privilege some areas at the expense of others; recognition that designation directs attention away from the value and potential of the wider countryside; the cluttered and overlapping pattern of designations and the need for rationalisation; the weakness of designations in connecting habitats and in expanding the stock of conservation resources; and because they reflect historic situations, designations may well be in the wrong place relative to future climatic conditions. In the absence of replacement landscape scale interventions though, Selman urges caution in promoting de-designation. A social critique is also evident and Gilg (1996) outlines two themes along which this type of criticism tends to align. The 'conspiracy theory' claims that land-owners and middle-class pressure groups have exerted disproportionate influence on planning and decision-making concerning what is good for the countryside, leading to an exclusive notion of rurality where middle-class recreation areas are protected while the rest of the population are encouraged to enjoy passive recreation away from high-value nature sites (Gilg, 1996; Allanson, 1996). Alternatively, some suggest that designations illustrate the hegemony of a scientific elite over how the countryside should be managed, and has led to an approach where fragile ecosystems are protected from what is judged to be recreational 'overuse' with the effect that the masses are herded to less sensitive areas. The present pattern of access management forms a public land-management spectrum: from high conservation value areas that are tightly controlled and where public access is restricted, to low conservation value areas where there is open access, often facilitated by the creation of country parks. Environmental conservation, therefore, is often seen as a motivation for exclusion or restriction of access (Gilg, 1996).

The Natural Environment White Paper 2011 and initiatives for integration

From many angles therefore, it seems that there is a case for a significant overhaul of approaches to planning and managing the rural environment of the UK. There are signs that this process may be starting. The Natural Environment White Paper 2011 (DEFRA, 2011) declares its aim of 'mainstreaming' the value of nature across our society by:

- facilitating greater local action to protect and improve nature;
- creating a green economy, in which economic growth and the health of our natural resources sustain each other, and markets, business and Government better reflect the value of nature;
- strengthening the connections between people and nature to the benefit of both; and
- showing leadership in the European Union and internationally, to protect and enhance natural assets globally.

Whilst the White Paper rules out the introduction of any new regulations, it pinpoints the Biodiversity 2020 strategy as a crucial blueprint that shows the way ahead. This advocates both a more integrated and democratic approach to nature conservation and environmental protection and the importance of engaging environmental values in local rural economies. It suggests that this can be achieved via a new partnership approach in the form of Local Nature Partnerships (LNPs) comprising people from local authorities, businesses, statutory authorities, civil society organisations, land managers and local environmental record centres, as well as people from communities themselves (see also Chapter 2). The Partnerships would be charged with raising awareness about the services and benefits of a healthy natural environment. Where the LNPs are supposed to show leadership and develop a shared local vision with a set of priorities, statutory bodies such as the Environment Agency, Natural England and the Forestry Commission will be expected to deliver more joined-up and consistent statutory and expert advice to local authorities and LNPs. LNPs are also encouraged to work with Local Enterprise Partnerships (LEPs – again, see Chapter 2).

Not only do these frameworks provide potential opportunities for integrating economic and environmental agendas at a sub-regional or local level (through a more participatory approach), but they also – as Tromans (2012) notes – resonate with the Natural Environment White Paper and the need to reconnect consumers with food production (Chapter 4). In Troman's words, measures in the 2011 White Paper help:

> [...] local authorities use their new duties and powers on public health, which include creating a 'Local Green Areas' designation, to allow local protection of green space, establishment of a Green Infrastructure Partnership to support development of green infrastructure, including allotments, and launching a new phase of volunteering opportunities – for example community food-growing projects.
>
> (Tromans, 2012: 20)

The Natural Environment White Paper also commits the government to encouraging further use of PES (Payments for Ecosystem Services) in safeguarding the natural environment (Dunn, 2011). As discussed above, PES constitute a potential solution for integrating environmental values and functions in the form of dynamic and evolving flows of services into rural economies. They also offer a potential means of addressing the elitist critique of nature and landscape conservation in the UK depending on how PES are designed and managed (Corbera *et al.*, 2007).

Concluding remarks

On the face of it, the Natural Environment White Paper seems to herald a new era of environmental planning in the UK, in which partnership working and more integrated environmental governance will come to the fore. Certainly, there are very good examples where some of the new arrangements set out in the White Paper are stimulating innovative bottom-up action. In Morecambe Bay, for example, stakeholders have seized upon the opportunity to establish a new Local Nature Partnership and associated Nature Improvement Area and are developing these as the focus of a range of collaborative initiatives that combine environmental enhancement with efforts to improve the economic and social wellbeing of local residents. However, it is interesting that so far, unlike many previous White Papers, the Natural Environment White Paper has not developed into a new act of parliament and a recent implementation update report (DEFRA, 2014e) offers no further commitment to supporting further action. The Local Nature Partnerships are a case in point. Although 48 LNPs have now been established – aided by an initial development fund of £1 million – this was a one-off allocation and little further support is indicated apart from the maintenance of a partnership database to allow sharing of good practice. Indeed of the 92 commitments set out on the White Paper, over half are reported as being completed and a further quarter have been completed but with some ongoing activity often of a limited nature. Of those items that are listed as 'in progress', many relate to international obligations, such as the commitment to environmental reporting and the 2020 mission to halt biodiversity loss. It is obviously important, through documents like the Natural Environment White Paper implementation update report, for the government to be able to show evidence that it is taking relevant action. However, the overall impression is of activity completed some time ago and little happening currently. The prospect of halting biodiversity loss in the UK by 2020, let alone enhancing landscape quality, achieving net gains in biodiversity and radically reducing greenhouse emissions therefore seems extremely remote. Indeed the Natural Environment White Paper can perhaps be viewed as another major step towards further 'hollowing out of the state', passing significant environmental responsibilities from an increasingly slim central government to equally depleted local authorities and others, with very few resources attached. Whether community-level action is capable of filling the gaps and providing the leadership necessary to address the very significant environmental challenges that are facing the UK is debatable. Overall, therefore, it is difficult to conclude a review of environmental planning in the UK at the moment on a positive note.

Summary

- While rural planning is increasingly connected to international environmental laws and principles, and has an obvious and significant role to play in influencing rural resource use and safeguarding environmental values through planning controls, it is clear that adequate resourcing is needed to deliver its potential. This includes access to specialist expertise and financial resources and increasingly space to collaborate with others to achieve positive environmental outcomes.
- In the UK – at least at the present time – many claim that cuts in central and local government funding and the policy approach adopted limits the role of land-use planning to mere 'policing' of environmental restrictions.
- Faced with continuing economic problems and rising demand for housing and transport, environmental and economic ends have become increasingly juxtaposed. Environmental policy is responding to this in the form of an ostensibly more integrated ecosystem approach, which conceptualises environmental values as ecosystem services, inherent to the functioning and sustainability of rural landscapes.
- Local Nature Partnerships (LNP) are a manifestation of this integrating agenda in the UK and they seem poised to engage a range of local stakeholders into networks working to operationalise new embedded visions of environmental quality and opportunities. However, without adequate

resourcing and support it seems that partnership approaches to environmental planning and management of this type will have difficulty in delivering these visions and grasping the opportunities for positive change that exist.

- The implications for the future of the countryside are not easy to judge, but in the UK at least it seems that the pace and direction of landscape change; habitat loss, degradation and fragmentation and changes in biodiversity; climate change; and resource scarcity are going to remain major challenges for some time to come.

Key readings

- Bell, S. and McGillivray, D. (2013) *Environmental Law*, Oxford University Press: Oxford.
  (Good overview of key areas of environmental law including climate change, water pollution, and waste management, plus separate chapters on town planning and conservation.)
- DEFRA (2005) *Securing the Future: Delivering the UK Sustainable Development Strategy*, HM Government: London.
  (Full text of the latest UK Sustainable Development Strategy, which informs many aspects of UK government activity.)

Key websites

- Convention on Biological Diversity Ecosystem Approach: www.cbd.int/ecosystem/
  (Provides access to wide range of EA resources including EA Operational Guidance, Case Studies and an EA Sourcebook.)
- Millennium Ecosystem Assessment: http://millenniumassessment.org/en/index.aspx
  (Provides access to all Millennium Ecosystem Assessment reports, associated statements and Green facts summary.)
- Ecosystems Knowledge Network: http://ecosystemsknowledge.net/about/events/nea2014
  (UK web portal to UK Ecosystem Services Assessment and associated research and case studies.)
- Natural England: www.gov.uk/government/organisations/natural-england
  (Provides access to all aspects of Natural England's work including land management, landscape and wildlife designations.)
- National Parks UK: www.nationalparks.gov.uk/
  (Web portal to all of the National Parks in the UK.)

Part 5

Future rural

11 Rural, urban or hybrid?

A retreating rural?

What does the future hold for rural planning in an era of accelerating global urbanisation? A great many issues, uncertainties and challenges have been identified and explored in the preceding chapters in relation to rural economies, communities and social equity, and the environment and landscapes of rural areas. But appropriate, sensitive and long-term planning and management policies and interventions need to be framed within a robust delineation of rural space (acknowledging the porosity of that delineation) and of course a broader strategy that is able to think across and integrate different domains. As Chapter 2 has shown, the territorial extent and the present and future purpose of rural space are culturally influenced, and particular national or regional cultural norms are in turn rooted in social history and, in relation to the countryside, a narrative of retreat from the rural and the growth of the industrial city. This is certainly the claim made in relation to the Western (and specifically British) experience of urbanisation that has followed a relatively linear trajectory of rural retreat, urbanisation and counter-urbanisation (Box 11.1), which seems to be a path familiar to many industrialised nations. Yet, as the next part of this chapter shows, this template is far from universal, and different cultural perspectives and expectations of the rural across different parts of the world relate to varying experiences of urbanisation and fundamentally different relationships between human societies and the natural environment.

In Western industrialised countries such as Britain, the decline of agriculture-related employment (Chapter 4) and the emergence of lowland mixed communities (Chapter 5) – which are increasingly disconnected from land-based economies – have become focal points of debate over the modern meaning of 'rurality' and the fundamental question of whether the histori- cally and culturally loaded term 'rural' retains currency. Contributing to this debate is the historic problem that the rural has often been presented in opposition to the urban: literally as 'other than urban' spaces; but this approach assumes a solid and broadly agreed definition of *urban* space. There

Box 11.1 Britain's rural trajectory

Britain's history of early urbanisation has been used to explain 'protectionist' attitudes towards the countryside that stem from nostalgia over a way of life that was quickly lost during the Industrial Revolution. The path of social change can be summarised as follows:

- 16th to early 19th century – successive acts of 'inclosure' driving land clearances across upland regions, but relative stability in the lowlands;
- 19th century – rural depopulation driven by combined effects of agricultural mechanisation and urban industrialisation;
- Late 19th century – rural celebration in art and literature perpetuating nostalgia for the lowland rural 'idyll' and the development of tourism to upland wild places;
- Early 20th century – development of conservation movements including the CPRE (1926) and the enactment of legislation to control outward urban sprawl;
- Mid 20th century (post-war period) – industrialisation of agriculture, protection of rural areas from development, inauguration of national parks and protected landscapes;
- Late 20th century – decline in agriculture's share of national GDP and local employment, emergence of lowland 'mixed communities' counter-urbanisation and second home ownership due to increasing car ownership and improved access with the development of national road infrastructure, but decline in rural amenities and public transport services as traditional local markets shrink;
- Today – existence of a rural housing crisis, with macro market affordability problems exacerbated by acute local housing supply problems, counter-urbanisation and gentrification, while a decline in public transport continues.

needs to be a relatively simple binary between the rural and the urban for oppositional definitions to make sense. It was noted in Chapter 1 that ideas of the rural are contested, being socially as well as practically constructed. And the same is true of the urban. The neat delineation of the 'urban' (as an entity separate and distinct from the non-urban) has been a subject of considerable debate, and as with the rural it is sometimes presented as a graded space, existing as a continuum from the most to least urban, but all nevertheless 'urban'. Taken to its extreme, there are today those who would argue that all space is essentially urban, either because it is obviously urbanised or it serves the needs of urban populations/economies, meaning that everyone and everything everywhere is urbanised. Those landscapes that might have been depicted as rural in the last eleven chapters are in fact 'extended urban landscapes' (Brenner, 2013). They are the products of interconnectedness (or 'connexity' – see Chapter 3) and inseparable parts of a global (or 'planetary') urban culture. Is this true? Has the label rural lost all meaning and currency? Or does the conceptual homogenisation of space, through claims of pervasive and universal urbanisation, in fact sidestep all of the analyses and critiques that point to a peculiar rural cultural and economic condition? This issue is picked up in Part 2 of this chapter.

Those who point to the existence of extended (or 'linked') urban landscapes and realities are not necessarily white-washing important spatial distinctions, or the need for tailored governance or planning approaches, but language is important. Ideas of the rural or the countryside are not merely historical constructions. They have currency and are imbued with sometimes deep cultural meaning. However, proponents of a graded urban prioritise the idea of interconnectedness over difference, although that is not to say that connectivity – in social, economic and physical terms (promoting it or dealing with its consequences) – is not a challenge in itself, being addressed in a variety of ways in different places. The challenge of connection illustrates the importance of geography, however places are badged. Practically, this challenge has often been central to governance and planning discourse, as demonstrated in Chapter 6 in relation to transport connectivity and the challenge of overcoming the jurisdictional barriers that often sit between cities and their hinterlands. City-regionalism has prevailed as a stable territorial tier in the Netherlands and Germany, while in Britain the first steps – following the revocation of regional strategies – towards a new urban–rural arrangement are perhaps expressed through the Local Enterprise Partnerships (LEPs). These appear to correspond, in many cases, with broad socio-economic geographies, following housing market areas (HMAs) and Travel-to-Work Areas (TTWAs). Increasingly, economic governance arrangements in England are acknowledging rural–urban connectivity and there is growing political interest in the creation of jurisdictional 'city regions' around some of the core cities. But there is sensitivity to such moves, not least from rural areas themselves, which often reject the idea of subservience to a city and feel that their local needs contrast markedly with those of the urban core. And this is despite clear economic connectivity, evidenced in daily commuting flows.

Why should this be so, and is Britain alone in its sensitivity to town and country distinctions? The next part of the chapter attempts to shed light on these questions by examining the trajectories, discourses and cultural norms elsewhere. By examining the experiences of urban–rural relations in different places, one aim is to interrogate the assumption of a receding rurality and expanding urbanity; another is to draw attention to the alternative notion of spatial 'hybridity' (Whatmore, 2002) and the idea that planning should concern itself primarily with the relationships and connections between places.

Part 1: Constructions of rural space

In much of Western society, the urban–rural dichotomy corresponds with a simple man–nature distinction. The urban is viewed as a product of human contrivance and the rural as something untamed, or at least closer in some respect to nature. Population growth and the intensification of urban use since the Industrial Revolution has accentuated this distinction, especially as economic activity is thought to have been concentrated in cities, leaving the countryside as a place of relative peace and quiet. Hebbert (1989) argues that 'deeper and endemic' tensions in advanced industrial societies between man and nature have been exacerbated by population growth and an intensification of resource use: urbanisation has come to sit in opposition to depopulating (or at least stagnant) rural areas. The denser the population and the higher the growth rate, the greater the potential for conflict and also for counter-urbanisation, viewed as a retreat from urban tensions. In general terms, this argument helps explain the long-running and deeper discussion over land-use and urban containment in the highly populated areas of southern England, where the 'defence' of the countryside is a regular battle cry. Yet the limits of this Western viewpoint concerning the relationship between man and nature needs to be acknowledged. Some oriental societies have developed radically different ontological stances grounded in particular cultivation practices, accepting that the rural is as much a product of human contrivance, and intensive use, as the urban. Hebbert (1989: 132) notes that the '[...] rice paddy is itself man-made and its irrigation channels require constant maintenance through collective human effort, just as the streets and canals of a town do'. Cultivated land in countries such as Japan has been fashioned by the assiduous efforts (contrivance) of man over several millennia, through land reclamation from the sea and from river estuaries and the terracing of mountain slopes that cover almost two thirds of the country, and maintained by cultivation channels, while rivers 'corseted in concrete and masonry' (Hebbert, 1989: 132) channel the excess rainfalls of the wet season, and recent afforestation schemes are as much about stabilising slopes from natural hazards and protecting river catchment basins, as they are about protecting scenic beauty and ecological diversity. Because so much of the Japanese landscape has been artificially constructed and maintained to support its rice economy, the urban and the rural appear equally man-made, diminishing any sense of a man–nature binary:

> 'The Rural' as a category embracing cultivated and wild areas but excluding urbanised areas, is a concept foreign to Japanese thinking, and so are the conditions of rural naturalness and vulnerability, and the consequent need for protection.
>
> (Hebbert, 1989: 132)

In contrast to the prevailing (if not exclusive) Western view of man as separate from nature, Hebbert describes the 'gregarious urban way' (Hebbert, 1989: 138) that many Japanese enjoy their mountain scenery – with adverts, souvenir shacks and so forth, while an 'absence of demand for artificially preserved sense of remoteness and facilities for private communion with nature' (Hebbert, 1989: 139) confirms the pervasive cultural viewpoint of man at the heart of – and central to – wider nature and landscape.

In the West, the rural has been romanticised, not least by the Romantic Movement itself, which emerged as a by-product of the Industrial Revolution. The rural came to be imbued with an interpretive cultural significance. It sat in opposition to what had been 'lost' because of urbanisation: the peace and quiet, and even the godliness of a past way of life. Romantic notions eventually gave way to a broader social construction of the countryside, which nevertheless continue to overpower physical and economic delimitations to a significant extent (see Chapter 1). The rural, as noted on several prior occasions, has been socially constructed or produced. It is differently perceived and experienced from one place to another. Building on Lefebvre's (1991) notion of the social *Production of Space*, Halfacree (2006) ascribes three elements to rurality: 'rural locality' (encapsulating the socio-spatial practices of production and consumption); 'representations of the rural' (which strictly relates to those expressed by capitalist interests, but which could be more liberally expanded to include cultural representations); and 'lives of the rural' (constructions tied to social practices and norms). Critically, Halfacree explains that Lefebvre himself did not distinguish between the urban and the rural, being opposed to 'dualisms' such as town and country which he viewed as being unified in an 'extended urban' space under capitalism (Brenner, 2013). Halfacree's three-fold model may therefore be regarded as an attempt to add resolution or shades to a palette; to re-insert the social-rural, while attempting to maintain Lefebvre's 'concrete universal' of *'production* that underlies capitalist spatiality' (Halfacree, 2006: 49 [author's original emphasis]). However, acknowledgement of a socially (and economically) distinct rural may be viewed as undermining, if not entirely contradicting, the 'extended urban' concept. It was shown in Chapter 5 that rural society has been transformed in the twentieth century, with those transformations underpinned by critical economic changes. A changed society has produced its own space: a space – in Britain at least – that is rooted in many of the ideas that emerged during the heyday of the Romantic Movement, and which is clearly defined in opposition to the urban. The English, in particular, are culturally unreceptive to the notion of spatial hybridity, under the guise of extended urbanism or some other model of graded urban. There is strong cultural attachment to the rural. But is this a uniquely English (or perhaps British, but see Chapter 1 on critical differences) distinction, born of a particular set of circumstances that other countries simply do not share?

The socially constructed rural

Western man's relationship with the land has long been regarded as symbiotic, as reflected in the language – the 'husbandmen' tending to a vulnerable 'mother earth' for example (Berry, 1977) – rather than intrinsic and inseparable as was the view that developed in the rice economies of the orient. The celebration of the rural that has developed in the West may be interpreted in part as a general reaction to a 'divorce' from the countryside that urbanisation has brought, but also, and perhaps critically, the pace and scale of change. The rural idyll was conjured from the imagery of folk tales and half memories of rural ways that had either disappeared or produced only a selective account of reality. Selective or not, real or imagined, cultural depictions of the rural undoubtedly became deeply embedded in the social psyche of industrial European societies: Britain with its lowland pastoral landscapes depicted in the works of Hardy and Vaughan-Williams, or Germany with its fascination for the fairytale forests recounted in the tales of the Grimm brothers, for example. The celebration of the bucolic tended to be limited to the landscapes that man could control, however. Beyond the farmland fields and friendly woods, the wild lands and deep forests stood dark and sinister as savage places to be feared and avoided. Even Britain's comparatively tame upland areas, which were almost entirely free from dangerous animals by the eighteenth century, were poorly regarded. In 1791, William Gilpen declared that '[...] there are a few who do not prefer the busy scenes of cultivation to the greatest of nature's rough productions' (Gilpen, 1791; quoted in MacFarlane, 2003: 14). A change of perception in the eighteenth and nineteenth centuries came partly with the conquest of Europe's high mountains and the emergence of a taste for the splendor of many uplands. Reflecting on this change, the author Robert MacFarlane notes that '[...] the qualities for which mountains were once reviled – steepness, desolation, perilousness – came to be numbered as their most prized aspects', continuing that '[...] to contemplate it now is to be reminded of a truth about landscapes: that our responses to them are for the most part culturally devised' (MacFarlane, 2003: 18). That is to say:

> We *read* landscapes, in other words, we interpret their own forms in the light of our own experience and memory, and that of our shared cultural memory [...] William Blake put his finger on this truth: 'The tree,' he wrote, 'which moves some to tears of joy is, in the eyes of others, only a green thing which stands in the way'.
>
> (MacFarlane, 2003: 18)

Only a rather general literature exists on the relationship between cultural celebration of the rural, land-use and planning; the linkages being diffuse and difficult to map out in terms of cause and effect. Despite the uncertainties surrounding causality, it can be said with considerable confidence that cultural depictions and celebrations have generated both the enthusiasm and the economic value to influence planning. Muir's writings on the Yosemite valley in California and Wordsworth's guide to England's Lake District (Figure 11.1), in which he described the area as '[...] a sort of national property in which every man has a right and interest who has an eye to perceive and a heart to enjoy' (Wordsworth, 1835: 92) were instrumental in the designation

Figure 11.1

Patterdale, Lake District National Park
Source: Jenny Post

of these areas as National Parks, and indeed the creation of an entire system of parks in the United States and Great Britain. In 1810   the same year that Wordsworth's guide established walking tourism in the Lake District, Walter Scott's 'Lady of the Lake' created a 'literary phenomenon' (Brown, 2012) that had an immediate impact in drawing hordes of tourists to Loch Katrine and a lasting legacy in the form of the Loch Lomond and Trossachs National Park (established in 2000).

The wild landscapes that later became national parks provided a point of common interest on both sides of the Atlantic and also a point of cultural divergence. In Britain, these were seen as places of romantic nostalgia: to preserve as reminders of a bygone times, in contrast to a rather more forward-looking North American view of the wilderness as places for man's betterment. The Romantic Movement of Europe and the Transcendentalists of North America had common roots in a concern for the wellbeing of urbanising societies, which had turned their backs on their rural past, and which were experiencing the rapidly accumulating loss of the countryside. But whilst the Romantics regarded the rural as a necessary anchoring for society to its past, the Transcendentalists saw it as the necessary ingredient to future fulfilment, allowing as it did, an 'occult relationship between man and the vegetable' (Emerson, 1985: 6). The higher state that man achieved in nature was widely depicted in the works of key writers – Emerson, Thoreau and Muir – and reverberated clearly in the landscape design and planning of Frederick Law Olmsted, a commissioner for Yosemite, creator of New York's Central Park and America's garden suburbs and, arguably, one of the greatest forces in the history of American planning. Escape into America's wild lands offered emancipation – escape from the oppression and tyranny associated with the 'old world' and its 'privatised' landscapes where '[t]he great mass of society, including those to whom it would be of greatest benefit, [are] excluded from it' (Olmsted, 1865: 22).

The mass-produced automobile became the means for emancipation from the early 1900s onwards. Indeed, as early as 1912 John Muir wrote that '[...] all the western mountains are still rich in wildness, and by means of good roads are being brought nearer civilization every year' (Muir, 1912: 71); a thought reflected in President Lyndon B. Johnson's address to the US congress in 1965, in which he declared that '[...] by making our roads highways to the enjoyment of nature and beauty we can greatly enrich the life of nearly all our people in city and countryside alike' (Johnson, 1965: 175). Moreover, Johnson poignantly declared that:

> [...] Concern is not with nature alone, but with the total relation between man and the world around him. Its object is not just man's welfare but the dignity of man's spirit [...] protection and enhancement of man's opportunity to be in contact with beauty must play a major role. This means that beauty must not be just a holiday treat, but a part of our daily life. It means not just easy physical access, but equal access for rich and poor, Negro and white, city dweller and farmer.
>
> (Johnson, 1965: 175)

In actuality the protected parks – and the automobiles necessary to access them – have produced rather different outcomes from those envisaged by the early pioneers. The North American wilderness exists as tracts – extensively so in parts – but counterpoised by industrial-grade agriculture across much of the lowlands and the outward sprawl of cities and development along major roads (Turner, 1996).

UK and US National Parks are illustrative of different approaches to planning and land management. One is highly protectionist and seeks to conserve landscape features, sometimes at considerable social and economic cost, whilst the other seeks to promote what it views as positive use of land. Green belts in the UK (see Chapter 10) seem also to be indicative of a particularly conservative brand of planning and can be contrasted with more laissez faire approaches, of the type that delivered sporadic speculative development in the Republic of Ireland and other European countries during the economic boom years of the early twenty-first century (Satsangi et al., 2010). But the former produces its own problems; its reliance on tight boundaries and presumption that all development should sit within them leads to a range of social and economic pressures. This was seen in Chapters 6, 7 and 8, with small village locations being starved of all sorts of development, producing undesirable patterns of back-commuting in support of services and a broad social reconfiguration (i.e. gentrification) of much of southern England. The defence of tight boundaries (which is sometimes simply a defence of the rural), either around growing towns or village envelopes, may create a range of conflicts when there are limited options for accommodating, for example, housing growth (Hamiduddin and Gallent, 2012; Gallent et al., 2013). The point is that these outcomes are socially-specific: US national parks are a product of a particular experience and discourse, as are their counterparts in Britain. Similarly, planning systems are calibrated to cultural and societal expectations: they deliver against a dominant discourse, delivering a majority countryside with its particular tyrannies and its regular failures in respect of minority needs.

Part 2: Towards planetary urbanisation?

Despite these more qualitative and cultural perspectives, which have delivered different policy and planning approaches in different parts of the world, the apparent convergence of the urban and the rural that is reflected by economic modes of production (Chapter 3) and by social habits and lifestyle patterns (Chapter 5) has recently been formalised in a theory of 'planetary urbanisation' (Brenner, 2013; Brenner and Schmidt, 2011). This outward projection of urban capitalism into the countryside is consistent with a long-standing Marxist view of unified space – not in itself 'inert' (Halfacree, 2006: 44) but a 'socially produced set of manifolds' (Crang and Thrift, 2000: 2) within which localities are formed 'by social processes or, less passively, both inscribed and used by social processes, products and means of production' (Halfacree, 2006:45). Halfacree acknowledges the difficulties in identifying rural space from a purely political economy perspective in today's developed world, where modes of production relate decreasingly to locality; a perspective reinforced by Amin and Thrift (2002: 1) who note that:

> The city is everywhere and in everything. If the urbanized world now is a chain of metropolitan areas connected by places/corridors of communication (airports and airways, stations and railways, parking lots and motorways, teleports and information highways), then what is not the urban? Is it the town, the village, the countryside? Maybe, but only to a limited degree. The footprints of the city are all over these places, in the form of city commuters, tourists, teleworking, the media, and the urbanization of lifestyles. The traditional divide between the city and the countryside has been perforated.

The above excerpt projects a very urban view of the 'non-urban', which rural theorists would strongly contest, and which will be unpacked later. Fundamentally, however, proponents of the extended urbanisation idea contend that capitalism is an essentially urban condition that shapes socio-spatial relations and therefore our own constructions of that space; rural space (as with all other forms) is 'imaginative' (and possibly romantic or transcendental) rather than 'immanent materially' (Mormont, 1990). Late capitalism, it is contended, has emerged from its urban 'command centres' to envelop surrounding regions and rural spaces with an 'urban tissue' (to use Lefebvre's own term) that has left very few parts of the world – the Amazon included – untouched by its physical imprint in the form of settlements, installations or transmission lines, communications and connectivity, production modes, and attitudes. In other words, Barbieri *et al.* (2009: 254) contend that:

> Extended urbanization carries within it the socio-spatial processes and forms that are proper to industrial capitalism, manifested both in its earliest expression, the industrial city, and its contemporary global multiple manifestations.

Yet, one of the key problems of this social representation of space is that it says little about the distinctiveness and identities of the localities and communities themselves – the reality of the physical world that individuals and groups hold and construct as being rural, rather than relying solely on generalised social and economic forms and functions. Bringing these two aspects together allows for localities to be '[...] *visualized* as islands of absolute space in a sea of relative space' (Smith, 1984: 87 quoted in Halfacree, 2006), with rural spaces identified as locations with 'distinctive causal forces' (Hoggart, 1990: 248). Bland descriptors are therefore unsatisfactory and greater resolution is required to identify the processes that have produced distinctive spaces. If all of this is accepted, then Brenner's depiction of 'extended urbanisation' (as a descriptor of the spaces beyond the core urban) seems too crude, being the product of a low-resolution examination of socio-spatial processes. As Cloke (2006: 18) observes:

> The concept of rurality lives on in the popular imagination and everyday processes of the contemporary world. The rural stands both as a significant imaginative space, connected with all kinds of cultural meanings ranging from the idyllic to the oppressive, and as a material object of lifestyle desire for some people.

A global survey of the terminology hints at the different ways in which non-urban spaces are viewed – rural, country, countryside, outback, wilderness, frontier and so forth (Cloke, 2006). Yet across the nomenclature, the rural – and its equivalents – are typically conceived in opposition to the urban; defined, to a large extent, by the *'other-than-urban'* (Cloke, 2006: 18) and rather less sharply shaped by strong and commonly agreed constructs of rurality (e.g. population density, economic activity, settlement pattern and so on: see Chapter 1). Indeed, matters are clouded further by perceptual or imaginary constructs of the rural that exist. These varying and contested perceptions of rurality give it a definitional 'slipperiness' that continues to be lubricated by the detachment of rural space and rural society. Indeed, Mormont (1990) believes that the changing relationship between society and the occupation of space means that 'rural society and space can no longer be seen as welded together'; a 'multiplicity of social spaces' therefore congregate at the same geographical point. This detachment of society and space has arisen from the well-documented physical outward migration of 'traditional' rural communities and the inward movement of affluent and highly mobile urban dwellers, while matters have been blurred further by processes of 'cultural urbanisation' that have shaped, and have been shaped by, economic change (Cloke, 2006: 19), leading to notions of 'post-ruralities' (Murdoch and Pratt, 1997). Such developments could point to the end of the rural (Hoggart, 1990) or at least a 'hollowing out' of distinct spatial delineations (Marsden, 2006: 3).

Given these developments and uncertainties, the concept of planetary urbanisation seems tenable *prima facie*, yet at least three significant doubts remain. First, the rural has always existed in the shadow of the urban. Why, therefore, should the increased global inter-connectivity facilitated by transport and communications technology be automatically cited as evidence of rural demise in the sense of the rural being dismissed as a historical construct lacking currency? Cities have always had an extensive footprint – the earliest acts of urbanisation were predicated on an agricultural surplus across a region, entailing transportation, trade and exchange. Except in the remotest, most disconnected regions of the world, left untouched by the process of urbanisation, urban and rural worlds have arguably been function-ally entwined since the birth of the city and the emergence of trade. Developments in transport and communications have undoubtedly caused an intensification of relations: everything *has become* more connected, but to a large extent this is because *everything is* connected. Therefore, when the complexity of social, spatial and physical relations is properly interrogated, supposed spatial divisions – lines on maps – appear to dissipate or are at least highly porous. But can porosity, or a blurring of urban and rural boundaries, be taken as evidence of diminished rurality? Returning to the first point, the rural has always existed in the shadow of the urban and as that shadow deepens and lengthens, the sense of the rural as distinct – and as something to be cherished – tends to grow. Urbanisation accentuates rurality. This seems to be particularly true in Britain, and especially England.

Second, the blurring of urban and rural boundaries seems, in some instances, to have resulted in a penetration of rural features into urban landscapes. Cloke (2006) argues that the 'ruralisation of the urban' is an under-researched phenomenon, but can be illustrated in two ways. *First,* recent models of housing, commercial and leisure developments in North America '[...] destabilize ideas about the city and country by producing city/country hybrids which owe as much to a bringing-nature-into-the-city as to a spreading-the-city-into-the-country' (Cloke, 2006: 19; citing Wilson, 1992). The examples cited to support this claim are fairly recent, all being from the late twentieth century. But earlier examples can also be found, particularly of great urban parks being introduced into cities in order to soften urban form and bring a sense of nature and wildness. In the US, the great landscape architect Frederick Law Olmsted created New York's Central Park for this reason and also attempted a union of city and nature with garden suburbs at Riverside, Illinois, and at Berkeley, California. Similar ideas were pursued in Britain, not least through the Garden Cities movement (see Chapter 2) led by Ebenezer Howard, who attempted to combine what he regarded as the best of town and country in a self-contained hybrid. The urban nature concept has been taken a step further in recent years with the advent of so-called 're-wilding' (Foreman, 2004): that is, the return of previously developed land to nature. Examples in London include the creation of wetland reserves from the site of a former sewage works at Barnes, a former gasworks on the Greenwich Peninsula and a former military range at Rainham (2020 Vision, 2010). The *second* illustration of ruralisation is the uptake of the 'village' concept in urban development and regeneration (Urbain, 2002; quoted in Cloke, 2006), which signals the search '[...] for a set of virtues in the city which are more commonly associated with the rural – seemingly fundamental and permanent virtues such as protection, solidarity, community spirit and identity' (Cloke, 2006: 19). It was noted in Chapter 5 that contrasting urban and rural patterns of sociability are viewed as a key marker of social difference, with some people drawn to the community life of villages for their perceived closeness and familiarity. Following the Urban Task Force (UTF, 1999), the creation of urban villages, which might retain and attract population, were viewed as a vehicle for delivering a broader urban renaissance. Cloke concludes that '[...] the blurring of rural-urban distinctions is bringing crucial changes to urbanity as well as to rurality' (Cloke, 2006: 19). Again, and as noted in Chapter 2, the fusion of town with country famously provided the basis for the Garden City model that in turn served as the springboard for future landscape-design-influenced urban planning models, including the British New Towns, Radburn and other American Garden Cities, Modernism and its preoccupation with buildings set amidst open parkland spaces, and, finally, suburbia itself – the most pervasive amalgamation of town and country.

The idea of extended urbanisation may, as noted in Chapter 1, merely be an attempt to reset the boundaries of urban studies: to engage in a form of 'intellectual imperialism', annexing the rural as a new frontier of the urban. If there is a crisis in urban studies, marked by a move to claim new territories, then that crisis seems remarkably similar to the one experienced in rural studies in the 1990s (Cloke, 2006). That crisis proved particularly fruitful, stimulating considerable debate as to the direction of the 'discipline' and the very nature of the rural: debates which have been summarised, all too briefly, in this Chapter and in Chapter 1. On the other hand, the idea of extended urbanism could morph into a broader debate on spatial hybridity, ultimately undermining the notion of a 'rural-urban dichotomy' (Hoggart, 1990).

Concluding remarks

Although the idea of rurality retains currency, social constructions of the rural will inevitably become more diffuse as a result of the global exchange of ideas and values facilitated by travel and telecommunications technology. The same processes that have given rise to greater individualism and social plurality and complexity could arguably have the opposite effect on urban and rural places, generating a muddled but ultimately ubiquitous sense of the rural, as functioning in a singular way and as subordinate to the urban. London is a case in hand: what does its population now think of the English countryside; how does it view its function and its relationship to the city? Does its population view the countryside within a shared cultural frame? The answer is of course no. The flow of people between European core cities has resulted in more mixed and fluid understandings of the rural, resulting in the dissolution of once distinct perceptions of the countryside. Yet these arguments privilege the urban viewpoint, offering a meta-narrative that ignores the lived reality of rural places.

The thematic chapters of this book have privileged the rural. They have each pointed to physical, cultural and economic features that are specific to the countryside and also to governance arrangements and policy interventions that acknowledge the existence of challenges that are differently constructed if not unique. It is of course true that the narrative of the countryside (recounted in this chapter, but set out in more detail in Chapter 2) links the rural to the urban: initially as a producer of urban goods and latterly as a place of urban consumption. But however important this relationship has been, and remains, public policy and planning is confronted by very particular challenges in the countryside. These were detailed in earlier chapters but can be summarised as (a) adaptation to the changing function of the countryside, through the cycle of productivist, consumption and neo-productivist episodes; (b) the increasing complexity of rural needs in an undeniably urbanising world; and (c) deep cultural attachment to the countryside, which results in particular development and landscape protection challenges (see Chapters 8 and 10).

All of these challenges are relational. The currently required adaptations (a) are those that would permit greater accommodation of development and renewed focus on food security and the procurement of ecosystem services. The consumption phase (which continues but now overlaps with a post- or neo-productivist period) is one of *in situ* demand for rural resources, largely land for new housing. But this now needs to be balanced, in the neo-produc-tivist period, with growing demands for increased food production in a world where food stability and security have again become important issues (Chapter 4). These are urban–rural challenges, generated by seemingly relentless urban-isation and also distant geo-political crises. The increasing complexity of rural needs (b) is also a relational issue, produced by the collision of rural and urban social structures during the period of consumption. This collision, and the import of hitherto urban wealth, has generated mismatches in political power, wealth and wellbeing. As noted earlier in this chapter, it has created a '[...] majority countryside with its particular tyrannies and its regular failures in respect of minority needs'. Its driver has been counter-urbanisation and the outcome has been gentrification in many rural areas, particularly but not exclusively in lowland England. And despite the social plurality of nearby cities, many rural areas are characterised by a shared and deep cultural attachment to the countryside (c). This may be nostalgic or transcendental in its origin, but it is now expressed in protectionist attitudes and in protection-ist outcomes realised through the exercise of growing political power. As is often remarked, the urban escapees utilise their capital and project their 'habitus' (an amalgam of tastes, values and behaviours) in order to transform places into a preferred reality (Bourdieu, 2005). The countryside has been socially reconstructed during the twentieth century by a returning population, which now holds entrenched views on future development.

The conclusion from this must be that the modern countryside, examined in the previous ten chapters, is a hybrid – produced in the shadow of the urban but presenting its own specific challenges to public policy and land-use planning. There are certainly strong connections between town and country and these connections have existed for thousands of years, intensifying through the twentieth century. Early intensification was social in nature, marked by the displacement and replacement of population. More recently, the environmental and ecological linkages between places have become increasingly apparent and today the rural finds itself on the front line of many global battles, relating to resource scarcity and climate change. This is an age in which the rural is increasingly defined by the challenges it faces and those challenges are found to be almost entirely a product of the inexorable link between town and country and the inter-connectivity of all human activity, wherever it occurs.

Summary

- By virtue of enhanced connectivity, all places can be seen as components of a global or planetary urban system. The rural label is a historical construct that has had its day. This is the view of some commentators, who argue that the concept of the 'extended urban' now adequately describes all spaces outside of core cities and other urban centres.

- But in this chapter, it is suggested that the rural has always – throughout the modern period, and even before the Industrial Revolution – existed in the shadow of the urban. Interdependencies have been apparent for millennia, but interdependence does not diminish the value of a label that signals distinct social, economic and landscape challenges.
- Equally, the idea of the social 'production of space' suggests how places are imbued with cultural meaning, which has important implications for how places are governed and planned. Social constructions of the rural are certainly expressed in public policy and especially in different planning approaches, which variably promote development or protect the landscape from 'urbanising' influences.
- However, a great many of the challenges faced by the countryside are relational, in the sense that they have been generated by rural–urban connections. The policy environment is a hybrid one, often dealing with urban pressures and anti-urban representations of the countryside. Hybridity is a feature of much of the countryside, even the remotest locations, as nowhere is entirely free from urban influence, or from external connection. That part of the 'extended urban' thesis has broad validity.

Key readings

- Halfacree, K. (2006) Rural space: constructing a three-fold architecture, in Cloke, P., Marsden, T., and Mooney, P. (eds) *Handbook of Rural Studies*, Sage: London.
  (This is a generally useful overview of how rural space/rurality can be understood.)
- Brenner, N. and Schmidt, C. (2011) Planetary urbanisation, in Gandy, M. (ed.) *Urban Constellations*, Jovis Verlag: Berlin, pp. 10–13.
  (This is a very short note by Brenner and Schmidt on the concept of extended urban landscapes, which is expanded upon in Brenner, 2013.)

Key websites

- The Urban Theory Lab at Harvard's Graduate School of Design (GSD): http://urbantheorylab.net/.
  (Led by Professor Neil Brenner, the pages of the Urban Theory Lab outline the work of Prof Brenner and others on planetary urbanisation.)
- RUC2011: www.ons.gov.uk/ons/guide-method/geography/products/area-classifications/2011-rural-urban/index.html.
  (This is the Office for National Statistics' page on the 2011 urban–rural classifications. Links to methods and data can be found here. There is also the suggestion of 'higher level geographies' to follow.)

12 The paths ahead

●●●

Introduction

There is no doubt that we are living in a time of increasing urbanisation and pervasive urban influence. As we have seen, faced with the growing interconnectedness of urban and rural localities (and indeed of all places across the globe) many commentators have questioned whether it still appropriate to distinguish between urban and rural areas from a planning and development perspective. We have, however, argued throughout this book that the distinction remains important. Rural areas continue to have their own particular representations, functions and political-economies that present planning with unique challenges. It also should not be forgotten that rural areas are still home to large numbers of people and are very extensive in their land coverage. Even in a country as highly urbanised as Britain, approximately 20 per cent of the population live in rural areas and nearly 80 per cent of the land area is not built-up and remains largely in agricultural use. It is important therefore that rural interests are not overlooked or simply merged with urban debates. Indeed in this final chapter as we look at the paths ahead we will argue that rural planning seems to be at a watershed and that a major reappraisal of planning approaches is needed. What form these approaches might take is perhaps debateable but there seems to be little doubt that careful and imaginative planning for rural areas will be increasingly important as we step further into the twenty-first century.

This final chapter starts by distilling from previous chapters the key trends and challenges that are likely to be the preoccupation of rural planning over the coming years. It then discusses what these trends and challenges might mean for a future rural development paradigm. In the second part of the chapter, the discussion focuses upon Britain and it revisits the scope of rural planning – its various components and functions – as introduced in Chapter 1, and considers what the paths ahead may look like here.

Part 1: Key trends and challenges

An important starting point in thinking about the future of rural planning is that the ongoing change and transformation of rural areas, which has been such a feature of the last 100 years, can be expected to continue for the foreseeable future. Equally, the increasing diversity and divergence in rural experience, which has also been a trend in recent times, is likely to remain a feature of rural life. Against this backcloth, the past chapters have identified a number of economic, social and environmental trends and challenges that are likely to be the focus of rural planning in years to come. These are summarised below.

Economy and land

Despite the increasing dominance of urban areas in the economic fortunes of nations all over the world, Chapters 3 and 4 revealed that there has been something of a reappraisal in recent times of the economic potential of rural regions and of the contribution they can make to national GDP. While there is continuing concern about the fortunes of some areas of 'production' countryside (often associated with falling population and employment and remoter rural regions), the economic growth being experienced in many areas of 'consumption' countryside (often associated with rising population and employment and rural regions with good connections to cities) has attracted increasing attention. So too have the growing number of rural regions with quietly successful diversified economies of a different kind, some being focused towards private service sector activity (in areas very close to cities) while others include strong elements of manufacturing. In looking to the future therefore, a whole range of rural development trajectories and economic profiles seem to be opening up, many of them with real potential to contribute to national economic growth. In all cases, however, there is little doubt that agriculture will remain the most extensive rural land-use and its development path will continue to be a significant contextual factor. In some areas, it can be envisaged that the sector will respond to global market pressures and demand through an intensification of activity bringing new farm-related built development to the countryside and marked changes in styles of land management. Such changes will challenge many people's preconceptions about what the countryside should look like. Elsewhere, agriculture may be at the centre of efforts to mobilise local assets, connect with local markets, shorten supply chains, and exploit the multi-functional potential of rural areas. In both instances, it will remain the backdrop to much rural economic activity and a distinctive feature of rural life. Another distinctive feature, particularly in poorer rural regions where economic drivers are weak, is likely to be the continuing growth of the social enterprise sector providing a more self-reliant development model that commoditises local assets and skills and retains benefits within local communities.

People and communities

Indeed, as discussed in Chapters 5, 6, 7 and 8, in all areas it can be envisaged that the strength of soft (social capital) assets will be increasingly central in determining the varied economic and social fortunes of the countryside. With the spread of neo-liberal philosophies and the further hollowing-out of the state anticipated, it seems inevitable that community action will become an even more important aspect of rural life, and a vital response to meeting a diversity of rural needs. Conscious development of social capital, harnessing the energy of communities to solve complex local challenges will be key, particularly in more remote and sparsely populated localities. Among these challenges, transport inevitably will continue to be one of the main issues facing rural dwellers. In many cases, community action can play a vital part in ensuring accessibility to employment, services and activities; providing some measure of personal mobility for the young and the old; and substituting declining public transport provision with more integrated/ community-based alternatives. Community-led initiatives are also likely to be critical in those areas experiencing ongoing decline in rural services. While ICT can be expected to play an increasing role here as well, it is only likely to offer partial solutions to sustaining acceptable levels of provision. Housing will also remain a central challenge both in areas facing rural decline and depopulation and areas facing counter-urbanisation and gentrification. These processes will continue to present a number of problems for rural communities, requiring either a general development and investment response or targeted assistance with the aim of widening housing access. Looking to the future, carefully targeted and locally tailored approaches to rural housing provision seem vital, with (again) community-based planning, perhaps being the appropriate scale and vehicle for addressing very local needs in many areas.

Landscape and environment

The long-standing environmental concerns discussed in Chapters 9 and 10 are also anticipated to continue to be central to the future of rural planning. For example, the scale and pace of landscape change in rural areas associated with increasing urbanisation and human activity is expected to remain a core concern. So too is ongoing habitat loss, degradation and fragmentation and changes in biodiversity, which reflect the ecological consequences of current patterns of human development in rural areas and the global reduction in areas of high biodiversity value. In both cases and in line with recent environmental thinking, it is anticipated that policy attention will shift from protection and conservation of designated sites and species to enhancement of all areas. Interest in re-wilding of some remote rural areas is expected to grow. As a result of international agreements it is envisaged that activity will increasingly make reference to the Ecosystem Approach and Ecosystem Services thinking and valuation of environmental assets will become a more prominent reference point in rural planning debates. Some rationalisation of established environmental designations is certainly needed and may also occur. Alongside these traditional rural planning issues, climate change associated with global warming and predicted impacts on sea-level rise, climate variability and water and food availability will provide a controversial setting for rural planning. The scale of mitigation activities will be highly dependent on political shifts but activity related to climate change adaptation and resilience can be expected to be the focus of increasing attention with significant impacts on rural land use envisaged, such as the development of natural flood-control solutions and increased woodland cover. The same can be envisaged for resource scarcity. Global population growth and increasing demands for food, timber, minerals and other resources that are mainly sourced from rural areas will inevitably result in intense competition for rural space and growing sensitivity to energy costs. As a consequence, levels of national, regional and local resource self-sufficiency are expected to receive greater political profile and this may see the development of an era of neo-productivism in rural policy. Local-level action to address environmental issues can also be expected to be of growing significance. However, developing understanding of the complexity and interrelatedness of environ-mental concerns and the scale and nature of the action needed mean that active involvement of international, national and local government and associated agencies is likely to be of continuing importance here.

Rural complexity and planning

The growing complexity of rural agendas is by no means confined to envi-ronmental matters. Indeed this has emerged as a common theme throughout this book. Although traditional perspectives that valued the 'simplicity' of rural life were never well-founded, they are certainly out of step with the countryside of today and tomorrow. In many areas there is rising appreciation of the complexity and contested nature of rural planning dilemmas. For example, conflicts between different environmental viewpoints and between pro-development and development-restraint perspectives do not seem to be ebbing and can only be expected to increase as global population growth and urban expansion intensify pressure on rural areas. There is also mounting complexity of institutional arrangements and power relations related to rural development and planning. Drivers for this are coming from many directions. For example the rising importance of international-level agreements and institutions in relation to environmental and agricultural matters is expanding the influence of institutions such as the UN and EU in a way that is very difficult for rural communities to grasp or connect with. Globalisation of market forces is having similar effects in relation to economic matters. At the same time social and political shifts in the wider global context including post-modernism and neo-liberalism are also bringing into question the appro-priateness of top-down national government interventions and promoting a systematic hollowing-out of the state in many countries. These trends in turn are adding to the complexity of the rural scene as traditional lines of responsibility are erased or blurred and leadership in rural planning, development and management becomes increasingly uncertain and unclear. Together with the growing appreciation of the 'wicked' nature of rural problems, this is resulting in a rise of partnership approaches to rural governance and community-scale action. However, there is little uniformity about the emerging arrangements, adding yet another layer of complexity to the picture. In this new 'composite' rural planning environment, traditional planning actors and mechanisms need careful recalibration if they are to play an effective role in the new scales and forms of governance that are emerging.

Early twenty-first-century rural development paradigm

Given these prospects, what might the rural development paradigm of the early twenty-first century look like? We have already seen in Chapter 2 the OECD's view on what this might entail, which also reflects much EU research. As shown in Table 12.1, the OECD envisage a broadening of rural development activities from a focus on agriculture and an ambition to equalise the economic performance of rural regions, to a much more diverse and multi-sectoral view of rural development aimed at realising the unique indigenous potential of rural regions. The OECD also sees a shift from state-led strategies based upon the provision of subsidies, to partnership action and strategically targeted investment. This certainly seems a very plausible summary of approaches to rural development in the early twenty-first century. However, it is perhaps only partial, reflecting the economic focus of the OECD and also of much EU activity in the post economic crash era. A rounder view of what the paradigm might and perhaps should entail can be gained by reference to the General Assembly Declaration that marked the end of the United Nations Conference on Sustainable Development held in Rio in 2012. This contains quite a number of references to rural development that are worth reflecting upon. The Declaration indicates that agriculture is unlikely to fade as a focus of rural development attention due to rising food security concerns. It also highlights ongoing social concerns about enhancing rural transport and access to rural services and perhaps a subtle, but notable, shift in emphasis in environmental attention to notions of supporting ecosystem resilience.

A different view of the key tools for rural development also emerges within the Declaration, highlighting new technology, innovative financing mechanisms, local capacity building and trade as key means of implementation. There is consistency though with the OECD on the idea that partnership working will continue to be important both at local and at international scales. Two major differences in emphasis are however apparent. The first relates to the Declaration's emphasis on strengthening urban–rural linkages for the benefit of rural regions, which perhaps surprisingly does not feature as a core element in the OECD picture. The second difference relates to

Table 12.1 Twenty-first-century rural development paradigm (developed from: OECD, 2006 and United Nations, 2012)

	Old approach	OECD new approach	Additional elements from United Nations
Objectives	Equalisation; farm income; farm competitiveness	Competitiveness of rural areas; valorisation of local assets; exploitation of unused resources	Enhancing rural transport, access to rural services and employment; increasing agricultural productivity and food security. Building ecosystem resilience. Urban–rural linkages. Intra- and inter-generational equity
Key target sector	Agriculture	Various sectors of rural economies (rural tourism, manufacturing, ICT industry, etc.)	Agriculture, ICT
Main tools	Subsidies	Investments	New technology; innovative financing mechanisms; capacity building; trade
Key actors	National governments, farmers	All levels of government (supra-national, national, regional and local); various local stakeholders (public, private, NGOs)	New local and global partnerships

questions of equity. While the OECD suggests that this will be of diminishing concern (recognising the very variable context and prospects of rural regions), the UN declaration continues to place the promotion of social equity at the heart of development debates. This includes both equity between developed and developing countries in the current era and also with future generations. As rising global population places increasing pressure on natural resources across the world, it does seem unlikely that equity considerations will disappear from rural development discussions; and in addition to more localised equity considerations, global perspectives can also be expected to feature here.

Part 2: The future of rural planning across the UK

How this rural development paradigm will play out in different countries over the coming years will of course vary considerably, reflecting their unique economic, social, environmental, geographical and above all historic, cultural and political contexts. We also must not forget the potential significance of dramatic events in changing what might seem well established trajectories and shifting activities in very different and unexpected directions. As we conclude our reflection on paths ahead for rural planning with some thoughts on Britain, this latter point seems particularly important to emphasise.

For example, in our discussions so far, it is evident that the post-war period was an important watershed in rural planning in Britain. Following the hardship and social upheaval of the war years, the election of a socialist government in 1945 brought profound change that reflected a broad consensus on the future direction of the nation and of rural areas in particular. The country not only saw the establishment of the modern planning system as a critical component of the new Welfare State, it also saw the introduction of a raft of special designations protecting the natural beauty and natural heritage of large areas of the British countryside and through green belts stopping the sprawl of many of Britain's largest towns and cities. At the same time special exemption from planning control was given to agriculture and forestry, which were seen as benign guardians of the countryside and fundamental to national food and resource security. In the first edition of this *Introduction to Rural Planning* we charted how the framework of rural planning that had been established in the post-war era, had evolved and been built upon over time, but remained recognisable with the system that was created post 1945. Indeed a central argument in the book was that the passing of the 2004 Planning and Compulsory Purchase Act, in fact saw a strengthening of the statutory planning system with the idea that it should play a more significant role within rural as well as urban affairs. Not only did the Act define for the first time the purpose of planning – sustainable development as an all-embracing remit – it also reflected spatial planning concerns of the day and saw the planning system as playing a key role in integrating the diverse planning agendas for rural areas, bringing place-based coherence to the many formal and informal planning regimes that shape the development of the countryside. In hindsight these were heady days for the British (and particularly English) planning system, and the modernist aspirations of the 2004 Act were very much out of step with underlying social and political trends. It now seems that scepticism about delivery of the cultural change that was envisaged both for the planning community and for others who were being encouraged to accept planning's integrating remit was well founded. The book was written in the early phases of the new planning regime and it was already evident that carrying forward the spirit of the 2004 Act was proving challenging. It is also interesting to note that the book was published in 2008, just as the global financial crisis was breaking.

Although the full implications of the 2008 financial crises are still unfolding, it is becoming increasingly apparent that we may be facing another watershed era in the UK and in many countries throughout the world. There can be little doubt that the economic crash and the subsequent election of the Coalition Government have significantly changed the context in which the planning systems are operating across the UK. The 2010 General Election brought to the fore once again long-standing social and political undercurrents of post-modernism and neo-liberalism and established the most anti-planning government in the UK since the Thatcher era of the 1980s. It also heralded substantial cuts in public-sector expenditure, which have affected all parts of central and local government. These are still ongoing as we conclude the second edition of the book and they are resulting in a major slimming down of public-sector activities. As part of this development, the scope of statutory planning along with other aspects of the Welfare State is increasingly being called into question. It also appears that the broad consensus that existed in the post war era about the future of the countryside is continuing to dissipate as political and public attention focuses more and more on economic matters. This is illustrated for example in the pro- and anti-fracking debates that potentially have great significance for the future of some rural areas. Here, employment and economic growth and resource security arguments are stridently juxtaposed against environmental protection and climate change concerns and it seems at present at least that the former are gaining ground. Post-2008 therefore a very different setting for rural planning appears to be emerging. Quite where this is taking rural planning in Britain is difficult to judge at the moment but we offer some suggestions with reference to the various components and functions of 'composite' rural planning that were outlined in the introduction to this second edition of the book.

Public land-use planning

Despite these changes and clear efforts to streamline the public land-use planning system since the 2010 elections, it does seem that key features of the system are now so embedded in the British public consciousness that proposals for a major withdrawal of requirements for planning permission or streamlining of rights of public consultation would be met with a public outcry, with fears about development in the countryside likely to be prominent in any ensuing debates. For the foreseeable future therefore it can only be envisaged that local planning authorities across the UK will continue to play an important role in regulating new development and changes in land use in the countryside through development control/development management, and that these activities will be undertaken with reference to national policy and local development plans. However, while the legal framework may remain intact, what is much less certain is whether local authorities will have the resources necessary to continue to deliver competent public planning services. The ability to produce local development plans in particular may prove challenging. This is because successive rounds of public spending cuts have taken a heavy toll and hollowed out the capacity of many local planning departments to such an extent that they now seem on the verge of being unworkable. It is therefore interesting to note recent developments in Wales, where a new Planning Bill includes a series of streamlining reforms including new ministerial powers to merge planning authorities (Welsh Government, 2014). In England, the 2004 Act has already established a duty for local planning authorities to co-operate and some rural authorities have been among those that have used this to produce joint development plans often driven by budgetary considerations and efforts to share resources and costs. Further rationalisation of public land-use planning functions in rural areas can therefore be expected over the coming years as efforts continue to recalibrate the system to the new political and budgetary realities. As part of this, a reduction in the number – and rise in the scale – of rural planning authorities and the production of more streamlined and strategic level development plans may be anticipated.

Spatial or territorial planning

Against this backdrop it is evident that the ambitious notion (reflected in the 2004 Planning and Compulsory Purchase Act) of the statutory spatial planning system playing a central role in co-ordinating public, private and third sector planning activities in rural areas is unlikely to be realised. Although there remains a strong commitment to statutory planning in Scotland and Wales, in England it seems that there is no longer the political or public support for this approach, let alone the resources needed to achieve it. In addition, it now seems rather utopian and unrealistic to think that the statutory planning system is well suited to this task. The legal requirements associated with the production of statutory development plans these days are highly procedural, time consuming and costly. This reflects their legal status in guiding planning decisions related to the many land and development interests involved. While, it is envisaged that statutory development plans will continue to play an important role in rural affairs (not least in relation to co-ordinating public investment and tying together various aspects of public sector policy), their appropriateness for taking on a wider role, leading area visioning and co-ordinating the initiatives of private and third sector partners does not seem well-founded. In an increasingly fast-moving and uncertain world, mechanisms that are lighter in touch and fleeter of foot seem to be needed to provide spatial integration of rural agendas. In looking for alternatives however, it seems that the suitable mechanisms to deliver integrated action will vary greatly across the UK. We have seen in Chapter 10 for example that in some cases National Park and AONB management plans are beginning to play this type of role. Elsewhere it has been noted that the development of action plans associated with EU rural LEADER funding has provided a focus for collaborative and joined-up working (Shucksmith, 2010), so future EU funding rounds may prove a useful co-ordinating stimulus in some areas. It will also be interesting to observe the progress of the devolution debate, which has reached new heights and urgency since the 2014 Scottish independence referendum. This may ultimately result in a return of regional governance in England and ideas for integrated regional strategies for Northern England are rapidly gaining ground (IPPR North, 2014). These are receiving support from all sectors and have the potential to play a valuable regional spatial planning role. However, from a rural perspective a note of caution is needed. To date these initiatives are being led by city regions and are very urban in their focus. Much more attention to important rural agendas and of developing urban–rural linkages will be required if they are to play a positive spatial planning role for rural areas as well and not lead to further sidelining of rural interests.

Community action and planning

The post-2010 political scene is also notable in its emphasis on local community action. This is advocated by the Coalition Government as a key part of the social shift that is needed to take up the reins from the new 'slimline' state. Past rural experience has perhaps been influential in this development as it has provided many examples of the potential value of community-based initiatives both in delivering critical rural services and also in wider community visioning and integrated place-based spatial planning at the local scale. However, many have expressed scepticism about whether efforts can be scaled up sufficiently and in a coherent way to address the withdrawal of state involvement (Bishop, 2010). The Coalition Government's Localism Act 2011, however, reflects belief in the place of community action and potentially has important implications for the future of rural planning across the UK. It is significant in that it establishes a new statutory tier of planning in England – neighbourhood planning involving the preparation of Neighbourhood Development Plans and Orders led by local communities. As discussed in Chapter 5, and at various other points, if passed by a local referendum these plans and orders form part of the statutory local development framework. So far, experience of the new arrangements has been mixed and it has been subject to much criticism, but there is no doubt that rural communities have been among the leaders in using the new powers. Difficulties with the arrangements include: the lack of resources and expertise available to support plan preparation; the complexity of the procedural requirements involved; and less than enthusiastic assistance from some local planning authorities faced with rapidly diminishing budgets. Concerns about the significant variation in community capacity and willingness to take on the new powers and the uneven coverage of plans under production have formed part of the critique of the new arrangements which some feel simply add another layer of unhelpful complexity, divergence and inequality to the rural scene (Parker *et al.*, 2014). There is no doubt that experience of the new powers has been problematic and it is entirely possible if we come to write a third edition of this book in years to come, like the 2004 Act ambitions, 'Neighborhood Planning' will be seen to be yet another passing phase.

However, at the present time it feels as though trends to localism (which neighbourhood planning reflects) are deep-seated in rural areas (part of a reaction to globalising forces and the shortcoming of market mechanisms) and growing in urban areas, and that any proposals to withdraw community powers would be met with strong opposition. Indeed the next step might be to make these powers more workable. So looking to the future, a strengthening of community campaigning and lobbying, community control of local services and community-led local development and planning in rural areas all seem quite possible. The picture is however likely to be very variable across the UK with clear hotspots of activity emerging but also areas where local-level action is limited. It is in these latter localities that state-led action and planning will be particularly important.

Countryside management

One area where community action and planning may become more prominent is in relation to countryside management. Although in the UK as elsewhere, countryside management agendas are increasingly influenced by international agreements related to landscape, biodiversity, climate change and agriculture, the capacity of public-sector actors to deliver international commitments and to continue to play a leading role in taking agendas forward, again seems very questionable. Reflecting such concerns, in 2014 the Green Alliance, a collection of key environmental NGOs published a document highlighting what it thought should be the political priorities for the manifestos of the main political parties for the May 2015 UK General Election. Their research indicated that local authorities will have faced a 40 per cent cut in their budgets by May 2015 since the 2010 austerity measures had started, and in order to protect statutory responsibilities in welfare (for example adult social care and elderly people), cuts to environmental management have been much more severe, probably at least 60 per cent. Such cuts have also affected the executive agencies of government and those with a strong rural component seem to have been among the hardest hit. Even in relation to areas potentially attracting the highest levels of national support – including the UK's national parks – major questions are now being raised about how they will be maintained and managed in the future (Green Alliance, 2014). New partnership approaches to countryside management are therefore very likely. In addition continuing restructuring of European support for agriculture towards broader multifunctional countryside management goals can also be envisaged particularly in relation to areas of most marginal agriculture – e.g. in the rural uplands. Here voices are mounting in favour of major re-wilding initiatives, underpinned by arguments that these may bring greater economic, social and environmental returns than the traditional hill farming that they might replace (Monbiot, 2013). Just as afforestation brought significant landscape change in the early twentieth century, the scene is now perhaps set for another era of more subtle and longer term transformation in these most iconic of the UK's rural landscapes. In other areas, it seems likely that countryside management objectives including ambitions for landscape and biodiversity enhancement and climate change mitigation and adaptation may be increasingly delivered in tandem with private sector development or major public infrastructure projects. The imaginative 'Greener Vision for HS2' put forward by The Wildlife Trusts (2014), illustrates how creative multifunctional as well as partnerhsip approaches to planning may have a valuable role to play in shaping future countryside management.

Other projects and programmes

The Wildlife Trusts' HS2 initiative is an example of what might be labelled a new 'opportunist turn' in rural affairs. The ruptures in the framework of rural planning that have occurred since the post-2008 crash have forced those involved in almost every sphere of rural life to reappraise their approach. It is apparent that past certainties and areas of consensus can no longer be taken for granted and a new realism about the need to work together to address the special challenges that rural areas face has emerged. Within this picture any complacency about the security of government funding even for what have seemed the bedrocks of the rural scene – the National Parks – has been removed. So too have any ideas that green belts and other areas that have received long-standing protection will remain unchallenged as the nation seeks to meet growing housing needs and search for economic growth opportunities. In this new era, government programmes are no longer the driving force of rural change and development, although EU funding is likely to remain significant. Instead, small-scale pots of money, available through for example the National Lottery and Coastal Communities Fund, are being targeted and imaginative collaborative projects are being developed as stakeholders pool their efforts to meet local needs and facilitate positive rural change. In the absence of government funding, small-scale project activity of this sort is taking on a new prominence and is providing opportunities for community capacity building that may grow to be important to the future of many rural areas. Local planning authorities often have a supportive role to play in such initiatives and they can use their plan-making expertise and statutory planning tools in ways that can add value to wider community efforts. For the foreseeable future therefore, project activity seems to be providing a vital area for rural experimentation in which stakeholders can co-operate to develop new ways of working appropriate to the uncertain and challenging opening of the twenty-first century. Our thoughts on the potential paths ahead for rural planning in the UK are summarised in Table 12.2.

Whatever the past reality, the modern countryside never lives up to the romantic myth of a timeless landscape. The transformations that have taken place during the last century are continuing at a relentless pace. And there are

Table 12.2 Rural planning in the UK: Possible paths ahead

Components	Prospects?
Public land-use planning	• continuing streamlining of the statutory planning system; • continuing land-use control and regulatory functions; • reducing number and growing scale of rural planning authorities; • evolution of more streamlined and strategic development plans.
Spatial or territorial planning	• emergence of new 'non statutory' mechanisms for area visioning and cross-sectoral collaboration; • much variability in these mechanisms is anticipated, reflecting differing local circumstances.
Community action and planning	• growing community capacity and support mechanisms; • growing community involvement in service delivery; • growing community involvement in neighbourhood planning (or equivalents); • role of public sector will remain important in areas of community action, e.g. through supportive actions.
Countryside management	• new partnership approaches will become important; • broader multi-functional countryside management goals; • diverse range of new initiatives including 're-wilding'.
Other projects and programmes	• declining importance of government programmes; • increasing importance of projects as vehicles for imaginative collaboration and community capacity building; • planning expertise and statutory tools have the potential to add value to project working.

even signs of acceleration, as societies grapple with the huge resource and environmental dilemmas that the planet now faces. It may transpire that the late twentieth century comes to be seen as a period of relative calm in the overall transformation of the British countryside, as national debates over land, housing and development pale against the global challenges of food security and climate change. It seems inevitable that the future countryside will be comprehensively re-shaped by these challenges and that rural planning will, therefore, remain a crucially important aspect of public policy and local action.

Summary

• Despite the dominant discourse and reality of global urbanisation, rural areas retain a functional distinctiveness that is also expressed through representations of the rural and particular political economies that present planning with key opportunities and challenges. It is vital that rural interests and agendas are not overlooked, or indeed merged with urban agendas.

- Ongoing economic, social and environmental change, increasing diversity and divergence in the trajectories of rural regions and the growing complexity of the rural scene will require thoughtful, locally tailored planning responses.
- Alongside traditional rural planning concerns centred on housing, transport and environmental protection, new imperatives concerning resource security and climate change are likely to feature prominently in future rural planning debates.
- In the context of a shrinking state and an inability, in many instances, of private enterprise to address the critical challenges of rural places, it seems likely that a range of local partnerships and community-level action will play an ever-larger part in addressing rural development and service needs, as well as engaging in local actions (for example, in the area of renewable community-based energy) that will, cumulatively, have global impacts.

References

2020 Vision (2010) *The Greater Thames Futurescapes Project: More Than Just an Urban Green Space*. Available at: www.2020v.org/assignments.asp?ref=14 (accessed 12 October 2014).

Abas, M., Punpuing, S., Jirapramulpitak, T., Guest, P., Tangchonlatip, K., Leese, M. and Prince, M. (2009) Rural–urban migration and depression in aging family members left behind, in *British Journal of Psychiatry*, 195(1), 54–60.

Abercrombie, P. (1959) *Town and Country Planning*, 3rd Edition, Oxford University Press: London.

ACORP (2014) *Association of Community Rail Partnerships:* www.acorp.uk.com (accessed 12 July 2014).

Action with Communities in Rural England (2009) *Services: A Position Paper*, ACRE: Cirencester.

Action with Communities in Rural England (2012) *Making a Difference: 25 Years of Community Action*. Available at: www.acre.org.uk/cms/resources/25-anniversary-doc-final.pdf

Age UK (2013) *Missed Opportunities: The Impact on Older People of Cuts to Rural Bus Services,* Age UK: London.

Agnitsch, K., Flora, J. and Ryan, V. (2006) Bonding and bridging social capital: the interactive effects on community action, in *Community Development*, 37(1), 36–51.

Albrechts, L. (2004) Strategic (spatial) planning re-examined, in *Environment and Planning B: Planning and Design*, 31(5), 743–758.

Allanson P. (1996) A sustainable rural economy, in Allanson P. and Whitby M. (eds) *The Rural Economy and the British Countryside*, Earthscan: London, pp. 119–131.

Allanson, P. and Whitby, M. (1996) Prologue: rural policy and the British countryside, in Allanson, P and Whitby, M. (eds) *The Rural Economy and the British Countryside*, Earthscan: London, pp. 1–18.

Allmendinger, P. and Tewdwr-Jones, M. (2006) Territory, identity and spatial planning, in Allmendinger, P. and Tewdwr-Jones, M. (eds) *Territory, Identity and Spatial Planning*, Routledge: London, pp. 3–21.

Amin, A. and Thrift, N. (2002) *Cities: Reimaging the Urban*, Polity Press: Cambridge.

Ansell, C. and Gash, A. (2007) Collaborative governance in theory and practice, in *Journal of Public Administration Research and Theory*, 18, 543–571.

ARHC (Affordable Rural Housing Commission) (2006) *Final Report*, DEFRA: London.

Bailey, N. (1990) Community development trusts: A radical third way, in Montgomery, J. and Thornley, A. (eds) *Radical Planning Initiatives: New Directions for Urban Planning in the 1990s*, pp. 150–163, Gower, Aldershot.

Banister, D. (2002) *Transport Planning*, Taylor and Francis: London.

Barbieri, A.F., Monte-Mór, R.L. and Bislborrow, R.E. (2009) Towns in the jungle: exploring linkages between rural-urban mobility, urbanization and development in the Amazon, in de Sherbiniin, A., Rahman, A., Barbieri, A., Fotso, J.C. and Zhu, Y. (eds) *Urban Population and Environment Dynamics in the Developing World: Case Studies and Lessons Learned*, CICRED: Paris.

Barker, K. (2004) *Review of Housing Supply: Final Report*, HMSO: London.

Bell, C. and Newby, H. (1971) *Community Studies: An Introduction to the Sociology of Local Community*, Allen and Unwin: London.

Bell, D. (2006) Variations of the rural idyll, in Cloke P., Marsden, T. and Mooney, P.H. (eds) *Handbook of Rural Studies*, Sage Publications: London, pp. 149–160.

Bell, S. and McGillivray, D. (2013) *Environmental Law*, Oxford University Press: Oxford.

Berry, W. (1977) *The Unsettling of America: Culture and Agriculture*, Sierra Club: San Francisco.

Bishop, K. and Phillips, A. (2004) Then and now: planning for countryside conservation, in Bishop, K. and Phillips, A. (eds) *Countryside Planning: New Approaches to Management and Conservation*, Earthscan: London, pp. 1–15.

Bishop, J. (2007) Plans without planners? in *Town and Country Planning*, 76, 340–344.

Bishop, J. (2010) From parish plans to localism in England, in *Planning Practice and Research*, 25(5), 611–624.

Blumer, H. (1948) Public opinion and public opinion polling, in *American Sociological Review*, 13(5), 542–554.

Booz and Company (2012) *This is for Everyone: The Case for Universal Digitisation*. Available at: www.go-on.co.uk/wp-content/uploads/2013/12/The-Booz-Report-Nov2012.pdf (accessed 17 July 2014): Booz and Co.

Bourdieu, P. (2005) Habitus, in Hillier, J. and Rooksby, E. (eds) *Habitus: A Sense of Place*, 2nd Edition, Ashgate: Aldershot, pp. 43–52.

Bramley, G. and Watkins, D. (2009) Affordability and supply: the rural dimension, in *Planning Practice and Research*, 24(2), 185–210.

Branca, G., Lipper, L., Neves, B., Lopa, D. and Mwanyoka I. (2009) *New Tools for Old Problems: Can Payments for Watershed Services Support Sustainable Agricultural Development in Africa?* ESA Working Paper No. 09-10, The Food and Agriculture Organization of the United Nations: Rome.

Brandt, J., Primdahl, J. and Reenberg, A. (1999) Rural land-use and dynamic forces: analysis of 'driving forces' in space and time, in Krönert, R., Baudry, J., Bowler, I.R. and Reenberg, A. (eds) *Land-use Changes and their Environmental Impact in Rural Areas in Europe*, UNESCO: Paris.

Brenner, N. (2013) Theses on urbanization, in *Public Culture*, 25(1), 85–114.

Brenner, N. and Schmidt, C. (2011) Planetary urbanisation, in Gandy, M. (ed.) *Urban Constellations*, Jovis Verlag: Berlin, pp. 10–13.

Briedenhann, J. and Wickens, E. (2004) Tourism routes as a tool for the economic development of rural areas: vibrant hope or impossible dream? in *Tourism Management*, 1, 71–79.

British Railways Board (1963) *The Reshaping of British Railways*. HMSO: London.

Broadland District Council (2011) *Broadland District Plan (Replacement) (update following the Joint Core Strategy Adoption – March 2011): Written Statement.* Available at: www.broadland.gov.uk/PDF/Chapter_5_to_Chapter_9_Local_Plan_(adopted_)_to_update_JCS_(Adopted)_March_2011.pdf (accessed 16 July 2014). Broadland District Council: Norfolk.

Brown, D.L. (2010) *Rethinking the OECD's New Rural Demography: Centre for Rural Economy* Discussion Paper Series No. 26, Newcastle University, Newcastle.

Brown, I. (2012) Literary pilgrimage as cultural imperialism and 'Scott-land', in Brown, I. (ed.) *Literary Tourism, The Trossachs and Walter Scott*, Scottish Literature International: Glasgow.

Bruton, M. (ed.) (1974) *The Spirit and Purpose of Planning*, Hutchinson and Co: London.

Buller, H. and Hoggart, K. (1994) *International Counter-Urbanisation: British Migrants in Rural France*, Avebury: Aldershot.

Buller, H., Morris, C. and Wright, E. (2003) *The Demography of Rural Areas: A Literature Review*, Department for Environment, Food and Rural Affairs: London

Cabinet Office (1999) *Rural Economies*, The Stationery Office, London.

Cabinet Office (2011) *Open Public Services White Paper*, HM Government: London.

Cadman, D. and Austin-Crowe, L. (1991) *Real Estate Development*, Spon Press: London.

Caffyn, A. (2004) Market town regeneration: challenges for policy and implementation, in *Local Economy*, 19(1), 8–24.

Cameron, D. (2009) *The Big Society* (Speech). Available at: www.conservatives.com/News/Speeches/2009/11/David_Cameron_The_Big_Society.aspx (accessed 23 February 2011).

Cameron, S. and Shucksmith, M. (2007) Market towns, housing and social inclusion, in Powe, N., Hart, T. and Shaw, T. (eds) *Market Towns: Roles, Challenges and Prospects*, Routledge: London, pp. 81–92.

Campaign for Better Transport (2013) *Buses in Crisis: A report on Bus Funding Across England and Wales*, Campaign for Better Transport: London.

Carmona, M., Carmona, S. and Gallent, N. (2003) *Delivering New Homes: Processes, Planners and Providers*, Routledge: London.

Carson, R. (1962) *Silent Spring*, Houghton Mifflin Co: Boston, MA.

Castells, M. (2004) *Informationalism, Networks, and the Network Society: A Theoretical Blueprinting, The Network Society: A Cross-Cultural Perspective*, Northampton, MA: Edward Elgar.

CBD COP (Convention on Biological Diversity, Conference of the Parties) (2000) *Fifth Meeting, Decision V/6 The Ecosystem Approach*, Secretariat to the Convention on Biological Diversity, Montreal.

CCRI, Rose Regeneration and the Rural Services Network. (2013) *Alternative Service Delivery Models in Rural Areas: Final Report.* Available at: http://sciencesearch.defra.gov.uk/Default.aspx?Menu=Menu&Module=More&Location=None&Completed=0&ProjectID=18554

Champion, A.G. (2000) Flight from the cities? In Bate, R., Best, R. and Holmans, A. (eds) *On the Move: The Housing Consequences of Migration*, York Publishing Services: York, pp. 10–19.

Chaskin, R. J., Brown, P., Venkatesh., S. and Vidal, A. (2001) *Building Community Capacity*, Aldine De Gruyter: New York.

Cherry, G. (ed.) (1976) *Rural Planning Problems*, International Textbook Co Ltd: London.

Cheshire, P. (1995) A new phase of urban development in western Europe? The evidence for the 1980s, in *Urban Studies*, 32, 1045–1063.

Christaller, W. (1933) *Central Places of Southern Germany*. Prentice Hall: London.

Ciaffi, D. and Gallent, N. (2014) Reflections on community action and planning, in Gallent, N. and Ciaffi, D. (eds) *Community Action and Planning: Contexts, Drivers and Outcomes*, Policy Press: Bristol.

Clark, D. (1984) Rural housing and countryside planning, in Blacksell, M. and Bowler, I. (eds) *Contemporary Issues in Rural Planning*, in SW Papers in Geography, pp. 93–104.

Clark, M. (2001) *Teleworking in the Countryside: Home-based Working in the Information Society*, Ashgate: Aldershot.

Cloke, P. and Edwards, G. (1986) Rurality in England and Wales 1981: a replication of the 1971 index, in *Regional Studies*, 20(4), 289–306.

Cloke, P. (1977) An index of rurality for England and Wales, in *Regional Studies*, 11, 31–46.

Cloke, P. (1979) *Key Settlements in Rural Areas*, Methuen: London.

Cloke, P. (1983) *An Introduction to Rural Settlement Planning*, London: Methuen.

Cloke, P. (2006) Conceptualising rurality, in Cloke, P., Marsden, T. and Mooney, P. (eds) *Handbook of Rural Studies*, Sage: London.

Cloke, P. and Goodwin, M. (1992) Conceptualizing countryside change: from post-Fordism to rural structured coherence, in *Transactions of the Institute of British Geographers*, 17, 321–336.

Cloke, P. and Little, J. (eds) (1997) *Contested Countryside Cultures: Otherness, Marginalisation and Rurality*, Routledge: London.

Cloke, P. and Shaw, D.P. (1983) Rural settlement policy in structure plans, in *Town Planning Review*, 54, 338–354.

Cloke, P., Goodwin, M. and Milbourne, P. (eds) (1997) *Rural Wales: Community and Marginalisation*, University of Wales Press: Cardiff.

Cloke, P., Marsden, T., and Mooney, P. (eds). (2006) *Handbook of Rural Studies*, Sage: London.

Clout, H. (1972) *Rural Geography: An Introductory Survey*, Pergamon Press: Oxford.

Collantes, F (2009) Rural Europe reshaped: the economic transformation of upland regions 1850 – 2000 in *The Economic History Review*, 62(2), 306–323.

Commission for Rural Communities (2006a) *Report of the Rural Advocate*, CRC: Cheltenham.

Commission for Rural Communities (2006b) *State of the Countryside 2006*, CRC: Cheltenham.

Commission for Rural Communities (2006c) *Rural Disadvantage: Reviewing the Evidence Base*, CRC: Cheltenham.

Commission for Rural Communities (2007) *A8 Migrant Workers in Rural Areas: Briefing Paper*, CRC: Cheltenham.

Commission for Rural Communities (2009) *Rural Proofing Guidance*, CRC: Cheltenham.

Commission for Rural Communities (2010) *Rural Economies Recession Intelligence*, Briefing note from CRC to Hilary Benn, CRC: Cheltenham.

Commission for Rural Communities (2012) *How Are Rural Interests Being Recognised Within Local Enterprise Partnerships?* CRC: Cheltenham.

Commission for Rural Communities (2013) *Rural Micro Business: What Makes Some Thrive in a Challenging Economic Climate?* CRC: Cheltenham.

Conservative Party (2010) *Open Source Planning: Green Paper 14*, Conservative Party: London.

Convention on Biological Diversity (2000) *Ecosystem Approach Principles, United Nations Environment Programme*, Available at: www.cbd.int/ecosystem/principles.shtml.

Cook, B. (2013) How we did it: allocating sites through a neighbourhood plan, *Planning Resource*, 14 June 2014.

Corbera E., Kosoy, N. and Martinez Tuna, M. (2007) Equity implications of marketing ecosystem services in protected areas and rural communities: case studies from Meso-America, in *Global Environmental Change*, 17, 365–380.

Council of Europe (2000) *The European Landscape Convention*, Council of Europe: Brussels.

Countryside Agency (1999) *The Rural Development Commission*, Countryside Agency: Cheltenham.

Countryside Agency (2001) *State of the Countryside*, Countryside Agency: Cheltenham.

Countryside Agency (2004) *What Makes a Good Parish Plan?* Countryside Agency: Cheltenham.

Countryside Agency (2006) *Landscape Beyond the View*, Countryside Agency: Cheltenham.

Countryside Agency, Department of Environment Food and Rural Affairs, Office of the Deputy Prime Minister, Office for National Statistics and the welsh Assembly Government (2004) *Rural and Urban Classification 2004: An Introductory Guide*, Countryside Agency: Cheltenham.

Courtney, P. and Errington, A. (2000) The role of small towns in the local economy and some implications for development policy, in *Local Economy*, 15(4), 280–301.

CPRE (Campaign to Protect Rural England) (2007) *Tranquillity Map: England*, CPRE: London. Available at: www.cpre.org.uk/resources/countryside/tranquil-places/item/1839- (accessed 19 October 2014).

CPRE (Campaign to Protect Rural England) (2012) *Green Belt: Under Renewed Threat?* CPRE: London.

Cracknell, J., Miller, F. and Williams, H. (2013) *Passionate Collaboration? Taking the Pulse of the UK Environmental Sector*, The Environment Funders Network: Maidenhead.

Crang, M. and Thrift, N. J. (2000) *Thinking Space*, 9. Psychology Press: Abingdon.

Crook, A.D.H, Monk, S., Rowley, S. and Whitehead, C.M.E. (2006) Planning gain and the supply of new affordable housing in England: understanding the numbers, in *Town Planning Review*, 77(3), 353–373.

Crouch D. (2006) Tourism, consumption and rurality, in Cloke, P., Marsden, T. and Mooney, P.H. (eds) *The Handbook of Rural Studies*, Sage Publications: London, pp. 355–364.

Cullingworth, B. and Nadin, V. (2006) *Town and Country Planning in the UK*, 14th Edition, Routledge: London.

Cullingworth, B., Nadin, V., Hart, T., Davoudi, S., Pendlebury, J., Vigar, G., Webb, D. and Townsend, T. (2014) *Town and Country Planning in the UK*, 15th Edition, Routledge: Abingdon.

Darling, E. (2005) The city in the country: wilderness gentrification and the rent gap, in *Environment and Planning A*, 37, 1015–1032.

DBIS (Department for Business, Innovation and Skills) (2010a) *Local Growth: Realising Every Place's Potential*, DBIS: London.

DBIS (Department for Business, Innovation and Skills) (2010b) *Manufacturing in the UK: An Economic Analysis of the Sector*, DBIS Occasional Paper No. 10A, DBIS: London.

DCLG (Department for Communities and Local Government) (2006a) *Strong and Prosperous Communities: Local Government White Paper*, DCLG: London.

DCLG (Department for Communities and Local Government) (2006b) *Planning Policy Statement 3: Housing*, DCLG: London.

DCLG (Department for Communities and Local Government) (2007) *Making Assets Work: The Quirk Review of Community Management and Ownership of Public*, DCLG: London.

DCLG (Department for Communities and Local Government) (2012) *National Planning Policy Framework*, DCLG: London.

DCLG (Department for Communities and Local Government) (2013) *You've Got the Power: A Quick and Easy Guide to Community Rights*, DCLG: London.

DCLG (Department for Communities and Local Government) (2014a) *Notes on Neighbourhood Planning*, 10th Edition, DCLG: London.

DCLG (Department for Communities and Local Government) (2014b) *Notes on Neighbourhood Planning: Edition 9*. Available at: www.gov.uk/government/uploads/system/uploads/attachment_data/file/287789/Notes_on_Neighbourhood_Planning_Edition_9.pdf (accessed 15 July 2014).

DCMS (Department for Culture, Media and Sport) (2009) *Digital Britain: Final Report*, DCMS: London.

DCMS (Department for Culture, Media and Sport) (2013) *Guidance Broadband Delivery UK*, DCMS: London.

DECC (Department of Energy and Climate Change) (2012a) *Energy Security Strategy*, DECC: London.

DECC (Department of Energy and Climate Change) (2012b) *The Unconventional Hydrocarbon Resources of Britain's Onshore Basins: Shale Gas*, DECC: London.

DECC (Department of Energy and Climate Change) (2013) *Developing Onshore Shale Gas and Oil: Facts about 'Fracking'*, DECC: London.

DECC (Department of Energy and Climate Change) (2014) *Energy Trends*, March, London: Department for Energy and Climate Change. Available Online at: www.gov.uk/government/collections/energy-trends.

DECC et al. (Department of Energy and Climate Change, Welsh Assembly Government, and Northern Ireland Assembly Government) (2009) *The Low Carbon Communities Challenge*, DECC: London.

DEFRA (Department for Environment, Food and Rural Affairs) (2005) *Securing the Future: Delivering the UK Sustainable Development Strategy*, HM Government: London.

DEFRA (Department for Environment, Food and Rural Affairs) (2006) *Sustainable Models of Community Retailing*, DEFRA: London.

DEFRA (Department for Environment, Food and Rural Affairs) (2007) *Attitudes and Behaviour in Relation to the Environment*, available at: www.defra.gov.uk/evidence/statistics/environment/pubatt.

DEFRA (2008) *Rural Challenges, Local Solutions: Building on the Rural Delivery Pathfinders in England*, DEFRA: London.

DEFRA (Department for Environment, Food and Rural Affairs) (2009) *Adapting to*

Climate Change: UK Climate Change Projections, DEFRA: London.

DEFRA (Department for Environment, Food and Rural Affairs) (2011) *The Natural Choice: Securing the Value of the Nature*, DEFRA: London.

DEFRA (Department for Environment, Food and Rural Affairs) (2012) *Rural Statement*, DEFRA: London.

DEFRA (Department for Environment, Food and Rural Affairs) (2013) *Guide to Rural Proofing: National Guidelines*, DEFRA: London.

DEFRA (Department for Environment, Food and Rural Affairs) (2014a) *Statistical Digest of Rural England*, DEFRA: London.

DEFRA (Department for Environment, Food and Rural Affairs) (2014b) *Food Statistics Pocket Book 2014*. DEFRA: London. Available at: www.gov.uk/government/collections/food-statistics-pocketbook

DEFRA (Department for Environment, Food and Rural Affairs) (2014c) *Stimulating Economic Growth in Rural Areas*. Available at: www.gov.uk/government/policies/stimulating-economic-growth-in-rural-areas/supporting-pages/rural-growth-network-pilots

DEFRA (Department for Environment, Food and Rural Affairs) (2014d) *Guidance: Diversifying Farm Businesses*. Available at: www.gov.uk/diversifying-farming-businesses

DEFRA (Department for Environment, Food and Rural Affairs) (2014e) *Natural Environment Whitepaper: Implementation Report*. DEFRA: London.

Delanty, G. (2003) *Community*, Routledge: London.

DfT (Department for Transport) (2006) *Social Inclusion: Transport Aspects*. Report compiled by the Centre for Transport Studies, Imperial College: London.

DfT (Department for Transport) (2011) *Local Frameworks for Funding Major Transport Schemes: Guidance for Local Transport Bodies*. Available at: www.gov.uk/government/uploads/system/uploads/attachment_data/file/15176/guidance-local-transport-bodies.pdf (accessed 1 October 2014) countryside agency 200.

DETR and MAFF (Department of Environment, Transport and the Regions and the Ministry of Agriculture, Fisheries and Food) (2000) *Our Countryside – The Future: A Fair Deal for Rural England*, Cmnd 4909, DETR and MAFF: London.

DoE and MAFF (Department of the Environment and the Ministry of Agriculture, Fisheries and Food) (1995) *Rural England: A Nation Committed to a Living Countryside*, Cmnd 3016, HMSO: London.

Dower, M. (1972) Amenity and tourism in the countryside, in Ashton, J. and Long, W. (eds) *The Remoter Rural Areas of Britain*, Oliver and Boyd: Edinburgh

Dühr, S., Colomb, C. and Nadin, V. (2010) *European Spatial Planning and Territorial Cooperation*, Routledge, Abingdon.

Dunn, H. (2011) *Payments for Ecosystem Services*, DEFRA Evidence and Analysis Series, Paper 4, Department for Environment, Food and Rural Affairs: London.

Eden District Council (2010) *Core Strategy Development Plan Document*, Eden District Council: Penrith.

EDORA (2010a) *European Development Opportunities for Rural Areas: Applied Research 2013/1/2: Final Report*, European Spatial Planning Observation Network, Brussels.

EDORA (2010b) *Country Profiles: United Kingdom*, Report No. 25.31. Available online at: www.espon.eu/main/Menu_Projects/Menu_AppliedResearch/edora.html.

Elson, M. (1986) *Green Belts: Conflict Mediation in the Rural Urban Fringe*, William Heinemann: London.

Emerson, R.W. (1985) *Nature*, Penguin: London.

Engels, F. (1872) *The Housing Question*, Progress Publishers: Moscow (first published in 1954 from Engels' articles for the Leipzig Volksstaat).

Escobar A. (2006) Difference and conflict in the struggle over natural resources: a political ecology framework, in *Development*, 49(3), 6–13.

European Commission (2003) *LEADER+ Programming and Implementation: Notice to Member States*, CEC: Brussels.

European Commission (2014) *Rural Development in the EU; Statistical and Economic Information Report 2013*, European Union: Brussels.

Evans, A.W. (1991) "Rabbit hutches on postage stamps": planning, development and political economy, in *Urban Studies*, 28(6), 853–870.

Evans, J.P. (2012) *Environmental Governance*, Routledge: London.

Fairbrother, N. (1970) *New Lives, New Landscapes*, The Architectural Press: London.

FAO (Food and Agriculture Organisation) (2009) *How to Feed the World in 2050*, FAO: Rome.

FAO (Food and Agriculture Organisation) (2013) *The State of Food and Agriculture 2013*, FAO: Rome.

Featherstone, D., Ince, A., MacKinnon, D., Strauss, K. and Cumbers, A. (2012) Progressive Localism and the construction of political alternatives, in *Transactions of the Institute of British Geographers*, 37, 177–182.

Ferraro, P. and Kiss, A. (2002). Direct payments to conserve biodiversity, in *Science*, 298, 1718–1719.

Fielding, A. (1982) Counter urbanisation in western Europe, in *Progress in Planning*, 17, 1–52.

Fischer, T. B. (2007) *The Theory and Practice of Strategic Environmental Assessment: Towards a More Systematic Approach*, Earthscan, London.

Fisher, F. (2005) *Citizens, Experts and the Environment: the Politics of Local Knowledge*, Duke Press: London.

Food Ethics Council (2006) *Sustainable Farming and Food: Emerging Challenges*, The Food Ethics Council: Brighton.

Ford, J., Quilgars, D. and Burrows, R. with Pleace, N. (1997) *Young People and Housing, Rural Research Report No. 31*, The Rural Development Commission: Salisbury.

Foreman, D. (2004) *Rewilding North America: A Vision for Conservation in the 21st Century*, Island Press: Washington.

Foster, C. D. and Beesley, M. E. (1963) Estimating the social benefit of constructing an underground railway in London, in *Journal of the Royal Statistical Society: Series A (General)*, pp. 46–93.

Foucault, M. (1991) Governmentality, in Burchell, G., Gordon, C., and Miller, P. (eds) *The Foucault Effect: Studies in Governmentality*, University of Chicago Press: Chicago, IL, pp. 87–104.

Fritsch O. and Newig J. (2012) Participatory governance and sustainability: findings of meta-analysis of stakeholder involvement in environmental decision making, in Brousseau, E., Dedeurwaerdere, T. and Siebenhuner, B. (eds) *Reflexive Governance for Global Public Goods*, The MIT Press: Cambridge, MA, pp. 181–204.

Gallent, N. (2000) Resource Allocation and Political Change in the UK in *Regional Studies*, 34(4), 389–394.

Gallent, N. (2009) Affordable housing in 'village England': towards a more systematic approach, in *Planning, Practice & Research*, 24(2), 263–283.

Gallent, N. (2013) Re-connecting 'people and planning': parish plans and the English localism agenda, in *Town Planning Review*, 84(3), 371–396.

Gallent, N. (2014) The social value of second homes in rural communities, in *Housing, Theory and Society*, 31, 2, 174–191.

Gallent, N. and Bell, P. (2000) Planning exceptions in rural England: past, present and future, in *Planning Practice and Research*, 15(4), 375–384.

Gallent, N. and Ciaffi, D. (eds) (2014a) *Community Action and Planning: Contexts, Drivers and Outcomes*, Policy Press: Bristol.

Gallent, N. and Ciaffi, N. (2014b) Communities, community action and planning, in Gallent, N. and Ciaffi, D. (eds) *Community Action and Planning: Contexts, Drivers and Outcomes*, Policy Press: Bristol.

Gallent, N., Morphet, J and Tewdwr-Jones, M. (2008) Parish plans and the spatial planning approach, in *Town Planning Review*, 79(1), 1–29.

Gallent, N. and Robinson, S. (2011) Local perspectives on rural housing affordability and implications for the localism agenda in England, in *Journal of Rural Studies*, 27(3), 297–307.

Gallent, N. and Robinson, S. (2012a) Community perspectives on localness and priority housing policies in rural England, in *Housing Studies*, 27(3), 360–380.

Gallent, N. and Robinson, S. (2012b) *Neighbourhood planning: communities, networks and Governance*, Policy Press: Bristol.

Gallent, N., Andersson, J. and Bianconi, M. (2006) *Planning on the Edge: The Context for Planning at the Rural-Urban Fringe*, Routledge: Abingdon.

Gallent, N., Hamiduddin, I. and Madeddu, M. (2013) Localism, down-scaling and the strategic dilemmas confronting planning in England, in *Town Planning Review*, 84(5), 563–582.

Gallent, N., Mace, A. and Tewdwr-Jones, M. (2002) *Second Homes in Rural Areas of England*, Countryside Agency: Cheltenham.

Gallent, N., Mace, A. and Tewdwr-Jones, M. (2005) *Second Homes: European Perspectives and UK Policies*, Ashgate: Aldershot.

Gallent, N., Shucksmith, M. and Tewdwr-Jones, M. (2003) *Housing in the European Countryside*, Routledge: London.

Garbett-Edwards D. (1972) The Establishment of New Industries (with Particular Reference to Recent Experience in Mid-Wales) in Ashton, J. and Long, W. (eds) *The Remoter Rural Areas of Britain*, Oliver and Boyd: Edinburgh.

Garcilazo, E. (2013) *Growth Trends and Characteristics of OECD Rural Regions*, OECD: Paris.

Garrod, B., Wornell, R. and Youell, R. (2006) Reconceptualising rural resources as countryside capital: the case of rural tourism, in *Journal of Rural Studies*, 22(2), 117–128.

Gibbs, D., Cocklin, C., and Dibden, J. (2008) Genetically modified organisms (GMOs) and the future of rural spaces, in *Geoforum*, 39, 145–147.

Gilg, A. (1976) 'Rural Employment' in Cherry, G. (ed.) *Rural Planning Problems*, International Textbook Co Ltd: London.

Gilg, A. (1996) *Countryside Planning: The First Half Century*, 2nd Edition, Routledge: Abingdon.

Gilg, A. (2005) *Planning in Britain: Understanding and Evaluating the Post-War System*, Sage Publications: London.

Gkartzios, M. and Norris, M. (2011) 'If you build it, they will come': governing property-led rural regeneration in Ireland, in *Land Use Policy*, 28(3), 486–494.

Glass, R. (1964) *London: Aspects of Change*, MacGibbon & Kee: London.

Glasson, J. and Marshall, T. (2007) *Regional Planning*, Routledge: London.

Glasson, J., Therivel, R., and Chadwick. A. (2013) *Introduction to Environmental Impact Assessment*, 4th Edition, Routledge: Abingdon.

Global Agriculture (2014) *Agriculture at a Cross Roads: Findings and Recommendations for Future Farming*. Retrieved 04 14, 2014, from Global Agriculture: www.globalagriculture.org/report-topics/industrial-agriculture-and-small-scale-farming.html.

Goodwin, P. (2012) Three views on 'peak car', in *World Transport Policy and Practice*, 17(4), 8–17.

Government of Ireland (2010) *Report on Rural Transport Provision*, Government of Ireland: Dublin.

Gray, D. (2014) Economic approaches to the rural, in Bosworth, G. and Somerville P. (eds) *Interpreting Rurality: Multidisciplinary approaches*, Routledge: London, pp. 32–53.

Gray, J. (2003) A rural sense of place: intimate experience in planning for countryside life, in *Interface*, 3(100), 93–96.

Green Alliance (2014) *Greener Britain: Practical Proposals for Party Manifestos from the Environment and Conservation Sector*, Green Alliance: London.

Green, C. and Hall, P. (2009) *Better Rail Stations*, Department for Transport: London.

Grossman, G.M. and Helpman, E. (1990) *Trade, Knowledge Spillovers, and Growth*, NBER Working Papers 3485, National Bureau of Economic Research, Inc: Cambridge, MA.

GSS (Government Statistical Service) (2013) *The 2011 Rural–Urban Classification For Small Area Geographies*, GSS: London.

Guy, C. (1998) Off-centre retailing in the UK: prospects for the future and implications for town centres, in *Built Environment*, 24(1), 16–30.

Habermas, J. (1984) *The Theory of Communicative Action: Part 1: Reason and the Rationalization of Society*, Beacon: Boston.

Halfacree, K. (1999) 'A new space or spatial effacement? Alternative futures for the post-productivist countryside, in Walford, N., Everitt, J. and Napton, D. (eds) *Reshaping the Countryside: Perceptions and Processes of Rural Change*, CABI Publishing: Wallingford.

Halfacree, K. (2006) Rural space: constructing a three-fold architecture, in Cloke, P., Marsden, T., and Mooney, P. (eds) *Handbook of Rural Studies*, Sage: London.

Halfacree, K. (2007) Trial by space for 'radical rural': introducing alternative localities, representations and lives, in *Journal of Rural Studies*, 23, 125–141.

Halfacree, K. (2010) Reading rural consumption practices for difference: Boltholes, castles and life-rafts, in *Culture Unbound*, 2, 241–263.

Halfacree, K. (2012) Heterolocal identities? Counter-urbanisation, second homes, and rural consumption in the era of mobilities, in *Population, Place & Space*, 18(2), 209–224.

Hall, P. (1974) The containment of rural England, in *The Geographical Journal*, 140(3), 386–408.

Hall, P. (1996) *Cities of Tomorrow*, Updated Edition, Blackwell: Oxford.

Hall, P. (2001) Sustainable cities or town cramming? in Layard, A., Davoudi, S. and Batty, S. (eds) *Planning for a Sustainable Future*, Spon Press: London.

Hall, P. (2011) The Big Society and the evolution of ideas, in *Town and Country Planning*, 80(2), 59–60.

Hall, P. (2013) *Good Cities, Better Lives: How Europe Discovered the Lost Art of Urbanism*, Routledge: London.

Hall, P. and Hay, D. (1980) *Growth Centres in the European Urban System*, Heinemann Educational: London.

Hamiduddin, I. and Gallent, N. (2012) Limits to growth: the challenge of housing delivery in England's 'under-bounded' districts, in *Planning Practice and Research*, 27(5), 513–530.

Hampshire County Council (2014) *Cango Bus Service*. Available at: www3.hants.gov.uk/passengertransport/communitytransport/cango/moreabout cango.htm.

Haskins, C. (2001) Rural recovery after foot-and-mouth disease, DEFRA: London. Available at: www.gov.uk/government/uploads/system/uploads/attachment_ data/file/86309/rural-recovery-after-fm-disease.pdf (accessed 11 February 2015).

Haughton, G. and Allmendinger, P. (2011) Moving on: from spatial planning to localism and beyond, in *Town and Country Planning*, 80(4), 184–187.

Haven-Tang, C. and Jones, E. (2014) Capitalising on rurality: tourism micro-businesses in rural tourism destinations, in Bosworth, G. and Somerville, P. (eds) *Interpreting Rurality: Multidisciplinary Approaches*, Routledge: London, pp. 237–250.

Headicar, P. (2009) *Transport Policy and Planning in Great Britain*, Routledge: Oxford

Healey, P. (1997) *Collaborative Planning: Shaping Places in Fragmented Societies*, Macmillan: Basingstoke.

Heart of the South West (2014) *Rural Growth Network*. Available at: www.heartofswlep.co.uk/rural-growth

Hebbert, M. (1989) Rural land use planning in Japan, in Cloke, P. (ed.) *Rural Land-Use Planning in Developed Nations*, Routledge: Abingdon.

Heidegger, M. (1971) Building, dwelling, thinking in *Poetry, Language, Thought*, Harper and Row: New York.

Henderson, J. and Heley, J. (2014) Governing beyond the metropolis: placing the rural in city-region development, in *Urban Studies*, published online May 2014.

Heseltine, M. and Leahy, T. (2011) *Rebalancing Britain: Policy or Slogan? Liverpool City Region – Building on its Strengths: An Independent Report*, Department for Business, Innovation and Skills: London.

Hill, R., Ketwell, K. and Slat, J. (2014) *Partnership Working in Small Rural Primary Schools: The Best of Both Worlds*, available from: http://cdn.cfbt.com/~/media/ cfbtcorporate/files/research/2014/r-partnership-working-small-rural-report-2014.pdf (accessed 17 July 2014). CfBT Education Trust.

Hindle, T., Spollen, M., and Dixon, P. (2004) *A Review of the Evidence on the Additional Costs of Delivering Services to Rural Communities*, SECTA: London.

HLPE (2012) *Food Security and Climate Change: A Report by the High Level Panel of Experts on Food Security and Nutrition*, The Committee on World Food Security: Rome.

HLPE (2013) *Biofuels and Food Security: A Report by the High Level Panel of Experts on Food Security and Nutrition*, Committee on World Food Security: Rome.

HM Government (1987) *Housing: The Government's Proposals (White Paper)*, HMSO: London.

HM Government (2010) *Food 2030*, DEFRA: London.

HM Treasury (2013) *The Green Book: Appraisal and Evaluation in Central Government*, The Stationery Office: London.

Hoggart, K, and Paniagua, A. (2001) What rural restructuring? in *Journal of Rural Studies*, 17, 41–62.

Hoggart, K. (1990) Let's do away with rural, in *Journal of Rural Studies*, 6, 245–257.

Hoggart, K. (2005) City hinterlands in European space, in Hoggart, K. (ed.) *The City's Hinterland: Dynamism and Divergence in Europe's Peri-Urban Territories*. Ashgate: Aldershot, pp. 1–18.

Hoggart, K. and Henderson, S. (2005) Excluding exceptions: housing non-affordability and the oppression of environmental sustainability? in *Journal of Rural Studies*, 21, 181–196.

Hoskins, W.G. (1955) *The Making of the English Landscape*, Hodder and Stoughton: London.

Hosszu, S. (2009) *Counterurbanisation: A Literature Study*, Danish Institute of Rural Research and Development, Working Paper No. 06/2009, University of Southern Denmark, Esbjerg.

House of Commons Select Committee: Environment, Food and Rural Affairs (2014) *Food Security*, House of Commons: London.

Howard, E. (1898) *To-Morrow, A Peaceful Path to Real Reform*, Swan Sonnenschein, London.

Hughes, A. and Reimer, S. (2004) Introduction, in Hughes, A. and Reimer, S. (eds) *Geographies of Commodity Chains*, Routledge: London, pp. 1–16.

Hutton, B. (2013) *Planning Sustainable Transport*, Routledge: Oxford.

Innes, J. and Booher, D. (2003) *The Impact of Collaborative Planning on Governance Capacity*, Working Paper 2003 – 3, Institute of Urban and Regional Development, University of California: Berkeley. Available at: http://escholarship.org/uc/item/98k72547

Innes, J. and Booher, D. (2004) Reframing public participation: strategies for the 21st Century, in *Planning Theory and Practice*, 5(4), 419–436.

Innes, J. and Booher, D. (2010) *Planning with Complexity: An Introduction to Collaborative Rationality for Public Policy*, Abingdon and New York: Routledge.

Institute of Rural Health (2012) *Rural Proofing Toolkit*. Available at: www.ruralproofingforhealth.org.uk/ (accessed 17 July 2014).

IGBP (International Geosphere-Biosphere Programme] (2001) *Amsterdam Declaration on Earth System Science*, IGBP: Stockholm.

IPCC [Intergovernmental Panel on Climate Change) (2013) *Climate Change 2013: The Physical Science Basis*, IPCC: Geneva.

IPCC (Intergovernmental Panel on Climate Change) (2014) *Climate Change 2014: Impacts, Adaptation and Vulnerability*, IPCC: Geneva.

IPPR North (2014) *The State of the North: Setting a Baseline for the Devolution Decade*, IPPR North: Manchester.

Jay, S., Jones, C., Slinn, P. and Wood, C. (2007) Environmental impact assessment: retrospect and prospect, in *Environmental Impact Assessment Review*, 27, 287–300.

Jessop, B. (1995) The regulation approach, governance and post-Fordism: alternative perspectives on economic and political change? In *Economy and Society*, 24(3), 307–333.

Jessop, B. (2002) Liberalism, neo-liberalism, and urban governance: a state theoretical perspective, in *Antipode*, 34(3), 452–472.

Jessop, B. (2003) Governance and metagovernance: on reflexivity, requisite variety, and requisite irony, in Bang, H. (ed.) *Governance, as Social and Political Communication*, Manchester University Press: Manchester.

Johnson, L.B. (1965) Beautification, in Nash, R. (1976) *The American Environment: Readings in the History of Conservation*, 2nd Edition, Addison-Wesley: Reading, MA.

JNCC (Joint Nature Conservation Committee) (2013) *UK Biodiversity Indicators in Your Pocket 2013: Measuring Progress Towards Halting Biodiversity Loss*, DEFRA: London. Available at: http://jncc.defra.gov.uk/page-4229

Jones, A. (2007) New wine in old bottles? England's parish and town councils and New Labour's neighbourhood experiment, in *Local Economy*, 22, 227–242.

Jones, C. and Murie, A. (2006) *The Right to Buy: Analysis and Evaluation of a Housing Policy*, Blackwell Publishing: London.

Joseph Rowntree Foundation (1994a) *Local Leadership and Decision Making*, JRF: York.

Joseph Rowntree Foundation (1994b) *The Governance Gap: Quangos and Accountability*, JRF: York.

Kenyon, S. (2010) The impacts of Internet use upon activity participation and travel: results from a longitudinal diary-based panel study, in *Transportation Research Part C: Emerging Technologies*, 18(1), 21–35.

Kidd, S., Plater, A. and Frid, C. (eds) (2011) *The Ecosystem Approach to Marine Planning and Management*, Earthscan, London.

Kilpatrick, S., Willis, K. and Lewis, S. (2014) Community action in Australian farming and fishing communities, in Gallent, N. and Ciaffi, D. (eds) *Community Action and Planning: Contexts, Drivers and Outcomes*, Policy Press: Bristol.

King, P. (2004) *Private Dwelling: Contemplating the Use of Housing*, Routledge: London

Kitchen, L. and Marsden, L. (2009) Creating sustainable rural development through stimulating the eco-economy: beyond the eco-economic paradox? In *Sociologia Ruralis*, 49(3), 274–294.

Krygsman, S., Dijst, M. and Arentze, T. (2004) Multimodal public transport: an analysis of travel time elements and the interconnectivity ratio, in *Transport Policy*, 11(3), 265–275.

Kunzmann, K. (2000) Strategic spatial development through information and communication, in Salet W. and Faludi A. (eds) *The Revival of Strategic Spatial Planning*, Royal Netherlands Academy of Arts and Sciences: Amsterdam, pp. 259–265.

Kurukulasuriya, L. and Robinson, N.A. (2006) *Training Manual on International Environmental Law*, UNEP, Paris.

Laconte, P. (2013) *Indirect Benefits of Public Transport on Land Use: Two Legal Innovations in Belgium,* Paper presented at the Sintropher Partnership Workshop, Bruges, 25 April, 2013.

Lange, A., Piorr, A., Siebert, R. and Zasada, I. (2013) Spatial differentiation of farm diversification: how rural attractiveness and vicinity to cities determine farm households' response to the CAP, in *Land Use Policy*, 31, 136–144.

LaRose, R., Gregg, J., Strover, S., Straubhaar, J. and Carpenter, S. (2007) Closing the rural broadband gap: promoting adoption of the internet in rural America, in *Telecommunications Policy*, 31, 359–373.

Leavett, A. (1985) *Role and Relationships of RCCs: Review on Behalf of the Development Commission*. Development Commission: London.

Lefebvre, H. (1970) *The Urban Revolution*, English Edition, University of Minnesota Press: Minneapolis, MN.

Lefebvre, H. (1991) *The Production of Space*, English Edition, Blackwell: Oxford.

Little, J. and Austin, P. (1996) Women and the rural idyll, in *Journal of Rural Studies*, 12(2), 101–111.

Lobley, M., Potter, C. and Butler, A. (2005) *The Wider Social Impacts of Changes in the Structure of Agricultural Businesses*, CRR Research Report No. 14, University of Exeter: Exeter.

Loft, C. (2014) *Last Trains: The Beeching Cuts and the Death of Rural England*, Biteback: London.

Longley, P., Webber, R., and Li, C. (2008) The UK geography of the e-society: a national classification, in *Environment and Planning A*, 40, 362–382.

Lorenzoni, I., Nicholson-Cole, S. and Whitmarsh L. (2007) Barriers perceived to engaging with climate change among the UK public and their policy implications, in *Global Environmental Change*, 17, 445–459.

Lovelock, J. (1979) *Gaia: A New Look at Life on Earth*, Oxford University Press: Oxford.

Lowe, P. and Ward, N. (2009) England's rural futures: a socio-geographical approach to scenario analysis, in *Regional Studies*, 43(10), 1319–1332.

Lowe, P. and Ward, N. (2007) *Rural Futures: A Socio-Geographical Approach to Scenario Analysis*, Paper presented at the Institute for Advanced Studies Annual Research Programme. 9–10 January 2007, Lancaster University.

Lowenthal, D. and Prince, H.C. (1965) English landscape tastes, in *Geographical Review*, 55(2), 186–222.

Lucas, K. and Jones, P. (2009) *The Car in British Society*, RAC Foundation: London.

Lyons, M. (2007) *Place Shaping: A shared Ambition for the Future of Local Government*, TSO: London.

Lyson, T. (2006) Global capital and the transformation of rural communities, in Cloke, P., Marsden, T. and Mooney, P. (eds) *Handbook of Rural Studies*, Sage: London, pp. 292–303.

MacFarlane, R. (2003) *Mountains of the Mind: A History of Fascination*, Granta: London.

MacGregor, B.D. (1976) Village life: facts and myths, in *Town and Country Planning*, 44(11), 524–527.

Marini, M.B. and Mooney, P. (2006) Rural economies, in Cloke, P. and Marsden, T. and Mooney, P. (eds) *Handbook of Rural Studies*, Sage: London, pp. 91–103.

Marsden T., Murdoch, J., Lowe, P., Munton, R. and Flynn, A. (1993) *Constructing the Countryside*, UCL Press: London.

Marsden, T. (1998) New rural territories: regulating the differentiated rural spaces, in *Journal of Rural Studies*, 14(1), 107–117.

Marsden, T. (1999) Rural futures: the consumption countryside and its regulation, in *Sociologia Ruralis*, 39, 501–520.

Marsden, T. (2006) Pathways in the sociology of rural knowledge, in Cloke, P., Marsden, T. and Mooney, P. (eds) *Handbook of Rural Studies*, Sage: London.

Martin, E.W. (1962) *The Book of the Village*, Phoenix House: London.

McCarthy, J., Lloyd, G. and Illsley, B. (2002) National parks in Scotland: balancing environment and economy, in *European Planning Studies*, 10(5), 665–670.

McGowan, M. (2010) The rise of computerised high frequency trading: use and controversy, in *Duke Law and Technology Review*, 8 November.

McNulty, R. (2011) *Realising the potential of GB Rail: Report of the Rail Value for Money Study*, Department for Transport: London.

Meen, G. (2005) On the economics of the Barker Review of Housing Supply, in *Housing Studies*, 20(6), 949–971.

Meinig, D.W. (1979) Introduction, in Meinig, D.W. (ed.) *The Interpretation of Ordinary Landscapes: Geographical Essays*, Oxford University Press: Oxford.

Mickwitz, P. (2003) A framework for evaluating environmental policy instruments: context and key concepts, in *Evaluation*, 9(4), 415–436.

MEA (Millennium Ecosystem Assessment) (2005) *Ecosystems and Human Well-being: Synthesis*, Island Press: Washington, DC.

Mitchell, C. (2004) Making sense of counter-urbanisation, in *Journal of Rural Studies*, 20, 15–34.

Mokhtarian, P.L. and Solomon, I. (1999) Travel for the fun of it, in *Access*, 15, 26–32.

Monbiot, G. (2006) 'Britain's most selfish people', in *The Guardian*, 23 May.

Monbiot, G. (2013) *Feral: Searching for Enchantment on the Frontiers of Rewilding*, Penguin UK: London.

Moore-Colyer, R. and Scott, A. (2005) What kind of landscape do we want? past, present and future perspectives, in *Landscape Research*, 30(4), 501–523.

Moran, A. (2007) *Community Land Trusts: Policy Seminar*, November, Housing Corporation: London.

Mormont, M. (1990) Who is rural? Or, how to be rural: towards a sociology of the rural, in Marsden, T., Lowe, P. and Whatmore, S. (eds) *Rural Restructuring: Global Processes and Their Responses*, David Fulton: London.

Morphet, J. (1998) *Rural Aspects of the Regional Agenda*, LGA Rural Commission: London.

Morphet, J. (2010) *Effective Practice in Spatial Planning*, Routledge: London.

Morphet, J., Tewdwr-Jones, M., Gallent, N., Spry, M., Hall, B., and Howard, R. (2007) *Shaping and Delivering Tomorrow's places-Reports Findings and Recommendations*, RTPI: London.

Morris, P. (1932) *Rural Housing*, CPRE: London.

Moseley, M. (1979) *Accessibility: The Rural Challenge*, Methuen: London.

Moseley, M. (2000) England's village services in the 1990s: entrepreneurialism, community involvement and the state, in *Town Planning Review*, 74(1), 415–433.

Moseley, M.J. (2002). Bottom-up 'village action plans': Some recent experience in rural England, in *Planning Practice and Research*, 17(4), 387–405.

Muir, J. (1912) The wild parks and forest reservations of the West, in Nash, R. (1976) *The American Environment: Readings in the History of Conservation*, 2nd Edition, Addison-Wesley: Reading (MA).

Mulgan, G. (1998) *Connexity: How to Live in a Connected World*, Harvard University Press, Cambridge, MA.

Murdoch, J. (1996) The Planning of Rural Britain, in Allanson, P. and Whitby, M. (eds) *The Rural Economy and the British Countryside*, Earthscan: London, pp. 132–149.

Murdoch, J. and Pratt, A.C. (1997) From the power of topography to the topography of power, in Cloke, P. and Little, J. (eds) *Contested Countryside Cultures: Otherness, Marginalisation and Rurality*, Routledge: London.

Nadin, V. (2007) The emergence of the spatial planning approach in England, in *Planning Practice and Research*, 22(1), 43–62.

Nathan, M. (2007) *A Question of Balance: Cities, Planning and the Barker Review*, Discussion paper no. 9, Institute for Public Policy Research, London.

National Institute of Adult Continuing Education (2012) *Community Learning in Rural Areas: A Report by NIACE to BIS and DEFRA*, NIACE: Leicester.

Natural England (2012) *Natural England Designations Strategy*, Natural England, Sheffield.

Natural England (2013) *National Character Area Map*, Natural England: Sheffield. Available at: http://webarchive.nationalarchives.gov.uk/20140605090108/ http://naturalengland.gov.uk/publications/nca/searchpage.aspx (accessed 19 October 2014).

Natural England (2013) *National Character Area Profile 111: The Northern Thames Basin*, Natural England: Sheffield.

Neal, S. and Agyeman, J. (eds) (2006) *The New Countryside? Ethnicity, Nation and Exclusion in Contemporary Rural Britain*, Policy Press: Bristol.

Newby, H. (1988) *The Countryside in Question*, Hutchinson: London.

Newby, H. (1979) *Green and Pleasant Land? Social Change in Rural England* Hutchinson: London.

Newhaven Research (2008) *All Pain, No Gain? Finding the Balance: Delivering Affordable Housing through the Planning System in Scotland*, Chartered Institute of Housing (Scotland): Edinburgh.

Newman, P. and Kenworthy, J. (1999) *Sustainability and Cities: Overcoming Automobile Dependence*, Island Press: Washington, DC.

Norris, M., Gkartzios, M. and Coates, D. (2013) Property-led urban, town and rural regeneration in Ireland: positive and perverse outcomes in different spatial and socio-economic contexts, in *European Planning Studies*, 22(9), 1841–1861.

OECD (Organisation for Economic and Co-operation and Development) (2006) *The New Rural Paradigm: Policies and Governance*, OECD Publishing: Paris.

OECD (Organisation for Economic and Co-operation and Development) (2011) *OECD-FAO Agricultural Outlook 2010–2020*, OECD Publishing: Paris.

Office of Fair Trading (2012) *Price and Choice in Remote Rural Communities*. Available at: http://webarchive.nationalarchives.gov.uk/20140402142426/ www.oft.gov.uk/OFTwork/consultations/remote-communities/;jsessionid= D1D7B8E71E951F29162F05F91D0668D8 (accessed 1 December 2014).

Office of Rail Regulation (2014) http://orr.gov.uk/about-orr/what-we-do/the-law/eu-law

ODPM (Office of the Deputy Prime Minister) (2003) *Sustainable Communities: Building for the Future*, HMSO: London.

ODPM (Office of the Deputy Prime Minister) (2005) *Affordability Targets: Implications for Housing Supply*, ODPM: London.

Olmsted, F.L. (1865) The Value and Care of Parks, In Nash, R. (1976) *The American Environment: Readings in the History of Conservation*, 2nd Edition, Addison-Wesley: Reading (MA).

Ostrom, E. (2008) Tragedy of the commons, in Durlauf, S.N. and Blume, L.E. (eds) *The New Palgrave Dictionary of Economics Online*, 2nd Edition, Palgrave Macmillan: London. Available at: www.dictionaryofeconomics.com/article?id= pde2008_T000193

Ostrom, E. (2010) Beyond markets and states: polycentric governance of complex economic systems, in *American Economic Review*, 100, 641–672.

Owen, S. (1998) The role of village design statements in fostering a locally responsive approach to village planning and design in the UK, in *Journal of Urban Design*, 3(3), 359–380.

Owen, S. (2002) Locality and community: towards a vehicle for community-based decision making in rural localities in England, *Town Planning Review*, 73, 1–21.

Owen S., Bishop, J. and O'Keeffe, B (2011) Lost in translation? Some issues encountered in transferring village design statements from England to Ireland, *Journal of Urban Design*, 16(3), 405–424.

Owen, S. and Moseley, M. (2003) Putting parish plans in their place: relationships between community based initiatives and development planning in English villages, in *Town Planning Review*, 74(4), 445–471.

Owen, S., Moseley, M. and Courtney, P. (2007) Bridging the gap: an attempt to reconcile strategic planning and very local community-based planning in rural England, in *Local Government Studies*, 33, 49–76.

Pahl, R.E. (1965) Urbs in Rure: *The Metropolitan Fringe in Hertfordshire*, Geographical Papers No. 2, London School of Economics and Political Science: London.

Pahl, R.E. (1975) *Whose City? And Further Essays on Urban Society*, Penguin: Harmondsworth.

Panelli, R. (2006) Rural society, in Cloke, P., Marsden, T. and Mooney, P. (eds) *Handbook of Rural Studies*, Sage: London, pp. 63–90.

Panelli, R. and Welch, W. (2005) 'Why community? Reading difference and singularity with community', in *Environment and Planning A: Environment and Planning*, 37, 1589–1611.

Parker, G. (2014) Engaging neighbourhoods: experiences with transactive planning with communities in England, in Gallent, N. and Ciaffi, D. (eds) *Community Action and Planning: Contexts, Drivers and Outcomes*, Policy Press: Bristol.

Parker, G., Lynn, T., Wargent, M. and Locality (2014) *User Experience of Neighbourhood Planning in England, Research*, University of Reading: Reading.

Pemberton, S. and Shaw, D. (2012) New forms of sub-regional governance and implications for rural areas: evidence from England, in *Planning Practice and Research*, 27, 441–458.

PIU (Performance and Innovation Unit) (1999) *Rural Economies*, The Stationery Office: London.

Pettigrew, P. (1987) A bias for action? Industrial development in Mid Wales, in Cloke, P. (ed.) *Rural Planning: Policy into Action?* Harper and Row: London.

Plunkett Foundation (2013) *A Better Form of Business 2013: Community-owned Village Shops*, Plunkett Foundation: Woodstock.

Potter, C. and Tilzey, M. (2005) Agricultural policy discourses in the European post-Fordist transition: neoliberalism, neo-mercantilism and multi-functionality, in *Progress in Human Geography*, pp. 581–600.

Potter, S. (2010) Transport integration-an impossible dream? Paper presented at *Universities' Transport Studies Group Annual Conference*, 5–7 January 2010, University of Plymouth.

Powe, N., Hart, T. and Shaw, T. (eds) (2007) *Market Towns: Roles, Challenges and Prospects*, Routledge: London.

Public Accounts Committee (2009) *Planning the Network Change Programme*. Available at: www.publications.parliament.uk/pa/cm200809/cmselect/cmpubacc/832/83205.htm (accessed 18 July 2014) UK Parliament: London.

Puga, D. (2010) The magnitude and causes of agglomeration economies, in *Journal of Regional Science*, 50(1), 203–219.

Pugh, R., Scharf, T., Williams, C. and Roberts, D. (2007) *Obstacles to Providing Rural Social Care: Research Briefing 22*, Social Care Institute for Excellence: London.

Putnam, R. (2000) *Bowling Alone: The Collapse and Revival of American Community*, Simon & Schuster: New York.

Rackham, O. (1986) *The History of the Countryside*, Phoenix: London.

Radaelli, C. (1999) The public policy of the European Union: whither politics of expertise? In *Journal of European Public Policy*, 6(5), 757–774.

Rayner, J. (2014). Why a supermarket price war is bad news for Britain's ability to feed itself, in *The Observer*, 8 June, p. 28.

Rees, P.W. (2014) 'Londoners "priced out" of housing market': interview with the *BBC*, 31 May 2014. Available at: www.bbc.co.uk/news/uk-england-london-27628579

Rhodes, D. (2007) Understanding governance: 10 years on, in *Organization Studies*, 28(8), 1243–1264.

Ribchester, C. and Edwards, W.J. (1999) The centre and the local: policy and practice in rural education provision, in *Journal of Rural studies*, 15(1), 49–63.

Riddlesden, D. and Singleton, A.D. (2014) Broadband speed equity: A new digital divide? In *Applied Geography*, 52, 25–33.

Rittel, H. and Webber, M. (1973) Dilemmas in a general theory of planning, in *Policy Sciences*, 4, 155–169.

Robertson, M.M. (2006) The nature that capital can see: science state and market in the commodification of ecosystem services, in *Environment and Planning D*, 46, 367–384.

Robertson, R.W. (1977) Second Home Decisions: The Australian Context, in Coppock, J.T. (ed.), *Second Homes: Curse or Blessing?* Pergamon Press: London.

Robinson, G.M. (1990) *Conflicts and Change in the Countryside*, Belhaven Press: London.

Rodgers, D. (2007) Proposed statutory definition of a Community Land Trust, in *Salford University (Initiative Leader) Community Land Trusts: Disposal Models for Affordable Housing*, Salford University: Salford.

RoSPA (2014) *Cycling Accidents: Facts and Figures*, RoSPA: London.

Rossi, A. and Lambrou, Y. (2008) *Gender and Equity Issues in Liquid Biofuels Production: Minimizing the Risks to Maximize the Opportunities*, Food and Agriculture Organization of the United Nations: New York.

RSPB (2012) *RSPB View: The Final National Planning Policy Framework*, RSPB, Sandy.

Rugg, J. and Jones, A. (1999) *Getting a Job, Finding a Home: Rural Youth Transitions*, The Policy Press: Bristol.

Rule, T.A. (2015) *Solar, Wind and Land: Conflicts in Renewable Energy Development*, Routledge: London.

Rural Coalition (2012) *The Rural Challenge*, ACRE: Cirencester.

Rural Services Network (2013) *State of Rural Service Provision 2013*, Rural Services Network: Tavistock, Devon.

Rydin, Y. (2011) *The Purpose of Planning*, Policy Press: Bristol.

Rydin, Y. (2014) Communities, networks and social capital, in Gallent, N. and Ciaffi, D. (eds) *Community Action and Planning: Contexts, Drivers and Outcomes*, Policy Press: Bristol.

Rye, J.F. (2011) Conflicts and contestations: rural populations' perspectives on the second home phenomenon, in *Journal of Rural Studies*, 27(3), 263–274.

Sage, D. (2012) A challenge to liberalism? The communitarianism of the Big Society and Blue Labour, in *Critical Social Policy*, 32(3), 365–382.

Sairinen, R. (2002) Environmental governmentality as basis for regulatory reform: the adaptation of new policy instruments in Finland, In Mol, A. and Buttel, F. (eds) *The Environmental State Under Pressure*, JAI/Elsevier: London, pp. 85–104.

Sands, P., and Peel, J. (2012). *Principles of International Environmental Law*, Cambridge University Press: Cambridge.

Satsangi, M. (2014) Communities, Land-ownership, housing and planning: reflections from the Scottish Experience, in Gallent, N. and Ciaffi, D. (eds) *Community Action and Planning, Contexts, Drivers and Outcomes*, Policy Press: London, pp. 117–130.

Satsangi, M., Gallent, N. and Bevan, M. (2010) *The Rural Housing Question: Communities and Planning in Britain's Countrysides*, Policy Press: Bristol.

Saunders, P. (1984) Beyond housing classes: the sociological significance of private property rights in the means of consumption, in *International Journal of Urban and Regional Research*, 8, 202–227.

Savage, W.G. (1919) *Rural Housing: With a Chapter on the After-war Problem*, T.F. Unwin: London.

SBSTTA (Subsidiary Body on Scientific, Technical and Technological Advice) (2007) *In-depth Review of the Application of the Ecosystem Approach*, Secretariat to the Convention on Biological Diversity, Montreal.

Schomers, S. and Matzdorf, B. (2013) Payments for ecosystem services: a review and comparison of developing and industrialised nations, in *Land Use Policy* (forthcoming).

Schoones, I. (2009) Livelihoods perspectives and rural development, in *Journal of Peasant Studies*, 36(1), 171–196.

Scott, M. (2012) Housing conflicts in the Irish countryside: uses and abuses of post-colonial narratives, in *Landscape Research*, 37(1), 91–114.

Scottish Natural Heritage (2011) *An Assessment of the Impacts of Climate Change on Scottish Landscapes and their Contribution to Quality of Life: Final Report*, SNH: Edinburgh.

Selman, P. (2009) Conservation designations: Are they fit for purpose in the 21st Century in *Land Use Policy*, 26(1), 142–153.

Sepulveda L. (2009) *Outsider, Missing Link or Panacea? Some Reflections about the Place of Social Enterprise (with) in and in Relation to the Third Sector*, Working Paper No. 15, The Third Sector Research Centre: Birmingham.

Sheller, M. and Urry, J. (eds) (2006) *Mobile Technologies of the City*, Routledge: Oxford.

Shiel, L., Richards, F., Robertson, M. and Innes, C. (2007) *Allocation of Land for Affordable Housing through the Planning System*, Scottish Government: Edinburgh.

Shoard, M. (2002) Edgelands, in Jenkins, J. (ed.) *Remaking the Landscape: The Changing Face of Britain*, Profile Books: London.

Short, B. (2006) Idyllic ruralities, in Cloke, P., Marsden, T. and Mooney P.H. (eds) *The Handbook of Rural Studies*, Sage Publications: London, pp. 133–148.

Shucksmith, M. (1981) *No homes for locals?* Gower Publishing: Hampshire.

Shucksmith, M. (1990) A theoretical perspective on rural housing: housing classes in rural Britain, in *Sociologia Ruralis*, 30(2), 210–229.

Shucksmith, M. (2010). Disintegrated rural development? neo endogenous rural development, planning and place shaping in diffused power contexts, in *Sociologia Ruralis*, 50(1), 1–14.

Shutt, J., Pugalis, L. and Bentley, G. (2012) LEPs – living up to the hype? The changing framework for regional economic development and localism in the UK, in Ward, M. and Hardy, S. (eds) *Changing Gear: Is Localism the New Regionalism?* Smith Institute: London, pp. 12–24.

Sillince, J.A.A. (1986) Why did Warwickshire Key Settlement Policy change in1982? An assessment of the political implications of cuts in rural services, in *The Geographical Journal*, 152(2), 176–192.

Simmie, J. (ed.) (1994) *Planning London*, UCL Press: London.

Sloman, L. (2003) *Rural Transport Futures: Transport Solutions for a Thriving Countryside*, Transport Research Laboratory: Wokingham.

Sloman, L., Cavill, N., Cope, A., Muller, L. and Kennedy, A. (2009) *Analysis and Synthesis of Evidence on the Effects of Investment in Six Cycling Demonstration Towns*, UK TRL: Wokingham.

Starkie, D. N. (1982) Road indivisibilities: some observations, in *Journal of Transport Economics and Policy*, 16(1), 259–266.

Steffen, W., Sanderson, A., Tyson, P.D., Jäger, J., Matson, P.A., Moore III, B., Oldfield, F., Richardson, K., Schellnhuber, H.J., Turner II, B.L. and Wasson, R.J. (2004) *Global Change and the Earth System: A Planet Under Pressure*, Springer: New York.

Stern, N. (2006) *Stern Review: The Economics of Climate Change*, Her Majesty's Treasury: London.

Stevenson, R.L. (1878) *Wanderings with a Donkey in the Cévennes*, Penguin: London.

Stoker, G. (2004) New localism, progressive politics and democracy, in Gamble, A. and Wright, T. (eds) *Restating the State?* Blackwell: London, pp. 1–25.

Stratford-upon-Avon District Council (2010) *Justification for the Preferred Development Strategy*, Available at: www.stratford.gov.uk (accessed 21 October 2010).

Sturzaker, J. (2011) Can community empowerment reduce opposition to housing? Evidence from rural England, in *Planning Practice and Research*, 26(5), 555–570.

Sturzaker, J. and Shaw, D. (2015) Localism in practice – lessons from a pioneer neighbourhood in England, in *Town Planning Review* (forthcoming).

Sutcliffe, R. and Holt, R. (2011) *Who is Ready for the Big Society?*, Consulting InPlace: Birmingham.

Swiss Agency for the Environment, Forests and Landscape, the Bureau of the Convention on Wetlands and the World Wide Fund for Nature (2002) *Sustainable Management of Water Resources: The Need for a Holistic Ecosystem Approach*, Ramsar COP8 DOC. 32, Secretariat to the Ramsar Convention, Glad, Switzerland.

Tansley, A.G. (1935) The use and abuse of vegetational terms and concepts, in *Ecology* 16(3), 284–307.

Taylor, M. (2008) *Living Working Countryside: The Taylor Review of Rural Economy and Affordable Housing*, DCLG: London.

Teasdale, S., Lyon, F. and Baldock, R. (2013) Playing with numbers: a methodological critique of the social enterprise growth myth, in *Journal of Social Entrepreneurship*, 4(2), 113–131.

Terluin, I.J. (2003) Differences in economic development in rural regions of advanced countries: an overview and critical analysis of theories in *Journal of Rural Studies*, 19, 327–344.

Tewdwr-Jones, M., Gallent, N. and Morphet, J. (2010). An anatomy of spatial planning: Coming to terms with the spatial element in UK planning, in *European Planning Studies*, 18(2), 239–257.

Thatcher, M. (1993) *The Downing Street Years*, Harper Press: London.

The Wildlife Trusts (2014) *A Greener Vision for HS2*, The Wildlife Trusts: Newark.

Tilzey, M. and Potter, C. (2008) Productivism versus post-productivism: modes of agri-environmental governance in post-Fordist agricultural transitions, in Robinson, G. (ed.), *Sustainable Rural Systems: Sustainable Agriculture and Rural Communities*, Ashgate: London, pp: 41–65.

Tönnies, F. (1887) *Gemeinschaft und Gesellschaft*, Fues's Verlag: Leipzig (translated as [1988] *Community and Society*, Library of Congress Publications: Washington, DC).

Townsend, L., Sathiaseelan, A., Fairhurst, G. and Wallace, C. (2013) Enhanced broadband access as a solution to the social and economic problems of the rural digital divide, in *Local Economy*, 28(6), 580–595.

Travis, A. (1972) 'Policy formulation and the planner' in Ashton, J. and Long, W. (eds) *The Remoter Rural Areas of Britain*, Oliver and Boyd: Edinburgh.

Tromans, S. (2012) Environmental and planning law: Uneasy bedfellows? Paper presented to the *40th Oxford Joint Planning Conference*, September 15, 2012, Available at: www.39essex.com/docs/articles/stroxfordplancon15sept12final.pdf

Troughton, M. (1999) Redefining 'rural' for the 21st Century, in Ramp, W., Kulig, J., Townsend, I. and McGowan, V. (eds) *Health in Rural Settings: Contexts for Action*, University of Lethbridge Press: Lethbridge, AB, pp. 21–38.

Turner, J. (1996) *The Abstract Wild*, University of Arizona Press: Phoenix.

United Nations (2012) *Future We Want*, General Assembly Declaration 66(2012).

United Nations (2013) *World Population Prospects: The 2012 Revision – Key Findings and Advance Tables*, UN: New York.

United Nations (2014) *World Urbanization Prospects: 2014 Revision*, United Nations, New York.

UNESCO (United Nations Educational, Scientific and Cultural Organization) (2012) *Florence Declaration on Landscape*, UNESCO: Paris.

UNEP (United Nations Environment Programme) (2001) *Ecosystems and Human Wellbeing: A Framework for Assessment*, Island Press, Washington, DC.

UNEP (United Nations Environment Programme) (2005) *Ecosystems and Human Wellbeing: Synthesis*, Island Press, Washington, DC.

UNEP (United Nations Environment Programme) (2007) *Global Environment Outlook 4*, UNEP: Nairobi.

Upper Eden Community Interest Company (2012) *Upper Eden NDP*, UECIC: Upper Eden.

Van den Berg, L., Drewett, R., Klaassen, L.H., Rossi, A. and Vijverberg, C.H.T. (1982) *Urban Europe: A Study of Growth and Decline*, Pergamon, Oxford.

Van der Pennen, T. and Schreuders, H. (2014) The fourth way of active citzenship: case studies from the Netherlands in Gallent, N. and Ciaffi, D. (eds) *Community Action and Planning, Context, Drivers and Outcomes*, Policy Press, Bristol. pp. 135–156.

Van der Ploeg, J. (2006) Agricultural production in crisis, in Cloke, P., Marsden, T. and Mooney, P. (eds) *The Handbook of Rural Studies*, Sage: London, pp. 258–277.

Van der Ploeg, J.D., Renting, H., Brunori, G., Knickel, K., Mannion, J., Marsden, T., De Roest, K., Sevilla Guzmán, E. and Ventura, F. (2000) Rural development: from practices and policies towards theory, in *Sociologia Ruralis*, 40(4), 391–408.

Van Hecken, G. and Bastiaensen J. (2010) Payments for ecosystem services: justified or not? A political view, in *Environmental Science and Policy*, 13, 785–792.

Van Huylenbroeck, G. and Durand, G. (eds) (2003) *Multifunctional Agriculture: A New Paradigm for European Agriculture and Rural Development*, Ashgate: Aldershot.

Veríssimo, D., Macmillan, D.C., Smith, R.J., Crees, J. and Davies, Z.D (2014) Has climate change taken prominence over biodiversity conservation? In *Bioscience*, 64(7), 625–629.

Walford, N. (1999) Geographical transition from productivism to post-productivism: agricultural production in England and Wales 1950s to 1990s, in Walford, N., Everitt, J. and Napton, D. (eds) (1999) *Reshaping the Countryside: Perceptions and Processes of Rural Change*, CABI Publishing: Wallingford.

Walker, A. (2014) *Public Houses: How Councils and Communities Can Save Public Houses*, LGiU: London.

Wang, H. (2004) An evaluation of the modular approach to the assessment and management of large marine ecosystems, in *Ocean Development and International Law*, 35, 267–286.

Ward, S. V. (2004) *Planning and Urban Change*, 2nd Edition, Sage: London.

Ward, N., Lowe, P. and Bridges, T. (2003) Rural and regional development: the role of the Regional Development Agencies in England, in *Regional Studies*, 37(2), 201–214.

WCED (World Commission for the Environment and Development) (1987) *Our Common Future*, Oxford University Press: Oxford.

Weber, M. (1968) *Economy and Society*, University of California Press: Berkeley.

Wellman, B. (2001) Physical place and cyber-place: The rise of personalized networking, in *International Journal of Urban and Regional Research*, 25(2), 227–252.

Welsh Government (2014) *The Planning (Wales) Bill*, Welsh Government, Cardiff.

Wenzell, T. (2009) Ecocriticism, early Irish nature writing, and the Irish landscape today, in *New Hibernia Review*, 13(1), 125–139.

Whatmore, S. (2002) *Hybrid Geographies: Natures, Cultures, Spaces*, Sage: London.

White, J. (1976) Recreation and tourism in the countryside, in Cherry, G. (ed.) *Rural Planning Problems*, Leonard Hill, London.

Williams, J. (2013) *Zero-carbon Homes: A Road Map*, Routledge: London.

Wilson G. (2008) From 'weak' to 'strong' multi-functionality: conceptualising farm-level multifunctional transitional pathways, in *Journal of Rural Studies*, 24, 367–383.

Winter, M. (1996) *Rural Politics: Policies for Agriculture, Forestry and the Environment*, Routledge: London.

Woods, M. (2005a) *Rural Geography: Processes, Responses and Experiences in Rural Restructuring*, Sage: London.

Woods, M. (2005b) *Contesting Rurality: Politics in the British Countryside*, Ashgate: Aldershot.

Woods, M. (2011) *Rural*, Routledge: London.

Woolcock, M. (1998) Social capital and economic development: towards a theoretical synthesis and policy framework, in *Theory and Society*, 27, 151–208.

Wragg, A. (2000) Towards sustainable landscape planning: experiences from the Wye Valley Area of Outstanding Natural Beauty, in *Landscape Research*, 25(2), 183–200.

Young, I.M. (1990) Justice and the politics of difference, Princeton University

Press: Princeton [selection republished in 2003 as 'City life and difference', in Campbell, S. and Fainstein, S. (eds) *Readings in Planning Theory*, 2nd Edition, Blackwell: Oxford, pp. 336–355].

Zonnefeld, I. (1990) Scope and contents of landscape ecology as an emerging science, in Zonnefeld, I. and Forman, R. (eds) *Changing Landscapes and Ecological Perspectives*, Springer: New York.

Index

Dower Report (1945) 214
drought 240
dynamic commuter areas 87
dynamic rural areas 87

EaRTH centre 112
Earth System Science 233–5; critical threshold and abrupt changes 235; exceeding natural variability 235; global change 235; human activities 235; single, self-regulating system 235
economy *see* rural economy
Ecosystem Approach (EA): barriers 272–3; definition of 270–1; ecosystem services 273–5; operational guidance 272; principles 271; promotion of 270; water and marine management 272
Ecosystem Services (ES): human wellbeing 274; and Millennium Ecosystem Assessment (MEA) 273–4; public goods 273; qualitative concepts 276; services and goods 104–5; valuation techniques 274–6
Eden District Council: Core Strategy 176, 177
Edgelands 245
EDORA 33; economic diversification of rural economies 73
education: free schools 189; further education providers 186; key services in rural areas 187–90; primary schools 186, 189; rationalisation of rural schools 189; rural population 178; village schools 187–9
e-learning 178
employment: changing structure in European uplands 33; convergence of urban and rural areas 32; impact of newcomers' wealth on low-paid workers 20; rural economy 80
employment centres 186
endogenous rural development 43, 91
energy consumption 234
energy security 255
Energy Security Strategy (2012) 255
Engels, Friedrich 201
English Heritage 114
entrepreneurialism: community 8; and development 21; lifestyle 105, 106, 117; local 90; social 10, 21, 22
environment: change at a local scale

232; climate change 239–40; collaborative governance 264; ecosystem services 273–5; environmental standards 264; governance issue 262–3; influence of human activities 235; international environmental law 261–2; key authorities with responsibility for 280–2; key international agreements 264–7; membership of conservation organisations 246; resource scarcity and security 240–2, 255–7; uncertainty in understanding and governance implications 263–4; *see also* climate change; habitat; international environmental law; landscape; UK environmental law
Environment Act (1995) 278
Environment Agency 28, 56, 281, 282, 290; remit 60
Environmental and Renewable Technologies Hub 112
The Environmental Assessment of Plans and Programs Regulations (2004) 278
Environmental Impact Assessment Directive (1985) 265
Environmental Impact Assessments (EIA) 157, 270, 277
Environmental Liability Directive (2004) 265
The Environmental Permitting Regulations (England and Wales) (2010) 278
Environmental Protection Act (1990) 278
escalator areas 136
Escobar, Andrés: subsistence livelihoods 106
EU (European Union): Directives 27; directives for conservation objectives 57; EDORA 33; emergence 43; encouraging competition between transport operators 155; funding for rural areas 56–7; international agreements 265; INTERREG (interregional) Structural Programmes 27
European Agricultural Fund for Rural Development (EAFRD) 56
European Agricultural Guarantee and Guidance Fund (EAGGF) 108

Romantic Movement 232, 301, 303
Royal Mail 166
Royal Town Planning Institute 185
Rural Advocate for England: Report
    (2006) 92–3
rural areas: and climate change 240;
    co-dependency with cities 16;
    community entrepreneurialism 8;
    declining employment during early
    20th century 45; delimitations of
    16; demand for property 136;
    drivers of economic growth 71;
    dynamic 87; economic
    diversification 104, 105; flight,
    quest and overflow of population
    movement 135–6; key services *see*
    key services; large property owners
    138; leisure destinations 52;
    multi-functionality 104; national
    policy frameworks 42; perceptions
    of 4–6; population replacement
    135; social and economic problems
    45–6; transient 88; *see also*
    localism; productivist era
    (1945-mid 1970s)
rural change: mix of means approach
    10; narrative of 19–21; perceptions
    of 4–6; shaped by political
    parameters 85
Rural Coalition 185
Rural Community Action Network
    (RCAN) 180
Rural Community Councils (RCC) 28,
    134, 180
Rural Development Commission
    (formerly Development
    Commission) 44–5
rural development funds 110–11
Rural Development Policy 110
Rural Development Programme 111
rural economy: ageing populations
    40–1; agrarian-based 33–4;
    agricultural diversification 113–16;
    business start-ups 82–3; capabilities
    77; consumption-countryside
    characteristics 35; consumptionist
    view 32; delineating 72–4;
    Development Areas 48;
    differentiation from the urban
    economies 72; distinctive
    tendencies 72–3; EC funding for
    non-agricultural development 111;
    economic diversification 73;
    ecosystem services and goods

104–5; extractive industries 73;
    future development 316–17; growth
    in GDP 38; high employment and
    low salaries 80–1; local economic
    governance 90–1; manufacturing
    companies 113; models of
    sustainability 102–4; part-time jobs
    81–2; performance differences in
    rural regions 38–9; poor
    performance of remote areas 38;
    prevalence of SMEs and
    micro-enterprises 82–4; producing
    high value goods and services 78–9;
    regional 84–5; regional economic
    governance 88–90; renewable
    energy production 118–19;
    restructuring post-1960s 32; rural
    tourism 117; secondary
    manufacturing industries 35; social
    enterprises 117–19; spatial diversity
    33; switch from primary production
    104; tertiary/private service sector
    35; theoretical understanding of
    74–8; tourism sector 83;
    unrecognised activities 105–6; *see
    also* localism
Rural Fair Share Campaign 173
Rural Growth Pilots Networks Fund
    111
Rural Growth Scheme 111
rural housing question *see* housing
rural idyll 13, 20, 52; commodifiable
    aspects of 106–7; commodification
    104; community spirit 126; imagery
    302
rurality: accentuated by urbanisation
    307; concept of 306; contrivance of
    human effort 300; cultural
    significance 301; debate about the
    focus of rural policy 53–4;
    definitions of 8–9; elements of 301;
    mapping of 14–16; perception of
    306–7; problems of 6; public policy
    and 7–8; romanticism of 232, 301;
    rural development 9–10; socially
    constructed 301–5; urban/rural
    dichotomy 299–301; views of 232
rural poor 139
rural proofing 178–9
Rural Services Network 173
Rural Strategy (2004) 54, 55, 174
Rural Task Force 89–90
rural typology 18
Rural White Paper 142, 173–4